ENVIRONMENTAL ORGANIC CHEMISTRY

Illustrative Examples, Problems, and Case Studies

RENÉ P. SCHWARZENBACH
Swiss Federal Institute of Technology (ETH)
Zürich, Switzerland
and
Swiss Federal Institute for Environmental Science
and Technology (EAWAG)
Dübendorf, Switzerland

PHILIP M. GSCHWEND
Department of Civil and Environmental Engineering
Massachusetts Institute of Technology
Cambridge, Massachusetts

DIETER M. IMBODEN
Swiss Federal Institute of Technology (ETH)
Zürich, Switzerland
and
Swiss Federal Institute for Environmental Science
and Technology (EAWAG)
Dübendorf, Switzerland

A Wiley-Interscience Publication
JOHN WILEY & SONS, INC.
New York / Chichester / Brisbane / Toronto / Singapore

This text is printed on acid-free paper.

Library of Congress Cataloging in Publication Data:

Schwarzenbach, René P., 1945–
Environmental organic chemistry : illustrative examples, problems, and case studies / by René P. Schwarzenbach, Philip M. Gschwend, Dieter Imboden.
p. cm.
"A Wiley-Interscience publication."
Includes biographical references and index.
ISBN 0-471-12588-1 (acid-free)
1. Organic compounds—Environmental aspects—Problems, exercises, etc. 2. Water chemistry—Problems, exercises, etc. 3. Organic compounds—Environmental aspects—Case studies. 4. Water chemistry--Case studies. I. Gschwend, P. M. II. Imboden, Dieter M., 1943– . III. Title.
TD196.073S79 1993 Suppl.
628.1'68—dc20 95-7375

Printed in the United States of America

10 9 8 7 6 5 4 3

ENVIRONMENTAL ORGANIC CHEMISTRY

Illustrative Examples, Problems, and Case Studies

CONTENTS

PREFACE

There is no question, a textbook without "Problems" has a problem. It was therefore an unexpected but welcome pleasure that our textbook "*Environmental Organic Chemistry*" was so well received, despite the fact that it contains no sections explicitly devoted to problems. Our initial intention in writing this book was just to compensate for this deficiency of the textbook by sharing the questions and problems that our students have loved and hated during various courses in environmental sciences over the past years. Yet, several of our readers, particularly those who have tried to use the book for independent study, have expressed their interest in simple illustrative calculations that help to understand and apply the theories presented in the different chapters of the textbook. In addition, we have been asked to extend the modeling concepts to other environmental systems such as rivers and porous media. Finally, we have taken this book as an ideal opportunity to elaborate on topics that were only briefly addressed in our textbook, and to include some of the current literature. All of these factors combined have led to a project that has kept us busy for much longer than initially planned. Nevertheless, we hope that this supplement to our textbook is still in time to help bring together theory and practice in this rapidly developing field.

A SHORT GUIDE TO THIS BOOK

This book is divided into two major parts, one that is mainly compound- or process-oriented (Chapters A to M), and one that is system-oriented (Chapters N to P). Each chapter of the first part refers to a chapter in the textbook. All of them, except Chapter F, contain "Illustrative Examples" as well as "Problems". The former serve to repeat and apply, step by step, the theoretical elements that are derived in the textbook. The latter are the reader's real challenge, his or her journey into the unknown, for which one is hopefully well-equipped with the necessary tools to survive the various problems. In order to make this book suitable for use in teaching courses, we do not include the answers to these problems. However, a short compilation of the solutions can be ordered by writing to the publisher (*John Wiley and Sons, Inc., 605 Third Avenue, New York, NY 10158-0012, USA*). To facilitate the reader's orientation, numerous cross-references are made to the equations, tables, and figures of the textbook. They are marked with *italics*.

Chapters N to P are devoted to case studies that focus on the integrative modeling of organic compounds in various aquatic systems. Since in the textbook such an approach was only attempted for lakes (*Chapter 15*), we felt it to be useful to present similar treatments for rivers (Chapter O) and for porous media, particularly groundwater systems (Chapter P). Again, the aim of these chapters is to familiarize the reader with basic concepts of modeling, rather than provide the state of the art in this field. We hope that with these chapters, we can provide the necessary background and insights that are required for the critical use of more sophisticated models.

ACKNOWLEDGMENTS

Lucky are those who can rely on collaborators, students, colleagues and friends who turn the writing of a book into a real pleasure. Above all, we are indebted to ***Béatrice Schwertfeger*** who was responsible for the final layout and for most of the typing of the camera-ready manuscript, and to ***Dieter Diem,*** who reviewed this book and also produced the index. We also thank *Vreni Graf Ammann* who has typed some of the chapters, and *Judith Perlinger* for proof-reading the text.

We are very grateful to *Josef Zeyer* for providing comments and materials to the biotransformation chapter, and to *Norman Brooks* for reviewing the chapter on river modeling. Valuable comments were made by *Stefan Haderlein* on the sorption chapter, and *Edi Höhn,* and *Fritz Stauffer* on the chapter dealing with transport in porous media.

Finally, with some guilty feelings, we would like to thank our families for their patience, particularly our wives Theres Schwarzenbach, Colleen Cavanaugh, and Sibyl Imboden, to whom we promised after finishing the textbook, that we would never write a book again!

RENÉ P. SCHWARZENBACH
Dübendorf, Zürich, Switzerland
PHILIP M. GSCHWEND
Cambridge, Massachusetts
DIETER M. IMBODEN
Dübendorf, Zürich, Switzerland

ENVIRONMENTAL ORGANIC CHEMISTRY

Illustrative Examples, Problems, and Case Studies

A. VAPOR PRESSURE

ILLUSTRATIVE EXAMPLES

Basic Vapor Pressure Calculations

Consider the chemical 1,2,4,5-tetramethylbenzene (abbreviated TeMB and also called durene). In the *CRC Handbook of Chemistry and Physics*, the following information on TeMB is found:

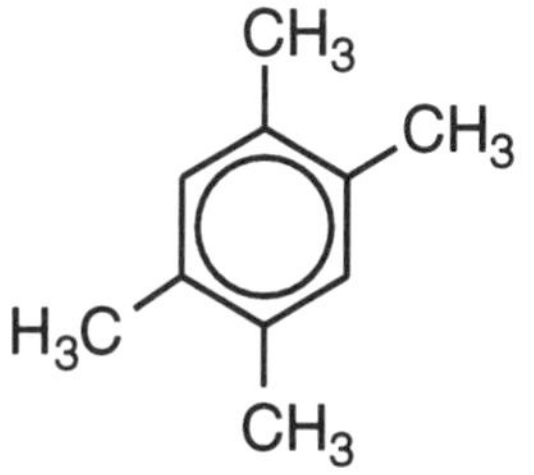

CH_3 CH_3 H_3C CH_3

1,2,4,5-tetramethylbenzene
TeMB

mw	134.2 $g \cdot mol^{-1}$
T_m	79.5°C
T_b	195.9°C

T (°C)	P^o (mm Hg)
45.0 s	1
74.6 s	10
104.2	40
128.1	100
172.1	400
195.9	760

Problem
Calculate the vapor pressure, P^o, in atmospheres and in pascals of TeMB at 20°C using the vapor pressure-temperature data given above. Also give the result in molar concentration and in mass concentration.

Answer
At 20°C, TeMB is a solid (T_m > 20°C). Thus, for calculating its P^o(s) at 20°C, only vapor pressure data for the solid (indicated by "s") may be used in *Eq. 4-7*. Convert temperatures in °C to K (add 273.2) and calculate 1/T values. Also, take the natural logarithms of the P^o-values:

1/T (K^{-1})	ln P^o (s) (mm Hg)
0.003143	0
0.002875	2.303

Perform a least squares fit of ln P^o versus 1/T (see Fig. A.1, note that often, one has more than two data pairs).

$$\ln P^o(s)\ (\text{mm Hg}) = -\frac{8609}{T} + 27.1$$

Set T = 293.2 K (= 20°C), calculate ln P^o, and get P^o(s):

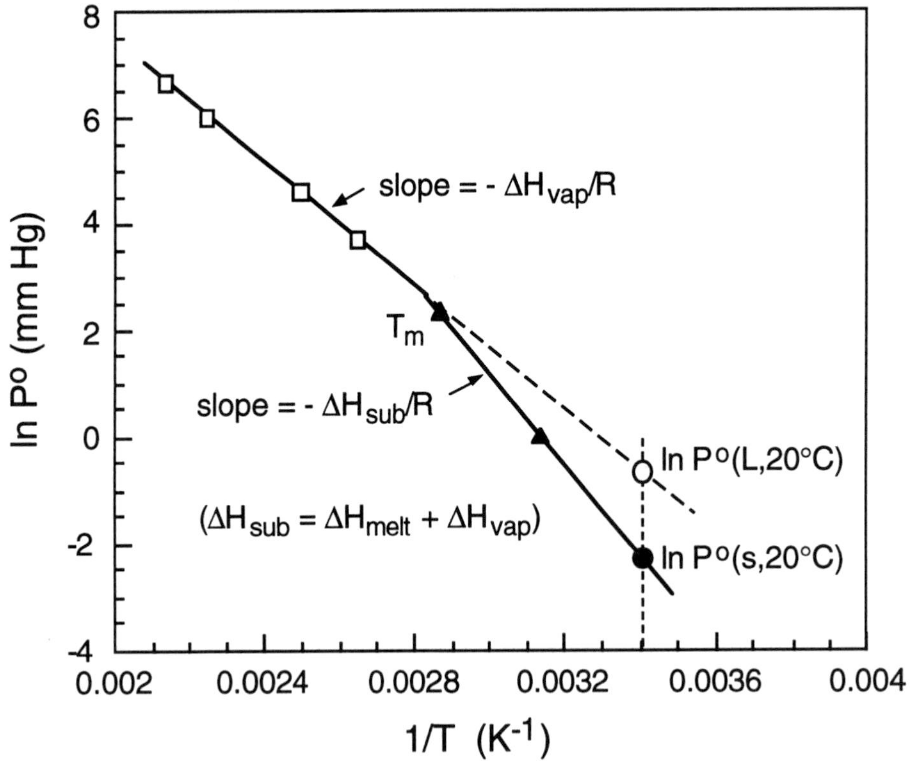

Figure A.1 Temperature dependence of the vapor pressure of TeMB: Plot of ln P^o (mm Hg) versus 1/T.

$$P^o(s) = 0.10 \text{ mm Hg} = 0.00013 \text{ atm} = 13 \text{ Pa}$$

For calculating the molar concentration, assume that TeMB behaves like an ideal gas (PV = nRT). Then the gas phase concentration, C_g, is given by

$$C_g = \frac{n_g}{V_g} = \frac{P^o}{RT}$$

With P^o = 0.00013 atm, R = 0.08206 L·atm·mol^{-1}·K^{-1}, and T = 293.2 K, the calculated concentration is (note that 1 mol TeMB corresponds to 134.2 g)

$$C_g = 5.4 \times 10^{-6} \text{ mol}\cdot\text{L}^{-1} = 0.73 \text{ mg}\cdot\text{L}^{-1}$$

Problem

Calculate the vapor pressure of the subcooled liquid of TeMB at 20°C using the data given above.

Answer

For calculating P^o(L) at 20°C, only vapor pressure data for the liquid (T entries without "s") may be used in *Eq. 4-7*. After converting the data:

1/T (K^{-1})	ln P^o (L) (mm Hg)
0.002650	3.689
0.002492	4.605
0.002246	5.991
0.002132	6.633

perform a least squares fit of ln P^o versus 1/T (see Fig. A.1):

$$\ln P^o(L) \text{ (mm Hg)} = -\frac{5676}{T} + 18.7$$

Set T = 293.2 K, calculate ln P^o, and get $P^o(L)$:

$$P^o(L) = 0.52 \text{ mm Hg} = 0.00068 \text{ atm} = 69 \text{ Pa}$$

Note that the use of Eq. 4-7 to extrapolate vapor pressure data hinges on the assumption that ΔH_{vap} for liquids or ΔH_{sub} (= $\Delta H_{vap} + \Delta H_{melt}$) for solids remains constant over the temperature range considered.

Problem

Estimate $P^o(s)$ of TeMB at 20°C using T_b and T_m data only. Compare the result with that calculated above.

Answer

Start with a rearrangement of *Eq. 4-19* where $P^o(s)$ is given as a function of $P^o(L)$, ΔS_{melt}, T_m, and T:

$$\ln P^o(s) = \ln P^o(L) - \frac{\Delta S_{melt}}{R}\left(\frac{T_m}{T} - 1\right) \qquad \text{(A-1)}$$

Note that you need to estimate $P^o(L)$ at 20°C and ΔS_{melt} at the melting point, T_m.

For estimating $P^o(L)$ (in atm), use

$$\ln P^o(L) \simeq -K_F (4.4 + \ln T_b) \left[1.8 \left(\frac{T_b}{T} - 1\right) - 0.8 \ln \left(\frac{T_b}{T}\right)\right] \qquad (4\text{-}18)$$

With $K_F = 1.00$ (see *Table 4.3* for substituted benzenes and $\mu = 0$), T_b = 273.2 + 195.9 = 469.1 K, and T = 293.2 K, you get

$$\ln P^o(L) \simeq -7.43$$

and

$$P^o(L) \simeq 0.00059 \text{ atm}$$

(This value compares reasonably with $P^o(L)$ calculated in question 2, i.e., 0.00068 atm.)

To estimate ΔS_{melt} in $J \cdot mol^{-1} \cdot K^{-1}$, use

$$\Delta S_{melt}(T_m) \simeq [56.5 + 10.5\,(n-5)] \qquad (4\text{-}20)$$

For TeMB, n = 1, and therefore set n = 5 in *Eq. 4-20*:

$$\Delta S_{melt}(T_m) \simeq 56.5 \ J \cdot mol^{-1} \cdot K^{-1}$$

Substitution of $P^o(L)$ and ΔS_{melt} in Eq. A-1 with T_m = 352.7 K and R = 8.315 $J \cdot mol^{-1} \cdot K^{-1}$, yields

$$\ln P^o(s) = -\,7.43 - 1.38 = -\,8.81$$

and therefore

$$P^o(s) = 0.00015 \text{ atm}$$

Comparing this result with that obtained by extrapolation of the vapor pressure data (0.00013 atm) shows that they "match" within about 20%. *Note that if one considers compounds with much higher T_m and T_b, the estimated result would typically be less accurate (see Table 4.4).*

Problem

Estimate the (molar) free energy ΔG_{melt}, enthalpy ΔH_{melt}, and entropy ΔS_{melt} of TeMB at 20°C using the vapor pressure-temperature data given above.

Answer

Recall from basic thermodynamics that ΔG_{melt} corresponds to the difference in chemical potential of a pure compound between its liquid and solid states (and is zero at T_m):

$$\Delta G_{melt} = \mu^o(L) - \mu^o(s) = [\mu^o(L)] - [\mu^o(L) + RT \ln \frac{P^o(s)}{P^o(L)}] = -\,RT \ln \frac{P^o(s)}{P^o(L)}$$

where the pure liquid is chosen as the reference state. Substituting the $P^o(s)$ and $P^o(L)$ values calculated above yields

$$\Delta G_{melt} = -(8.315\ J \cdot mol^{-1} \cdot K^{-1})(293.2\ K) \ln \frac{(0.00013\ atm)}{(0.00068\ atm)}$$

$$\Delta G_{melt} = +4.0\ kJ \cdot mol^{-1}$$

The ΔH_{melt} value can be calculated from ΔH_{sub} and ΔH_{vap}:

$$\Delta H_{melt} = \Delta H_{sub} - \Delta H_{vap}$$

ΔH_{sub} can be derived from the slope of ln $P^o(s)$ versus 1/T (Fig. A.1):

$$\Delta H_{sub} = -R \cdot slope = 8.315\ J \cdot mol^{-1} \cdot K^{-1} \cdot 8625\ K = 71.6\ kJ \cdot mol^{-1}$$

Similarly, ΔH_{vap} is obtained from the slope of ln $P^o(L)$ versus 1/T (Fig. A.1):

$$\Delta H_{vap} = -R \cdot slope = 8.315\ J \cdot mol^{-1} \cdot K^{-1} \cdot 5609\ K = 47.2\ kJ \cdot mol^{-1}$$

Therefore,

$$\Delta H_{melt} = +24.4\ kJ \cdot mol^{-1}$$

Finally, since $\Delta G_{melt} = \Delta H_{melt} - T\Delta S_{melt}$, one obtains

$$\Delta S_{melt} = (24.4\ kJ \cdot mol^{-1} - 4.0\ kJ \cdot mol^{-1}) / 293.2\ K$$

$$\Delta S_{melt} = 69.6\ J \cdot mol^{-1} \cdot K^{-1}$$

This assumes a constant ΔH_{melt} over the temperature range considered, and therefore attributes all of the ΔG_{melt} change to a change in ΔS_{melt} with changing temperature.

Problem

Calculate the mole fraction of TeMB in an ideal liquid mixture of aromatic hydrocarbons if, at 20°C, this compound's equilibrium concentration in the gas phase above the mixture is 35 μg TeMB· L_{air}^{-1}.

Answer

Recall from *Chapter 3* that at equilibrium the fugacity of the compound in the gas phase is equal to its fugacity in the liquid mixture:

(for a gas)

$$f_{gas} = \theta_i \, x_i \, P_{tot} = P_i \text{ (ideal gas)} \qquad (3\text{-}11)$$

(for a liquid mixture)

$$f_{liquid} = \gamma_i \, x_i \, P_i^o{}_{\text{pure liquid}} \qquad (3\text{-}13)$$

Hence, at equilibrium, the partial pressure of the compound above the liquid mixture is given by

$$P_i = \gamma_i \, x_i \, P_i^o{}_{\text{pure liquid}} = \gamma_i \, x_i \, P_i^o(L)$$

When assuming an ideal liquid mixture, γ_i is set to 1. (In reality γ_i is greater than 1.) Thus,

$$x_i = P_i \, / \, P_i^o(L)$$

Use the ideal gas law

$$P_i = \frac{n_{i,g}}{V_g} \cdot RT = C_{i,g} \cdot RT$$

where $C_{i,g}$ is the gas phase concentration on a mole per liter basis and R is 0.08206 L·atm· mol^{-1}· K^{-1}. With

$$C_{i,g} = \frac{35 \times 10^{-6} \text{ g}\cdot\text{L}^{-1}}{134.2 \text{ g} \cdot \text{mol}^{-1}} = 2.6 \times 10^{-7} \text{ mol}\cdot\text{L}^{-1}$$

one obtains

$$P_i = (2.6 \times 10^{-7}\ mol \cdot L^{-1})\ (0.08206\ L \cdot atm \cdot mol^{-1} \cdot K^{-1})\ (293.2\ K)$$
$$= 6.3 \times 10^{-6}\ atm$$

Using P_i^o (L) as calculated above (6.8×10^{-4} atm), one gets

$$x_i \simeq 0.01$$

PROBLEMS

● *A-1 A Public Toilet Problem*

Pure 1,4-dichlorobenzene (1,4-DCB) is still used as a disinfectant and air-refreshener in some public toilets. As an employee of the health department of a large city you are asked to evaluate whether the 1,4-DCB present in the air in such toilets may pose a health problem to the toilet personnel who are exposed to this compound for several hours every day. In this context you are interested in the maximum possible 1,4-DCB concentration in the toilet air at 20°C. Calculate this concentration in g per m^3 air assuming that

(a) you go to the library and get the vapor pressure data given below (*CRC Handbook of Chemistry and Physics*), or

(b) you have no time to look for vapor pressure data, but you know the boiling point (T_b) and the melting point (T_m) of 1,4-DCB (see below).

Compare the two results. What would be the maximum 1,4-DCB concentration in the air of a public toilet located in Death Valley (Temperature 60°C)? Any comments?

Cl–C₆H₄–Cl

1,4-DCB

mw 147.0 $g \cdot mol^{-1}$
T_m 53.0°C
T_b 173.9°C

T (°C)	P^o (mm Hg)
29.1s	1
44.4 s	4
54.8	10
84.8	40
108.4	100
150.2	400

● A-2 *How Much Freon Is Left in the Old Pressure Bottle?*

In a dump site, you find an old 3-liter pressure bottle with a pressure gauge that indicates a pressure of 2.7 atm. The temperature is 10°C. From the label you can see that the bottle contains Freon 12 (i.e., dichlorodifluoromethane, CCl_2F_2). You wonder how much Freon 12 is still left in the bottle. Try to answer this question. In the *CRC Handbook of Chemistry and Physics* you find the following data on CCl_2F_2:

CCl_2F_2 (structure: Cl–C–F with Cl above and F below)

Freon 12

mw 120.91 $g \cdot mol^{-1}$
T_m -158°C
T_b -29.8°C

T (°C)	P^o (atm)
-29.8	1
-12.2	2
16.1	5
42.4	10
74.0	20

● *A-3 True or False?*

Somebody bets you that at 60°C, the vapor pressure of 1,2-dichlorobenzene (1,2-DCB) is smaller than that of 1,4-dichlorobenzene (1,4-DCB), but that at 20°C, the opposite is true, that is, P^o(1,2-DCB, 20°C) > P^o(1,4-DCB, 20°C). Is this person right? If yes, at what temperature do both compounds exhibit the same vapor pressure? Try to answer these questions by using only the T_m and T_b values given below for the two compounds.

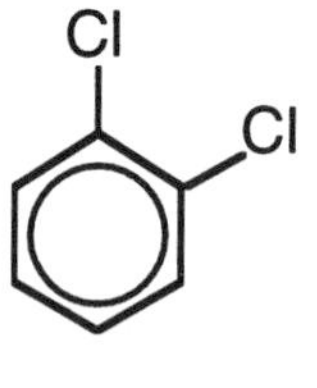

Compound	T_m	T_b
1,2-DCB	-17.0°C	180.0°C
1,4-DCB	53.1°C	174.0°C

● *A-4 Speciation of Organic Pollutants in the Atmosphere*

Ligocki and Pankow (1989) have determined the concentrations of a variety of organic pollutants in the air above the city of Portland (Oregon). Using a special sampling device, they were able to distinguish between the concentrations of the compounds present in gaseous (C_g) and in particulate forms (C_p) respectively. At an air temperature of 7°C, they got the experimental data given below for the three aromatic hydrocarbons anthracene, phenanthrene, and benz(a)anthracene. Note that C_g and C_p are the concentrations of the compounds in the respective form per total air volume.

Junge (1977) suggested a simple model for estimating the fraction of a given compound present in particulate form (Φ_p):

$$\Phi_p = \frac{C_p}{C_g + C_p} = \frac{c^* S_T}{P^o(l,L) + c^* S_T}$$

where

$P^o(l,L)$ is the vapor pressure of the (subcooled) liquid compound (in atm),

S_T is the total particle surface area (in $cm^2 \cdot cm^{-3}$ air). For urban air, *Bidleman (1988)* suggests an average value for S_T of 10^{-5} $cm^2 \cdot cm^{-3}$ air, and

c^* is a bulk parameter that describes the interactions of the compound with the particle surface. For the compounds considered here, *Junge (1977)* proposes an average c^* value of 1.7×10^{-4} atm · cm.

Calculate Φ_p for the three compounds using the model suggested by Junge and compare the results with the experimental data. For more details about the model, see *Bidleman (1988)*.

compound	T_m(°C)	T_b(°C)	C_g (ng·m^{-3} air)	C_p (ng·m^{-3} air)
anthracene	217.5	342.2	3.4±2.2	0.035±0.22
phenanthrene	99.5	340.2	26±10	0.28±0.25
benz(a)anthracene	162.0	435.0	0.32±0.14	1.2±0.8

● *A-5 A Solvent Spill*

In a factory in which metal parts are degreased with organic solvents, somebody drops a full glass bottle containing a mixture of 5 liters of tetrachloroethene (PER) and 10 liters of trichloroethene (TCE) in a closed room at 20°C. The room has a total volume of 50 m^3. The bottle breaks and the solvent mixture is spilled on the floor. Soon one can smell the solvent vapors in the air. Try to answer the following questions by assuming that PER and TCE form an ideal mixture in the liquid phase.

(a) What was the composition (in mole fractions) of the liquid mixture in the bottle, and what is it at equilibrium in the remaining liquid on the floor of the room?

(b) What are the maximum (= equilibrium) concentrations of PER and TCE in the air of the room? How much of each solvent component has evaporated?

(c) If the same accident happened in your sauna (volume = 15 m^3, T = 80°C), what maximum PER and TCE concentrations would you and your friends be exposed to there?

In the *CRC Handbook of Chemistry and Physics* you find the following data for PER and TCE:

$Cl_2C=CCl_2$

PER

$Cl_2C=CHCl$

TCE

PER	
mw	165.8 $g \cdot mol^{-1}$
ρ (20°C)	1.62 $g \cdot cm^{-3}$
T_m	- 19°C
T_b	120.8°C

TCE	
mw	131.4 $g \cdot mol^{-1}$
ρ (20°C)	1.46 $g \cdot cm^{-3}$
T_m	- 73°C
T_b	86.7°C

T (°C)	P^o (mm Hg)
- 20.6s	1
13.8	10
40.1	40
61.3	100
100.0	400
120.8	760

T (°C)	P^o (mm Hg)
- 43.8	1
- 12.4	10
11.9	40
31.4	100
67.0	400
86.7	760

B. SOLUBILITY AND ACTIVITY COEFFICIENT IN WATER

ILLUSTRATIVE EXAMPLES

Calculating Aqueous Activity Coefficients (γ_w^{sat}) from Solubilities (C_w^{sat})

Problem
Calculate the aqueous activity coefficients for (a) chloroform (a liquid), (b) lindane (a solid), and (c) vinyl chloride (a gas) at 25°C using the data provided.

Answer (a)

$CHCl_3$ (H–C(Cl)(Cl)–Cl)

trichloromethane
(also chloroform)

mw	119.4 g·mol^{-1}
T_m	– 63.5 °C
T_b	61.7 °C
P^o(25°C)	2.6 x 10^{-1} atm
C_w^{sat}(25°C)	6.5 x 10^{-2} mol · L^{-1}

For chemicals that are liquids at the temperature of interest, rearrangement of *Eq. 5-8* gives:

$$\gamma_w^{sat} = \frac{1}{x_w^{sat}}$$

From the measured molar solubility data provided, the mole fraction solubility (x_w^{sat}) can be substituted by (see *Eq. 5-9*):

$$x_w^{sat} = C_w^{sat} \cdot V_w$$

with V_w = 1.8×10^{-2} L·mol^{-1}. Thus, chloroform's aqueous activity coefficient at saturation may be calculated:

$$\gamma_w^{sat}(\text{chloroform, 25 °C}) = \frac{1}{(6.5 \times 10^{-2}\ \text{mol·L}^{-1})(1.8 \times 10^{-2}\ \text{L·mol}^{-1})} = 860$$

Answer (b)

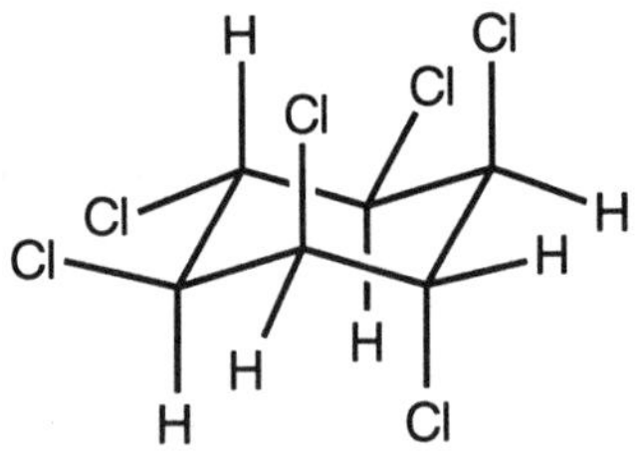

γ-1,2,3,4,5,6-hexachlorocyclohexane (also lindane)

mw	290.8 g·mol^{-1}
T_m	112.9 °C
T_b	323.4 °C
P^o(s,25°C)	8.3×10^{-8} atm
C_w^{sat}(s,25°C)	2.6×10^{-5} mol·L^{-1}

For chemicals that are solids at the temperature of interest, rearrangement of *Eq. 5-8* gives:

$$\gamma_w^{sat} = \frac{1}{x_w^{sat}} \cdot \frac{P^o(s)}{P^o(L)}$$

If P^o(l) data at higher temperatures were available, one could extrapolate these data to estimate P^o(L) at 25°C. If P^o(L) is not known, recall that *Eq. 4-19* approximates the ratio of solid and hypothetical liquid vapor pressures at a single temperature:

$$\frac{P^o(s)}{P^o(L)} = e^{-(\frac{\Delta S_{melt}(T_m)}{R})(\frac{T_m}{T}-1)}$$

Since lindane does not have a flexing chain of greater than 5 units, the entropy of melting for this compound can be calculated using *Eq. 4-20* with n = 5:

$$\Delta S_{melt}(T_m) \simeq 56.5 \ J\cdot mol^{-1}\cdot K^{-1}$$

Thus the aqueous activity coefficient is

γ_w^{sat} (lindane, 25°C) ≃

$$\frac{1}{(2.6\cdot 10^{-5} mol\cdot L^{-1})(1.8\cdot 10^{-2} L\cdot mol^{-1})}\cdot e^{-(\frac{56.5 J\cdot mol^{-1}\cdot K^{-1}}{8.31 J\cdot mol^{-1}\cdot K^{-1}})(\frac{112.9K+273.2K}{298.2K}-1)}$$

$$\gamma_w^{sat} \text{(lindane, 25 °C)} \simeq 290{,}000$$

Answer (c)

mw	62.4 $g\cdot mol^{-1}$
T_m	−153.8 °C
T_b	−13.4 °C
P^o (L,25°C)	3.9 atm
C_w^{sat}(1atm,25°C)	4.5 x 10^{-2} $mol\cdot L^{-1}$

chloroethene
(also vinyl chloride)

A rearrangement of *Eq. 5-8* also gives the means to calculate aqueous activity coefficients for chemicals that are gases at the temperature of interest:

$$\gamma_w^{sat} = \frac{1}{x_w^{sat}}\cdot\frac{1 \ atm}{P^o(L)}$$

Thus, vinyl chloride's aqueous activity coefficient at saturation may be calculated:

γ_w^{sat}(vinyl chloride, 25°C) =

$$\frac{1}{(4.5 \times 10^{-2}\ \text{mol}\cdot\text{L}^{-1})(1.8\times10^{-2}\ \text{L}\cdot\text{mol}^{-1})} \cdot \frac{1\ \text{atm}}{3.9\ \text{atm}}$$

$$\gamma_w^{sat}(\text{vinyl chloride, 25°C}) = 320$$

Evaluating the Concentration Dependence of Aqueous Activity Coefficients

Problem

Are the aqueous activity coefficients (γ_w^{sat}) calculated above for saturated solution conditions suitable for less concentrated conditions?

Answer

In order for γ_w^{sat} to be invariable for all concentrations (i.e., $\gamma_w^{sat} = \gamma_w^{\infty}$), the organic solute of interest must be present at sufficiently low levels in solution such that the hydration shells of individual molecules do not contact one another on average, even at the highest possible concentration (i.e., saturation). From experience this means that the solute must occupy less than about 0.1 % of the solution volume (see *Fig. 5.8*).

For chloroform (density at 25°C : $\rho^{25} = 1.57\ \text{g}\cdot\text{mL}^{-1}$), the volume fraction occupied at saturation is

volume fraction (chloroform, 25°C)

$$= (6.5 \times 10^{-2}\ \text{mol}\cdot\text{L}^{-1})\ (119.4\ \text{g}\cdot\text{mol}^{-1})\ (1570\ \text{g}\cdot\text{L}^{-1})^{-1}$$

$$= 0.0049\ \text{L}_{\text{chloroform}} \cdot \text{L}^{-1}_{\text{solution}}$$

$$= 0.49\ \%$$

Thus, molecules of chloroform may "feel one another" just a little at saturation, and γ_w^{sat}(chloroform, 25°C) may be a little lower than γ_w^{∞} (chloroform, 25°C) – on the order of 10% (see also *Munz and Roberts, 1986*).

For lindane (ρ^{25} = 1.9 g·mL^{-1}),

volume fraction (lindane, 25°C)

$$= (2.6 \times 10^{-5}\ \text{mol}\cdot\text{L}^{-1})\ (290.8\ \text{g}\cdot\text{mol}^{-1})\ (1900\ \text{g}\cdot\text{L}^{-1})^{-1}$$

$$= 3.9 \times 10^{-6}\ \text{L}_{\text{lindane}} \cdot \text{L}^{-1}_{\text{solution}}$$

$$= 0.00039\%$$

Consequently, this compound is sufficiently diluted even at saturation that individual lindane solute molecules are virtually never interacting in aqueous solution (γ_w^{sat} (lindane, 25°C) = γ_w^{∞} (lindane, 25°C)).

Finally, for vinyl chloride (ρ^{25} = 0.965 g·mL^{-1}), the fraction of solution occupied by this compound at saturation is about

volume fraction (vinyl chloride, 25°C)

$$= (4.5 \times 10^{-2}\ \text{mol}\cdot\text{L}^{-1})\ (62.5\ \text{g}\cdot\text{mol}^{-1})\ (965\ \text{g}\cdot\text{L}^{-1})^{-1}$$

$$= 0.0029\ \text{L}_{\text{vinyl chloride}} \cdot \text{L}^{-1}_{\text{solution}}$$

$$= 0.29\ \%$$

Therefore, as in the case of chloroform, the vinyl chloride γ_w^{sat} is probably somewhat lower than γ_w^{∞}. *For practical purposes, this effect can generally be neglected for nonpolar compounds. For polar compounds exhibiting much higher solubilities, however, their aqueous activity coefficients will vary substantially at solution concentrations greater than ~1 %.*

Estimating Aqueous Solubilities from Compound Size

Problem
Estimate the aqueous solubility of hexachloroethane (HCE) at 25°C using molar volume as a measure of its size.

Answer
Using data for a series of chemicals from the same compound class (i.e., chloroalkanes) so as to reflect consistent intermolecular interactions with water, find a correlation between liquid solubilities and size (e.g., molar volumes) similar to those shown in *Fig. 5.5.*

compound	mw ($g \cdot mol^{-1}$)	ρ^{25} ($g \cdot mL^{-1}$)	V_{org} ($mL \cdot mol^{-1}$)	log C_w^{sat}(l,L) (in $mol \cdot L^{-1}$)
chloromethane (L)	50.5	0.93	54.5	- 0.22
dichloromethane	84.9	1.317	64.5	- 0.64
trichloromethane	119.4	~ 1.4	~85	- 1.19
tetrachloromethane	153.8	~ 1.5	~100	- 2.20
1,1-dichloroethane	99.0	1.168	84.8	- 1.30
1,2-dichloroethane	99.0	~ 1.2	~ 85	- 1.07
1,1,1-trichloroethane	133.4	~ 1.3	~ 110	- 2.07
1,1,2,2-tetrachloroethane	167.9	1.587	105.8	- 1.74
γ-1,2,3,4,5,6-hexachlorocyclohexane	290.8	~ 1.5	~ 200	- 4.90

A plot of these data shows a strong inverse correlation between compound size and aqueous solubility (of liquids) as expected from unfavorable intermolecular interactions between these generally nonpolar organic chemicals and water (see Fig. B.1).

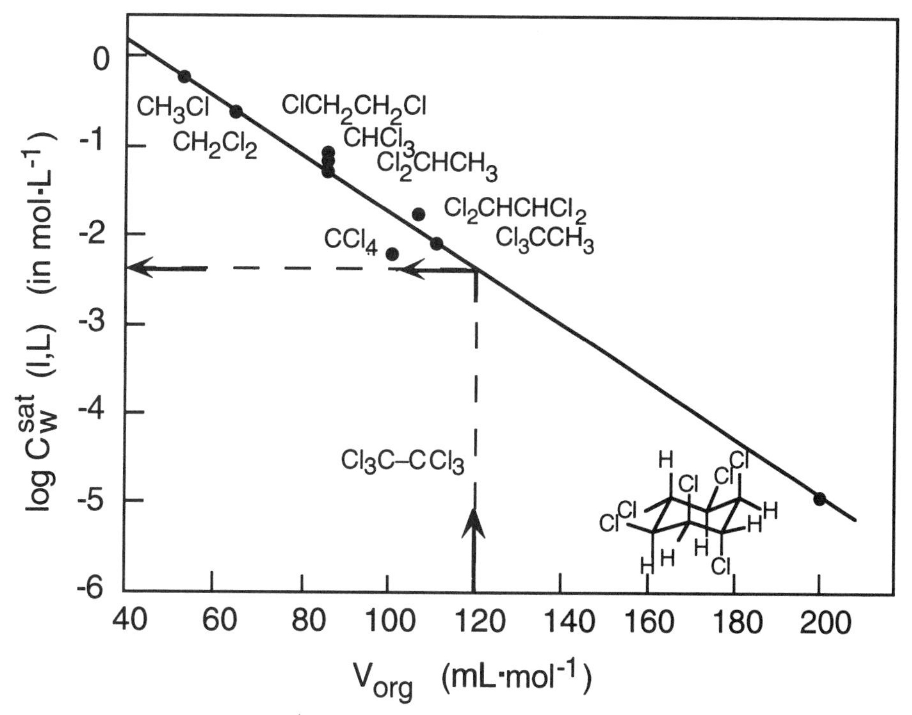

Figure B.1 Plot of log C_w^{sat}(l,L) versus molar volume (V_{org}) for a series of polyhalogenated aliphatic hydrocarbons.

Fitting these data by linear least squares yields

$$\log C_w^{sat}(\mathrm{l,L}) = -\ 0.032 \cdot V_{org} + 1.43 \qquad (r^2 = 0.99\ ,\ N = 9) \quad \text{(B-1)}$$

For HCE, find its molar volume using the ratio of its molecular weight ($236.7\ \mathrm{g \cdot mL^{-1}}$) and its liquid density at 25°C (~ $2.0\ \mathrm{g \cdot mL^{-1}}$):

$$V(\mathrm{HCE}) \simeq (236.7\ \mathrm{g \cdot mol^{-1}})\ (2.0\ \mathrm{g \cdot mL^{-1}})^{-1}$$
$$\simeq 120\ \mathrm{mL \cdot mol^{-1}}$$

Inserting this result in the correlation equation derived above yields

$$\log C_w^{sat}\ (\mathrm{HCE,\ L,\ 25°C}) \simeq -\ 2.4$$

or

$$C_w^{sat}\ (\text{HCE, L, 25°C}) \simeq 4.0 \times 10^{-3}\ \text{mol·L}^{-1}$$

Since hexachloroethane is a solid at 25°C (T_m = 186°C), its solubility at this temperature is a function of both its compatibility with water (C_w^{sat}(L)) and the free energy required to melt the solid:

$$x_w^{sat}(s) = \frac{1}{\gamma_w^{sat}} \cdot \frac{P^o(s)}{P^o(L)} \qquad (5\text{-}8)$$

Since for liquids, $\gamma_w^{sat} = x_w^{sat}(l,L)^{-1}$, and because $x_w^{sat} = C_w^{sat} \cdot V_w$, this expression may be rewritten:

$$C_w^{sat}(s) = C_w^{sat}(L) \cdot \frac{P^o(s)}{P^o(L)}$$

Using the relation

$$\frac{P^o(s)}{P^o(L)} = e^{-\left(\frac{\Delta S_{melt}(T_m)}{R}\right)\left(\frac{T_m}{T}-1\right)}$$

where ΔS_{melt}(HCE) with n = 5 is estimated as 56.5 J·mol^{-1}· K^{-1}, one finds

$$C_w^{sat}(\text{HCE, s, 25°C}) \simeq C_w^{sat}(\text{HCE, L, 25°C}) \cdot e^{-\left(\frac{\Delta S_m}{R}\right)\left(\frac{T_m}{T}-1\right)}$$

$$\simeq (4.0 \times 10^{-3}\ \text{mol·L}^{-1})\ e^{-\left(\frac{56.5\text{J·mol}^{-1}\text{·K}^{-1}}{8.31\text{J·mol}^{-1}\text{·K}^{-1}}\right)\left(\frac{186°\text{C}+273.2\text{K}}{298.2\text{K}}-1\right)}$$

$$C_w^{sat}\ (\text{HCE, s, 25°C}) \simeq 1.0 \times 10^{-4}\ \text{mol·L}^{-1}$$

The experimental value at 22.3°C is 2.1 x 10^{-4} mol·L^{-1} (*Horvath, 1982*).

Estimating the Effect of Temperature on Aqueous Solubilities and Aqueous Activity Coefficients

Problem
Estimate the aqueous solubilities and activity coefficients of (a) chloroform, (b) lindane, and (c) vinyl chloride at 5°C.

Answer (a)
Since chloroform (trichloromethane) is a liquid at both 25°C and 5°C, the magnitude of change in its solubility with solution temperature is dictated by its excess enthalpy of solution. *Assuming this* ΔH_s^e *is constant over the temperature range of interest*, solubility change with temperature is described by

$$\log C_w^{sat}(l,L) = -\frac{\Delta H_s^e}{2.303\,RT} + \text{constant} \qquad (5\text{-}19)$$

Generally, the value of ΔH_s^e for low molecular weight nonpolar compounds is near zero (e.g., benzene's $\Delta H_s^e \simeq +2\ \text{kJ}\cdot\text{mol}^{-1}$ at 25°C; see *Table 5.4*). *Horvath (1982)* reports the following chloroform mole fraction solubilities:

T(°C)	x_w^{sat}
0	0.001514
10	0.001366
20	0.001249
25	0.001203
30	0.001168

Converting the temperatures to K, calculating T^{-1}-values, converting mole fraction solubilities to molar ones ($C_w^{sat}(T) = x_w^{sat}(T)/\bar{V}_w(T)$) and taking logarithms, one obtains

1/T (K^{-1})	$\log C_w^{sat}$ (in $mol \cdot L^{-1}$)
0.003660	– 1.076
0.003531	– 1.120
0.003411	– 1.160
0.003353	– 1.177
0.003298	– 1.190

These data exhibit a linear trend (Fig. B.2), which can be fit to *Eq. 5-19* yielding:

$$\log C_w^{sat}\text{(chloroform)} = +\frac{318}{T} - 2.24$$

This result implies the assumption of constant ΔH_s^e is reasonable over this temperature range, and

$$\Delta H_s^e\text{(chloroform, 0-30°C)} \simeq -(2.303)(8.315\ J \cdot mol^{-1} \cdot K^{-1})(318\ K)$$
$$\simeq -6\ kJ \cdot mol^{-1}$$

To find C_w^{sat}(chloroform, 5°C), set T = 278.2 K and calculate $\log C_w^{sat}$:

$$\log C_w^{sat}\text{(chloroform, 5°C)} = +\frac{318}{278.2} - 2.24$$
$$= -1.10$$

$$C_w^{sat}\text{(chloroform, 5°C)} = 8.0 \times 10^{-2}\ mol \cdot L^{-1}$$

Since chloroform is a liquid, its aqueous activity coefficient at 5°C may now be found;

$$\gamma_w^{sat}\text{(chloroform, 5°C)} = \frac{1}{x_w^{sat}\text{(chloroform, 5°C)}}$$

$$= \frac{1}{C_w^{sat}(\text{chloroform, 5°C}) \cdot V_w\,(5°C)}$$

$$\approx \frac{1}{(7.9 \times 10^{-2}\ \text{mol·L}^{-1})(1.8 \times 10^{-2}\ \text{L·mol}^{-1})}$$

$$\gamma_w^{sat}\,(\text{chloroform, 5°C}) = 700$$

as compared to 850 at 25°C (see above).

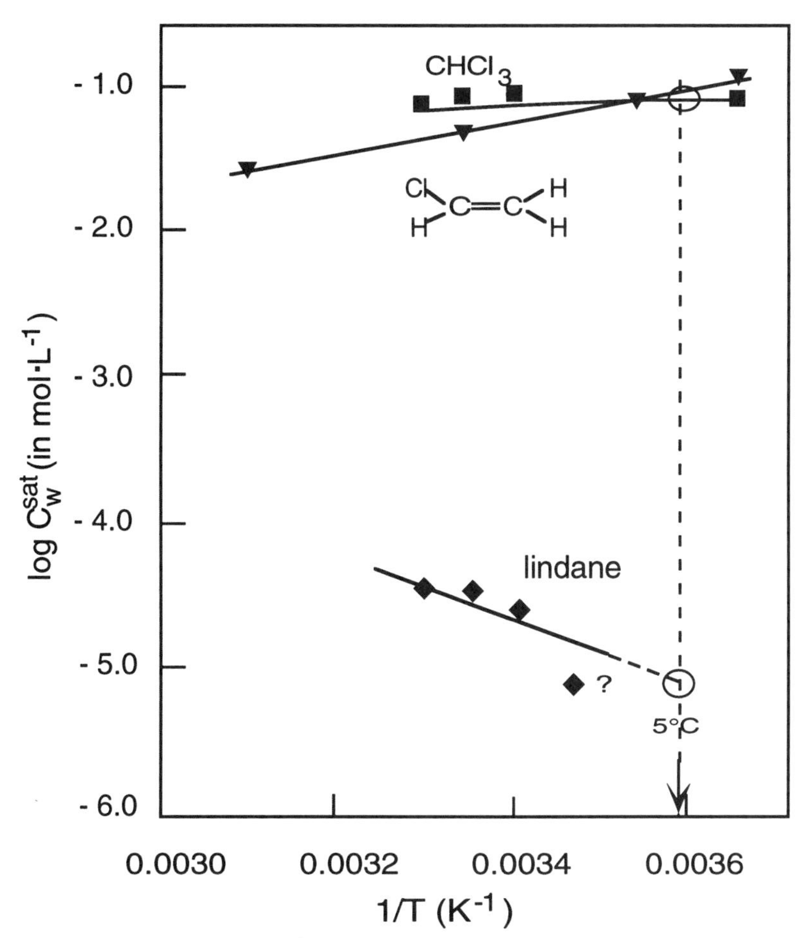

Figure B.2 Plot of log C_w^{sat} versus 1/T for chloroform, vinyl chloride, and lindane.

Answer (b)
At 5°C, lindane is a solid (T_m = 112.8°C), therefore use *Eq. 5-21* to relate aqueous solubilities at various temperatures:

$$\log C_w^{sat} = -\frac{(\Delta H_{melt} + \Delta H_s^e)}{2.303\ RT} + \text{constant} \qquad (5\text{-}21)$$

Horvath (1982) gives the following temperature-solubility data, which after conversion to reciprocal T and log C_w^{sat}, can be examined for their linear relation (Fig. B.2):

T(°C)	1/T (K^{-1})	x_w^{sat}	log C_w^{sat} (in $mol \cdot L^{-1}$)
15	0.003470	0.00000012	– 5.177
20	0.003411	0.00000035	– 4.712
25	0.003353	0.00000048	– 4.575
30	0.003298	0.00000057	– 4.500

It appears there is something wrong with the 15°C solubility, so one might choose to exclude it from the regression (as we do here):

$$\log C_w^{sat}(\text{lindane, s}) = -\frac{1876}{T} + 1.696$$

This result implies a total enthalpy of solution:

$$\Delta H_s = \Delta H_{melt} + \Delta H_s^e = (2.303)(8.315\ J \cdot mol^{-1}K^{-1})(1876\ K)$$

$$\simeq 36\ kJ \cdot mol^{-1}$$

If ΔH_{melt} is estimated as $T_m \cdot \Delta S_{melt} \simeq (112.8 + 273.2)(56.5)$

$$\simeq 22\ kJ \cdot mol^{-1},$$

the value of ΔH_s^e(lindane) must, therefore, be about +14 $kJ \cdot mol^{-1}$, a similar magnitude as exhibited by polycyclic aromatic hydrocarbons of similar size *(Table 5.4)*.

Finally to estimate C_w^{sat}(at 5°C), first assume ΔH_s remains constant down to 5°C, and setting T = 278.2 K, calculated

$$\log C_w^{sat}(\text{lindane, } 5°C) = -\frac{1876}{278.2} + 1.696$$

$$= -5.048$$

$$C_w^{sat}(\text{lindane, } 5°C) = 9.0 \times 10^{-6}\ \text{mol}\cdot\text{L}^{-1}$$

Since lindane is a solid, to find its aqueous activity coefficient at 5°C use

$$\gamma_w^{sat}(\text{lindane, } 5°C) = \frac{1}{x_w^{sat}(\text{lindane, } 5°C)} \cdot \frac{P^o(s)}{P^o(L)}$$

$$= \frac{1}{C_w^{sat}(\text{lindane, } 5°C) \cdot V_w(5°C)} \cdot \frac{P^o(s)}{P^o(L)}$$

As done before, except now at 5°C

$$\frac{P^o(s)}{P^o(L)} \simeq e^{-\left(\frac{56.5\,\text{J}\cdot\text{mol}^{-1}\cdot\text{K}^{-1}}{8.31\,\text{J}\cdot\text{mol}^{-1}\cdot\text{K}^{-1}}\right)\left(\frac{112.9°\text{C}+273.2\,\text{K}}{278.2\,\text{K}}-1\right)}$$

$$\simeq 0.072$$

Therefore,

$$\gamma_w^{sat}(\text{lindane, } 5°C) = \frac{1}{(9.0 \times 10^{-6}\ \text{mol}\cdot\text{L}^{-1})(1.8 \times 10^{-2}\ \text{L}\cdot\text{mol}^{-1})} \cdot (0.072)$$

$$\gamma_w^{sat}(\text{lindane, } 5°C) = 450,000$$

an increased value relative to 25°C (i.e., 290,000), which is consistent with the positive ΔH_s^e.

Answer (c)

At 5°C, vinyl chloride (chloroethene) is a gas ($T_b = -13.4$°C); therefore use *Eq. 5-21* to relate aqueous solubilities at various temperatures:

$$\log C_w^{sat} = -\frac{(\Delta H_{condense} + \Delta H_s^e)}{2.303\ RT} + \text{constant} \qquad (5\text{-}21)$$

Horvath (1982) gives the following temperature-solubility data, which after conversion to reciprocal T and $\log C_w^{sat}$ can be plotted (see Fig. B.2) and fit to *Eq. 5-21*.

T(°C)	1/T (K^{-1})	x_w^{sat}	$\log C_w^{sat}$(in mol·L^{-1})
0	0.003660	0.00158	– 1.057
25	0.003353	0.000798	– 1.355
50	0.003094	0.000410	– 1.648

This fit again shows good linearity over the temperature range involved:

$$\log C_w^{sat}(\text{vinyl chloride, g}) = +\frac{1042}{T} - 4.86$$

This result implies a total enthalpy of solution:

$$\Delta H_s = \Delta H_{condense} + \Delta H_s^e \simeq (2.303)(8.315\ \text{J}\cdot\text{mol}^{-1}\text{K}^{-1})(-1042\ \text{K})$$

$$\simeq -20\ \text{kJ}\cdot\text{mol}^{-1}$$

If $\Delta H_{condense} = -\Delta H_{vap}$ is estimated as

$$-T_b \cdot \Delta S_{vap} \simeq -(-13.4 + 273.2)(85\ \text{J}\cdot\text{mol}^{-1})$$

$$\simeq -22\ \text{kJ}\cdot\text{mol}^{-1}$$

the value of ΔH_s^e (vinyl chloride) must, therefore, be about + 2 kJ·mol^{-1}, a reasonable near-zero value for such a small nonpolar compound.

To estimate C_w^{sat} at 5°C, set T = 278.2 K and calculate

$$\log C_w^{sat} \text{ (vinyl chloride, 5°C)} = + \frac{1042}{278.2} - 4.86$$

$$= - 1.12$$

$$C_w^{sat} \text{ (vinyl chloride, 5°C)} = 7.6 \times 10^{-2} \text{ mol·L}^{-1}$$

Since vinyl chloride is a gas, to find its aqueous activity coefficient at 5°C, use:

$$\gamma_w^{sat} \text{ (vinyl chloride, 5°C)} = \frac{1}{x_w^{sat} \text{ (vinyl chloride, 5°C)}} \cdot \frac{1 \text{ atm}}{P^o(L)}$$

$$= \frac{1}{C_w^{sat}\text{(vinyl chloride, 5°C)} \cdot V_w\text{(5°C)}} \cdot \frac{1 \text{ atm}}{P^o(L)}$$

Using vapor pressure data from the *CRC Handbook of Chemistry and Physics*, find for liquid vinyl chloride

$$\ln P(\text{atm}) = -\frac{2944}{T} + 11.35$$

so

$$P^o(\text{L, at 5°C}) = 2.2 \text{ atm}$$

and γ_w^{sat} (vinyl chloride, 5°C)

$$= \frac{1}{(7.6 \times 10^{-2} \text{ mol·L}^{-1})(1.8 \times 10^{-2} \text{ L·mol}^{-1})} \cdot \frac{1 \text{ atm}}{2.2 \text{ atm}}$$

$$\gamma_w^{sat} \text{ (vinyl chloride, 5°C)} = 330$$

a value not too different from the 25°C result, (i.e., 320) which is consistent with the small ΔH_s^e.

Estimating the Effect of Dissolved Salts on Aqueous Solubilities and Activity Coefficients

Problem
Estimate the solubility of (a) chloroform, (b) lindane, and (c) vinyl chloride in seawater at 25°C and 30‰ salinity.

Answer
To assess the influence of dissolved salts on solubility, use the Setschenow relation (*Eq. 5-22*):

$$\log C_{w\ salt}^{sat} = \log C_{w}^{sat} - K^{s} \cdot [salt]$$

or

$$C_{w\ salt}^{sat} = 10^{-K^{s} \cdot [salt]} \cdot C_{w}^{sat}$$

The value of $K^s(M^{-1})$ for an organic compound is a function of its hydrophobic surface area (see *Table 5.6)*.

In the absence of specific information, estimate:

K^s(chloroform)	$\simeq$ 0.2	(like benzene)
K^s(lindane)	$\simeq$ 0.3	(like phenanthrene, pyrene)
K^s (vinyl chloride)	$\simeq$ 0.1	

Seawater contains a mixture of salts, which generally remain in fixed proportions. At a total salt concentration of 35‰ (35 parts per thousand), the molar abundance of various major ions in seawater are *(Millero and Sohn, 1992)*:

Na^+	0.469 M	Cl^-	0.546 M
K^+	0.0102 M	HCO_3^-	0.00186 M
Ca^{2+}	0.0103 M	Br^-	0.000841 M
Mg^{2+}	0.0528 M	SO_4^{2-}	0.0282 M

To estimate such a mixture at 35‰ one could dissolve the following combination of salts in water:

		for 35‰	for 30‰
$NaCl$	@	0.411 M	0.352 M
KCl	@	0.00936 M	0.00802 M
$CaCl_2$	@	0.0103 M	0.00883 M
$MgCl_2$	@	0.0528 M	0.0453 M
Na_2SO_4	@	0.0282 M	0.0242 M
$NaHCO_3$	@	0.00186 M	0.00159 M
Total		0.514 M	0.440 M

In the case of seawater at 30‰, all of these concentrations would be reduced by the factor 30/35. Now to estimate K^s precisely, one could then sum the contributions of each salt according to *Eq. 5-23*:

$$K^s(\text{chloroform}) \simeq K^s_{NaCl} \cdot \frac{[NaCl]}{\text{Total}} + K^s_{KCl} \cdot \frac{[KCl]}{\text{Total}} + K^s_{CaCl_2} \cdot \frac{[CaCl_2]}{\text{Total}}$$

$$+ K^s_{MgCl_2} \cdot \frac{[MgCl_2]}{\text{Total}} + K^s_{Na_2SO_4} \cdot \frac{[Na_2SO_4]}{\text{Total}}$$

$$+ K^s_{NaHCO_3} \cdot \frac{[NaHCO_3]}{\text{Total}}$$

and judging from values for benzene and naphthalene in *Table 5.7*

$$K^s(\text{chloroform}) \simeq (0.2)\frac{(0.352)}{(0.440)} + (0.2)\frac{(0.00842)}{(0.440)} + (0.3)\frac{(0.00883)}{(0.440)}$$

$$+ (0.3)\frac{(0.0453)}{(0.440)} + (0.5)\frac{(0.0242)}{(0.440)}$$

$$+ (0.3)\frac{(0.00159)}{(0.440)}$$

$$\simeq 0.2$$

similar to the initial estimate above.

Finally, the table below shows the estimated solubilities for chloroform, lindane, and vinyl chloride in 30‰ seawater (0.44 M):

compound	C_w^{sat} (M)	K^s (M^{-1})	$K^s \cdot [0.44\ M]$	$10^{-K^s[0.44M]}$	$C_{w,salt}^{sat}$ (M)
chloroform	6.5×10^{-2}	0.2	0.088	0.82	5.3×10^{-2}
lindane	2.6×10^{-5}	0.3	0.13	0.74	1.9×10^{-5}
vinyl chloride	4.5×10^{-2}	0.1	0.044	0.90	4.1×10^{-2}

These results demonstrate that seawater salinities typically reduce aqueous solubilities by less than a factor of 2.

Note: Because the free energy contributions of phase change to the overall free energy of solution (condensation or melting) are not affected by salts in the solution, it is the aqueous activities coefficients that are increased as salt concentration increases:

compound	$\gamma_w^{sat}(25°C)$	$\gamma_w^{sat}(25°C, 30‰) = \gamma_w^{sat} \cdot 10^{+K^s \cdot [0.44\ M]}$
chloroform	860	1000
lindane	290,000	390,000
vinyl chloride	320	360

Estimating the Concentration in Water of Organic Compounds Dissolving from an Organic Liquid Mixture

Problem
What is the concentration of chloroform, lindane, and vinyl chloride in groundwater (subscript "gw") at 5°C at equilibrium with a large volume of methylene chloride (CH_2Cl_2; mw = 84.9 $g \cdot mol^{-1}$) containing these substances at the weight percents: 1%, 0.1%, and 0.01%, respectively?

Answer
Use *Eq. 5-37* to calculate the resultant groundwater concentrations;

$$x_{gw} = \gamma_{CH_2Cl_2mix} \cdot x_{CH_2Cl_2mix} \cdot \gamma_{gw}^{-1}$$

or recalling that $C_w = x_w \cdot V_w^{-1}$ (V_w in $L \cdot mol^{-1}$):

$$C_{gw}\,(mol \cdot L^{-1}) = \gamma_{CH_2Cl_2mix} \cdot x_{CH_2Cl_2mix} \cdot \gamma_{gw}^{-1} \cdot V_w^{-1}$$

Since the chemicals of interest are chemically similar to the methylene chloride solvent, values of $\gamma_{CH_2Cl_2mix}$ should be near 1, though probably higher for lindane than for the others (see *Table 7.1* for examples of some hydrocarbons dissolved in hexane.) Here we take $\gamma_{CH_2Cl_2mix}$ to be 2 for chloroform and vinyl chloride and 5 for lindane. Next calculate the mole fraction in the CH_2Cl_2mix. Generally, this is done as

$$x_{mix}(i) = \frac{(\text{wt fraction } i)(\text{mw } i)^{-1}}{(\text{wt fraction } i)(\text{mw } i)^{-1} + (\text{wt fraction } j)(\text{mw } j)^{-1} + \ldots (\text{wt fraction solvent})(\text{mw solvent})^{-1}}$$

Collecting all these calculation inputs in the table below and using V_w (5°C) = 0.018 $L \cdot mol^{-1}$ in the modified form of *Eq. 5-37*, the groundwater concentrations at equilibrium with the spilled solvent are estimated:

compound	$\gamma_{CH_2Cl_2mix}$	wt fraction in CH_2Cl_2mix	$x_{CH_2Cl_2mix}$	$\gamma_w(5°C)$	C_{gw} (mol·L^{-1})
chloroform	~2	0.01	0.0071	690	$1x10^{-3}$
lindane	~5	0.001	0.00029	450000	$2x10^{-7}$
vinyl chloride	~2	0.0001	0.00014	330	$5x10^{-5}$

The greatest uncertainty in the calculation is probably the choice for $\gamma_{CH_2Cl_2mix}$ for the various compounds.

PROBLEMS

● *B-1 The Missing Information*

For assessing the environmental partitioning behavior of the four compounds (I-IV) indicated below, you need to know their vapor pressures and aqueous solubilities at 25°C. Since you have only found the melting points, T_m, of the compounds, you call a friend at EPA and you ask for the missing data. While she is speaking to you over the phone, you write down the numbers:

P^o (atm)	C_w^{sat} (mol L^{-1})	compound
1.9×10^{-4}	1.6×10^{-5}	
1.4×10^{-3}	1.0×10^{-4}	
2.6×10^{-4}	6.3×10^{-1}	
3.0×10^{-5}	2.5×10^{-6}	

After you have hung up the phone, you realize that you did not indicate which data set belongs to which compound. Since you are a little bit embarrassed, you try to make the assignment yourself. Comment on your choices!

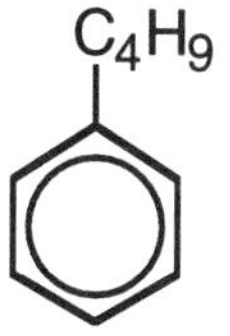

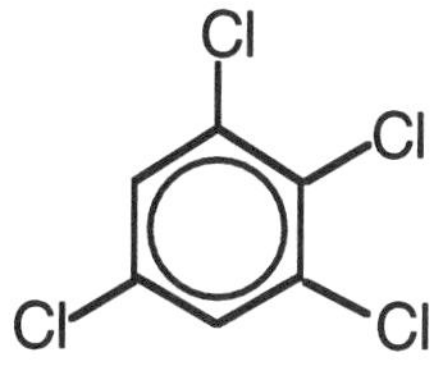

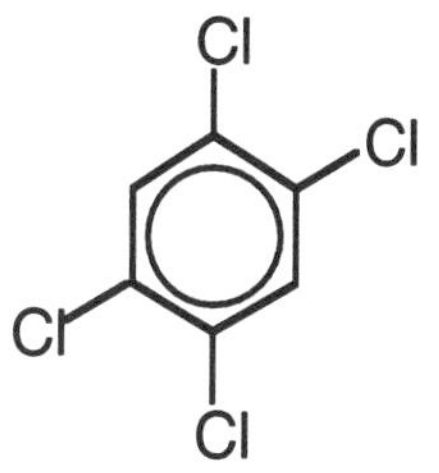

	phenol	n-butyl benzene	1,2,3,5-tetra-chlorobenzene	1,2,4,5-tetra-chlorobenzene
	I	II	III	IV
T_m(°C)	43.0	- 88.0	54.5	140.0

● *B-2 Is It Possible to Prepare These Aqueous Solutions?*

Somebody wants to prepare the following (aqueous) solutions at 25°C and 1 atm total pressure: (a) 0.1 g·L^{-1} 1,3-butadiene in distilled water, (b) 0.5 g·L^{-1} styrene in seawater (35‰ salinity), and (c) 25 μg·L^{-1} benzo[a]pyrene in a 40% methanol/60% water (v/v) mixture (see *Fig. 5.9*). The person wants to know from you whether this is possible or whether the solubility of the compounds is too low. Because you do not find any aqueous solubility data for these compounds, you use the UNIFAC program to estimate the "UNIFAC aqueous activity coefficients", $\gamma_{w\ UNIFAC}^{\infty}$, for the componds at infinite dilution. From these values, you then derive the aqueous solubilities of the liquid compounds at 25°C by using the relationship given by *Banerjee (1985)*:

$$\log C_w^{sat}(l,L)\ (\text{in mol L}^{-1}) \simeq 1.20 - 0.78 \log \gamma_{w\ UNIFAC}^{\infty}$$

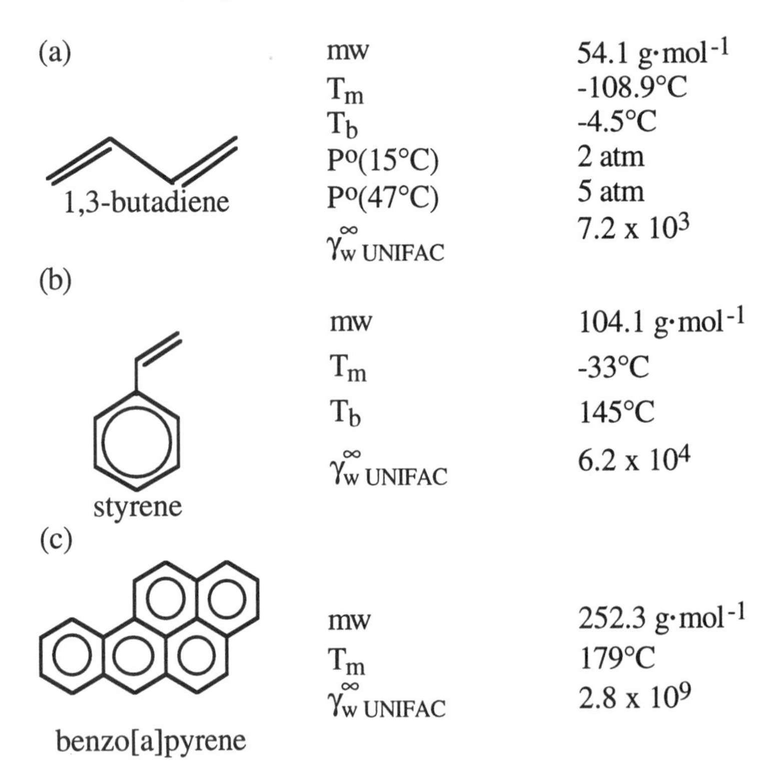

(a)
1,3-butadiene
mw 54.1 g·mol-1
Tm -108.9°C
Tb -4.5°C
Po(15°C) 2 atm
Po(47°C) 5 atm
γw∞ UNIFAC 7.2 x 103
(b)
styrene
mw 104.1 g·mol-1
Tm -33°C
Tb 145°C
γw∞ UNIFAC 6.2 x 104
(c)
benzo[a]pyrene
mw 252.3 g·mol-1
Tm 179°C
γw∞ UNIFAC 2.8 x 109

● *B-3 A Tricky Stock Solution*

You work in an analytical laboratory and you are asked to prepare 250 mL of a 0.2 M stock solution of anthracene in toluene (ρ^{20} (toluene) = 0.87 g·cm^{-3}) as solvent. You look up the molecular weight of anthracene, go to the balance, weigh out 8.91g of this compound, put it into a 250 mL volumetric flask, and then you fill the flask with toluene. To your surprise (or not?), even after several hours of intensive shaking, there is still a substantial portion of undissolved anthracene present in the flask, although your intuitive assumption that anthracene and toluene form an almost ideal liquid mixture is not that incorrect. What is the problem? Try to give a rough estimate of how much anthracene has actually been dissolved, and what it's concentration is in the stock solution (at 20°C). The necessary data can be found in the *Appendix* of the textbook.

● *B-4 How Much Methyl Bromide Is Present in the Water Droplets on the Lettuce in a Greenhouse?*

In a greenhouse (25°C), gaseous methyl bromide (CH_3Br) is applied as a fumigant to protect the vegetables against a variety of bugs. The partial pressure of CH_3Br in the air of the greenhouse is 0.01 atm. Curious as you are, you wonder what the concentration of CH_3Br is in the water droplets on the lettuce leaves. Being a well-trained environmental organic chemist, you are aware that CH_3Br hydrolyzes to CH_3OH and Br^- (see *Chapter 12*). However, you assume that gas exchange with the air is fast as compared to hydrolysis, and, therefore, you assume air-water equilibrium. Calculate the saturation-concentration of CH_3Br in the water at a partial pressure of 0.01 atm. Unfortunately, your search for aqueous solubility data of CH_3Br is unsuccessful. In the *CRC Handbook of Chemistry and Physics* you only find the following data for CH_3Br:

H–C(H)(H)–Br methyl bromide	mw	94.9 g·mol^{-1}
	T_m	- 95.0°C
	density of liquid CH_3Br(0°C)	1.73 g·cm^{-2}

T(°C)	P°(atm)
3.6	1
23.3	2
54.8	5
84.0	10

In addition, solubility data (25°C) are available for the following compounds:

compound	mw g·mol^{-1}	density °C (g·cm^{-3})	log C_w^{sat} (l,L)
chloromethane(L)	50.5	0.96^0, 0.93^{20}	- 0.22
dichloromethane	84.9	1.33^{20}	- 0.64
trichloromethane	119.4	1.48^{20}	- 1.19
tetrachloromethane	153.8	1.60^{20}	- 2.20
tribromomethane	252.7	2.89^{20}	-1.91

● *B-5 A Small Bet with an Oceanographer*

A colleague of yours who works in oceanography bets you that both the solubility as well as the activity coefficient of naphthalene are larger in seawater at 25°C than in distilled water at 5°C. Is this not a contradiction? How much money do you bet? Estimate C_w^{sat} and γ_w^{sat} for naphthalene in seawater at 25°C and in distilled water at 5°C using the data given in *Figs. 5.3, 5.6, 5.7, Tables 5.4 and 5.6* , and in the *Appendix.*

C. AIR-WATER PARTITIONING: THE HENRY'S LAW CONSTANT

ILLUSTRATIVE EXAMPLES

Estimating Henry's Law Constants (K_H) from Vapor Pressure and Aqueous Solubility

Problem

Estimate K_H (in atm·L·mol^{-1}) and K_H' ("dimensionless" Henry's Law constant) at 25°C for (a) chlorobenzene (a liquid), (b) fluorene (a solid), and (c) bromomethane (a gas) using the data provided.

Answer (a)

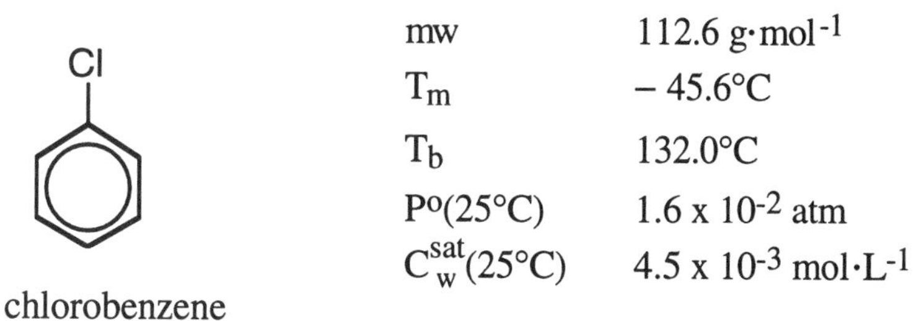

mw	112.6 g·mol^{-1}
T_m	– 45.6°C
T_b	132.0°C
P^o(25°C)	1.6 x 10^{-2} atm
C_w^{sat}(25°C)	4.5 x 10^{-3} mol·L^{-1}

chlorobenzene

Approximate K_H by the ratio of vapor pressure and aqueous solubility:

$$K_H \simeq K_H^{sat} = \frac{P^o}{C_w^{sat}} \qquad (6\text{-}9)$$

Inserting the data given above into *Eq. 6-9* yields

$$K_H \simeq \frac{1.6 \times 10^{-2}\ \text{atm}}{4.5 \times 10^{-3}\ \text{mol}\cdot\text{L}^{-1}} \simeq 3.6\ \text{atm}\cdot\text{L}\cdot\text{mol}^{-1}$$

which, in this case, matches exactly the experimental value (see *Appendix*). The dimensionless Henry's Law constant is then obtained by

$$K_H' = \frac{K_H}{RT} \qquad (6\text{-}3)$$

that is,

$$K_H' \simeq \frac{3.6\ \text{atm}\cdot\text{L}\cdot\text{mol}^{-1}}{(0.08206\ \text{atm}\cdot\text{L}\cdot\text{mol}^{-1}\cdot\text{K}^{-1})\ (298.2\text{K})} = 0.15$$

Answer (b)

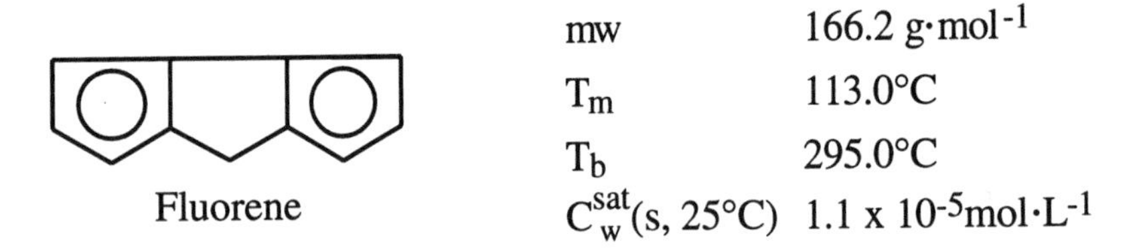

Fluorene

mw	166.2 $g\cdot mol^{-1}$
T_m	113.0°C
T_b	295.0°C
C_w^{sat}(s, 25°C)	1.1 x $10^{-5} mol\cdot L^{-1}$

Fluorene is a solid at 25°C. Make the same assumptions as for chlorobenzene and approximate K_H by *Eq. 6-9* using either the combination $P^o(s)$, $C_w^{sat}(s)$, or of $P^o(L)$, $C_w^{sat}(L)$ values.

$$K_H \simeq \frac{P^o(s)}{C_w^{sat}(s)} = \frac{P^o(L)}{C_w^{sat}(L)}$$

Since $P^o(s)$ of fluorene is not given, estimate it from T_b ($\rightarrow P^o(L)$) and T_m ($P^o(L) \rightarrow P^o(s)$) using *Eq. 4-18* with K_F = 1.00 and *Eq. 4-21* with n = 5, respectively (see illustrative calculations in the vapor pressure section). The calculated values are 6.3 x 10^{-6} atm for $P^o(L)$ and 8.5 x 10^{-7} atm for $P^o(s)$. Insertion of $P^o(s)$ and $C_w^{sat}(s)$ in *Eq. 6-9* yields

$$K_H \simeq \frac{8.5 \times 10^{-7}\ \text{atm}}{1.1 \times 10^{-5}\ \text{mol}\cdot\text{L}^{-1}} = 0.078\ \text{atm}\cdot\text{L}\cdot\text{mol}^{-1}$$

and

$$K'_H = 0.0032$$

Answer (c)

$H-C(H)(H)-Br$

bromomethane
CH_3Br

mw	94.9 $g\cdot mol^{-1}$
T_m	−93.6°C
T_b	3.6°C
P^o(L, 25°C)	1.8 atm
C_w^{sat}(1 atm, 25°C)	1.6 x 10^{-1} $mol\cdot L^{-1}$

Bromomethane is a gas at 25°C. Because its solubility is given at 1 atm partial pressure, the K_H-value can be directly approximated by

$$K_H \simeq \frac{1\ \text{atm}}{C_w^{sat}(1\text{atm})} = \frac{1\ \text{atm}}{1.6 \cdot 10^{-1}\text{mol}\cdot\text{L}^{-1}} = 6.3\ \text{atm}\cdot\text{L}\cdot\text{mol}^{-1}$$

and

$$K'_H = 0.26$$

Note that C_w^{sat} (L, 25°C) of CH_3Br is equal to (C_w^{sat}(1atm, 25°C) · [P^o(L, 25°C) / 1 atm]); i.e., C_w^{sat}(L, 25°C) = 2.9 x $10^{-1}$$mol\cdot L^{-1}$.

Estimating the Effect of Temperature on K_H

Problem
Above, you have estimated the K_H-values of chlorobenzene, fluorene, and bromomethane at 25°C. Calculate the K_H's of these compounds at 5°C.

Answer
Over small temperature ranges for which the enthalpy change of air-water transfer, ΔH_{Henry}, is assumed to be constant, the temperature dependence of K_H can be expressed as

$$\ln K_H(T) = -\left(\frac{\Delta H_{Henry}}{R}\right)\frac{1}{T} + \text{const}$$

If $K_H \simeq K_H^{sat}$, ΔH_{Henry} may be approximated by the difference between ΔH_{vap} and ΔH_s^e:

$$\ln K_H(T) \simeq \ln K_H^{sat}(T) = -\left(\frac{\Delta H_{vap} - \Delta H_s^e}{R}\right)\frac{1}{T} + \text{const} \quad (6\text{-}11)$$

Recall that for solids, $\Delta H_{vap} = \Delta H_{sub} - \Delta H_{melt}$ and $H_s^e = \Delta H_s - \Delta H_{melt}$.

To calculate K_H at a temperature T other than 298.2K (25°C), insert each 298.2K and T in *Eq. 6-11*. Subtraction of the two equations obtained and rearrangement yields:

$$\ln K_H(T) = \ln K_H(298.2) + \left(\frac{\Delta H_{vap} - \Delta H_s^e}{R}\right)\left(\frac{1}{298.2} - \frac{1}{T}\right)$$

or

$$K_H(T) = K_H(298.2) \cdot e^{+\left(\frac{\Delta H_{vap} - \Delta H_s^e}{R}\right)\left(\frac{1}{298.2} - \frac{1}{T}\right)}$$

As discussed extensively in Chapter A, ΔH_{vap} can be derived from vapor pressure data, which in the case of chlorobenzene, fluorene, and bromomethane can be found in the *CRC Handbook of Chemistry and Physics*.

In contrast to ΔH_{vap}, H_s^e-values are more difficult to obtain, since measurements of the temperature dependence of aqueous solubilities are rather scarce (see also comments in *Section 5.4*). Note that for nonpolar compounds as the ones considered here, a rough estimate of H_s^e may be made from the size of the molecule (e.g., from molar volume or total surface area) using H_s^e-values of appropriate reference compounds.

Using the ΔH_{vap}-values derived from vapor pressure data, and H_s^e -values taken form *Table 5.4* (fluorene) or estimated from the size of the molecules (chlorobenzene, bromomethane), the following results are obtained:

	ΔH_{vap} (kJ·mol^{-1})	ΔH_s^e (kJ·mol^{-1})	ΔH_{Henry} (kJ·mol^{-1})	K_H(25°C) (atm·L·mol^{-1})	K_H(5°C) (atm·L·mol^{-1})
chlorobenzene	41	~ 6[a]	~ 35	3.6	1.3
fluorene	56.5[b]	15.5	41	0.078	0.024
bromomethane	23.5	~ 0[c]	~ 23.5	6.3	3.1

[a] Using the values in *Table 5.4* and assuming that size (chlorobenzene) = 1/2 (size of benzene + size of naphthalene). Note that using H_s^e-values of chlorinated benzenes would be more appropriate.

[b] Extrapolated from vapor pressure data above T_m.

[c] See *Fig. 5.6*, temperature dependence of the solubility of the superheated liquid. H_s^e is probably slightly negative.

Thus, for the three compounds at 5°C, the K_H-values are smaller than the K_H-values at 25°C by a factor of 2 to 3.

Estimating the Effect of Salt on Air-Water Partitioning

Problem
Calculate the equilibrium distribution ratio, $K_H(sw)$, of fluorene between air and seawater at 25°C.

Answer
Use the Setschenow relationship to relate the solubility (and thus the activity coefficient) of fluorene in distilled water and in seawater:

$$\log C_{w.salt}^{sat} = \log C_w^{sat} - K^s[salt]_t \qquad (5\text{-}22)$$

or

$$C_{w.salt}^{sat} = C_w^{sat} \cdot 10^{-(K^s[salt]_t)}$$

Approximate $K_H(sw)$ by

$$K_H(sw) = \frac{p^o}{C_{w.salt}^{sat}} = \frac{p^o}{C_w^{sat} \cdot 10^{-(K^s[salt]_t)}} = K_H^{sat} \cdot 10^{+(K^s[salt]_t)}$$

Since the K^s-values of fluorene for seawater is not known, use the K^s-value for anthracene ($K^s \simeq 0.3$, see *Table 5.6*). With a total salt concentration of about 0.5 M and using the K_H-value derived above, one then gets

$$K_H(sw) = 0.078 \text{ atm·L·mol}^{-1} \text{ x } 10^{(0.3)(0.5)} = 0.11 \text{ atm·L·mol}^{-1}$$

Thus, for fluorene, due to the salting-out effect, the equilibrium distribution ratio between air and seawater is 1.4 times greater than its distribution ratio between air and distilled water.

Calculating the Equilibrium Partitioning of a Compound between an Air Volume and a Water Volume

Problem
Consider a well-sealed 1-L flask containing 100 mL of distilled water, 900 mL of air, and 10 µg total (M_{tot}) of either chlorobenzene, fluorene, or bromomethane.

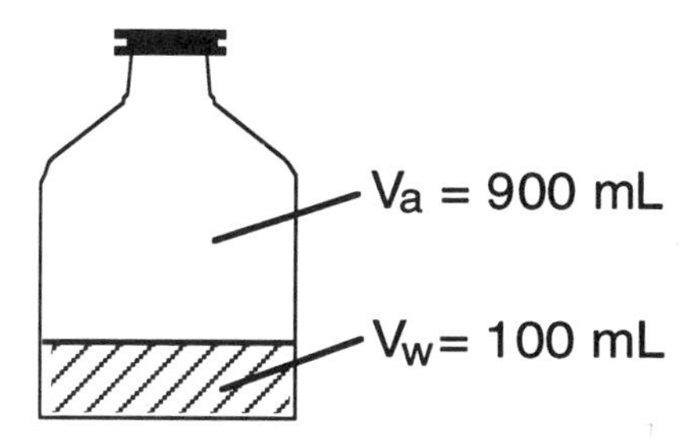

Calculate the concentration of each of the three compounds in the water and in the air at equilibrium at 25°C.

Answer
Calculate the fraction, f_w, of the compound in the aqueous phase:

$$f_w = \frac{\text{mass in aqueous phase}}{\text{total mass}}$$

$$= \frac{C_w V_w}{C_w V_w + C_a V_a} = \frac{1}{1 + \frac{C_a V_a}{C_w V_w}} = \frac{1}{1 + K_H' \frac{V_a}{V_w}}$$

where C_a and C_w are the concentrations of the compound in the air and in the water, respectively.

The concentrations in the aqueous phase and in the air are then obtained by

$$C_w = \frac{f_w M_{tot}}{V_w}$$

and

$$C_a = \frac{(1-f_w)M_{tot}}{V_a}$$

where M_{tot} = 10 μg. With $V_a/V_w = 9$ and using the K_H'-values estimated above, one gets

compound	f_w (–)	C_w ($\mu g \cdot L^{-1}$)	C_a ($\mu g \cdot L^{-1}$)
chlorobenzene	0.43	43	6.3
fluorene	0.97	97	0.31
bromomethane	0.30	30	7.7

PROBLEMS

● *C-1 A Small Ranking Exercise*

Rank the four compounds (I-IV) indicated below in the order of increasing tendency to distribute from water to air: (a) by using your chemical intuition (comment on your choice), and (b) by using the structural unit contributions given in *Table 6.4*.

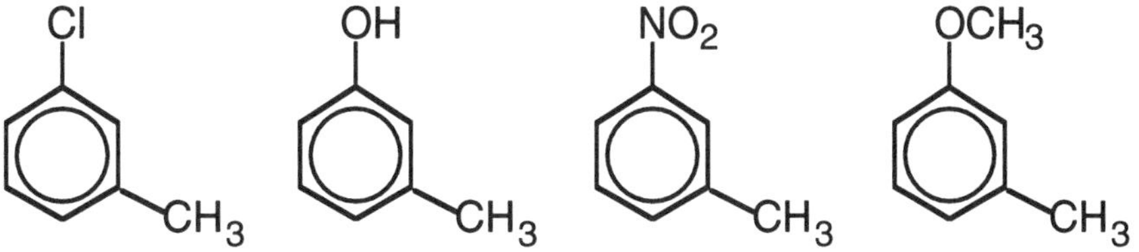

Cl — CH_3 (I); OH — CH_3 (II); NO_2 — CH_3 (III); OCH_3 — CH_3 (IV)

I II III IV

● *C-2 Raining Out*

Because of the increasing contamination of the atmosphere by organic pollutants, there is also a growing concern about the quality of rainwater. In this context, it is interesting to know how well a given compound is scavenged from the atmosphere by rainfall. Although for a quantitative description of this process, more sophisticated models are required, some simple equilibrium calculations are quite helpful.

Assume that the three compounds (V-VII) indicated below are present in the atmosphere at low concentrations. Consider now a drop of water (volume - 0.1 mL, pH = 6.0) in a volume of 100 L of air (corresponds about to the air-water ratio of a cloud (*Seinfeld, 1986*)). Calculate the fraction of the total amount of each compound present in the water drop at 25°C and at 5°C assuming equilibrium between the two phases. Use the data given in *Table 6.3* and in the *Appendix*, and comment on any assumption that you make.

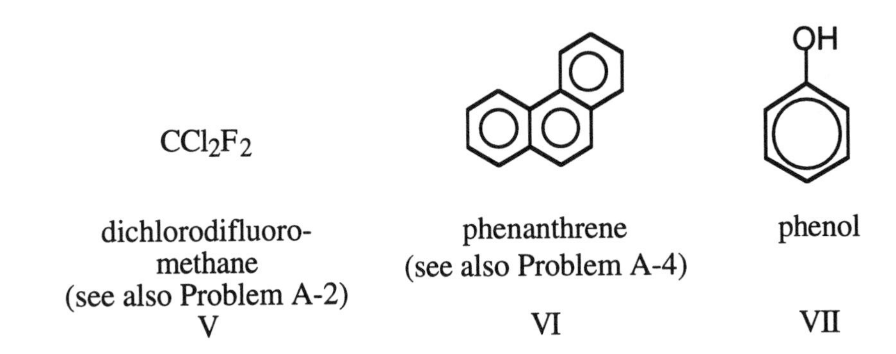

● *C-3 Watch Out When Analyzing for Volatile Organic Compounds (VOCs)*

You are the boss of a commercial analytical laboratory, and your job is to check all results before they are sent to the customers. One day, you look at the numbers from the analysis of benzene in water samples of very different origins, namely

(a) Uncontaminated groundwater,
(b) Seawater (ionic strength $\simeq$ 0.5 M),
(c) Water from the Dead Sea (ionic strength $\simeq$ 5 M), and
(d) Leachate of a hazardous waste site containing 40% (v:v) methanol.

For all samples, your laboratory reports a benzene concentration of 1 µg L^{-1}. Knowing the problems associated with the analysis of volatile organic compounds, you inquire immediately about the handling of the samples. Here we go! The samples (100 mL) were put into 1 L flasks which were then sealed and stored at 5°C for several days. Then, an aliquot of the water was withdrawn and analyzed for benzene. What were the original concentrations of benzene in the 4 samples? (The data required to answer this question can be found in *Chapters 5* and *6* and in the *Appendix*).

D. ORGANIC SOLVENT-WATER PARTITIONING; BIOCONCENTRATION

ILLUSTRATIVE EXAMPLES

Estimating the Activity Coefficients of Organic Compounds in Organic Phases

Problem

Calculate the activity coefficients for naphthalene and aniline in water-saturated hexane (γ_h), in water-saturated octanol (γ_o), and in water (γ_w) at 25°C using the data given below. Compare and discuss the results.

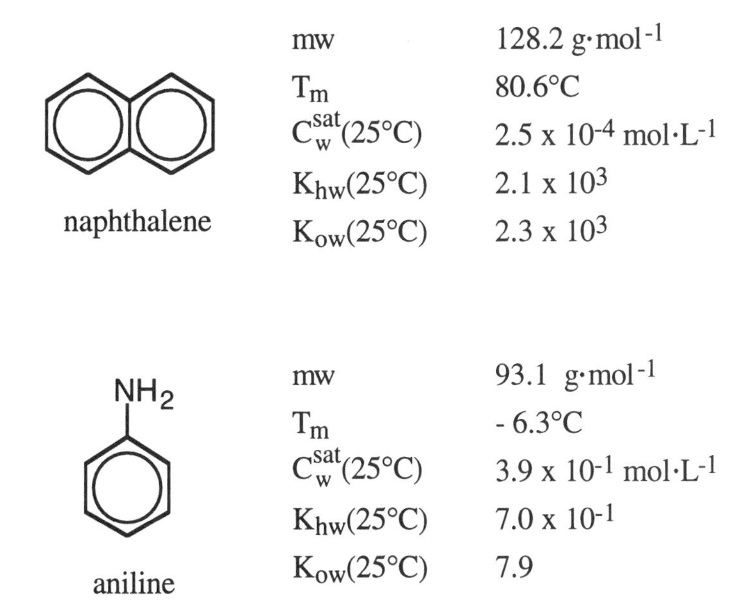

naphthalene		
	mw	128.2 g·mol^{-1}
	T_m	80.6°C
	C_w^{sat}(25°C)	2.5 x 10^{-4} mol·L^{-1}
	K_{hw}(25°C)	2.1 x 10^3
	K_{ow}(25°C)	2.3 x 10^3

aniline		
	mw	93.1 g·mol^{-1}
	T_m	- 6.3°C
	C_w^{sat}(25°C)	3.9 x 10^{-1} mol·L^{-1}
	K_{hw}(25°C)	7.0 x 10^{-1}
	K_{ow}(25°C)	7.9

Answer

Since immiscible phases partition organic chemicals according to the relative incompatibilities of those chemicals in each phase, partition constants simply reflect the ratios of the corresponding activity coefficients and molar volumes in dilute solution:

$$K_{sw} = \frac{C_s}{C_w} = \frac{\gamma_w V_w}{\gamma_s V_s} \tag{7-4}$$

Thus one may solve for the activity coefficients of solutes in *water-saturated organic media* if K_{sw} is known:

$$\gamma_s = \gamma_w \cdot \frac{1}{K_{sw}} \cdot \frac{V_w}{V_s} \tag{D-1}$$

As a first approximation, γ_w may be expressed by γ_w^{sat} which is related to the aqueous solubility of the liquid compound (see *Eqs. 5-8* and *5-9*):

$$\gamma_w \simeq \gamma_w^{sat} = \frac{1}{C_w^{sat}(1,L) \cdot V_w} \tag{D-2}$$

Since naphthalene is a solid at 25°C, its liquid aqueous solubility may be estimated from its melting point (*Eqs. 5-8*, *5-9,* and *4-21*):

$$C_w^{sat}(L) = C_w^{sat}(s) \cdot e^{+(6.8)(\frac{T_m}{T}-1)} \tag{D-3}$$

Hence, a $C_w^{sat}(L)$ value of 8.9×10^{-4} mol·L^{-1} is calculated from Eq. D-3 for naphthalene. Inserting the liquid aqueous solubilities of the two compounds into Eq. D-2 yields:

$$\gamma_w^{sat}\text{(naphthalene)} = 62{,}500$$
$$\gamma_w^{sat}\text{(aniline)} = 140$$

Consider now whether γ_w at dilute solution is equal to γ_w^{sat}. Assuming a liquid density of approximately 1 g·mL^{-1} for naphthalene, the volume fraction, f_v, occupied by the compound in the water at saturation is about

$$f_v \simeq (2.5 \times 10^{-4}\ \text{mol}\cdot L_w^{-1})(128.2\ \text{g}\cdot\text{mol}^{-1})(1 \times 10^{-3}\ L_n\cdot g^{-1})$$

$$\simeq 3.2 \times 10^{-5}\ L_n\cdot L_w^{-1} \simeq 0.003\ \%$$

Hence, the nonpolar naphthalene is sufficiently diluted even at saturation, and, therefore, it can be assumed that $\gamma_w \simeq \gamma_w^{sat}$. Note that some organic chemicals, for example, surface active compounds, may still self-associate significantly, though present at quite low volume fractions; an example is oleic acid *(Jung et al., 1987)*.

In the case of aniline (density $\simeq$ 1 $g\cdot mL^{-1}$), the calculated volume fraction is substantially higher:

$$f_v \simeq (3.9 \times 10^{-1}\ \text{mol}\cdot L_w^{-1})\ (93.1\ \text{g}\cdot\text{mol}^{-1})(1 \times 10^{-3}\ L_a\cdot g^{-1})$$

$$\simeq 3.6 \times 10^{-2} \simeq 3.6\%$$

Hence, for aniline and other compounds that may self-associate by hydrogen bonding, one cannot *a priori* assume that $\gamma_w = \gamma_w^{sat}$. This has been demonstrated by *Schwarzenbach et al. (1988)* who found self-association of 4-nitrophenols at even lower concentrations. Nevertheless, for the following calculations, neglect the error introduced by the assumption made in Eq. D-2.

Using $V_h \simeq 0.13\ L\cdot mol^{-1}$ and $V_o \simeq 0.13\ L\cdot mol^{-1}$ for the water-saturated molar volumes of hexane and octanol, respectively, calculate now the γ_s values using Eq. D-1 with $\gamma_w \simeq \gamma_w^{sat}$. The resulting values are

γ_h (naphthalene)	$\simeq$ 4	γ_o (naphthalene)	$\simeq$ 3.8
γ_h (aniline)	$\simeq$ 28	γ_o (aniline)	$\simeq$ 2.5

A comparison of these results shows that the nonpolar compound, naphthalene, is accommodated similarly well in both solvents, while the more polar compound, aniline, is much more incompatible with hexane as compared to octanol (note that the hexane molecules cannot hydrogen bond with polar groups such as the amino group of aniline). Finally, the results also demonstrate again the ability of octanol to accommodate both nonpolar as well as polar solutes (see also *Fig. 7.2*).

Estimating K_{ow} by Various Methods

Problem
Estimate the octanol-water partition constants of naphthalene and aniline from (a) their aqueous solubilities, (b) their hexane-water partition constants, and (c) structural group contributions.

Answer (a)
Use the linear free-energy relationship (LFER) *Eq. 7-14* to calculate log K_{ow}:

$$\log K_{ow} = -a \cdot \log C_w^{sat}(l,L) + b \qquad (7\text{-}14)$$

For naphthalene, take the a and b values given in *Table 7.2* for polycyclic aromatic hydrocarbons (with $C_w^{sat}(l,L) = 8.9 \times 10^{-4}\ mol \cdot L^{-1}$, see above):

$$\log K_{ow} = -0.87 \log (8.9 \times 10^{-4}) + 0.68$$

$$= 3.33$$

Hence,

$$K_{ow}\ (\text{naphthalene}) = 2.2 \times 10^3$$

which compares very well with the experimental value of 2.3×10^3.

For aniline, use the a and b values for substituted benzenes including polar substitutents:

$$\log K_{ow} = -0.72 \log (3.9 \times 10^{-1}) + 1.18$$

$$= 1.47$$

and, therefore,

$$K_{ow}\text{ (aniline)} = 30$$

This is not a very good estimate (K_{ow}(exp) = 7.9), which may not be too surprising when considering the rather poor correlation coefficient (R^2 = 0.87) for this equation.

Answer (b)
Use the LFER given in *Figure 7.6* for nonpolar compounds to estimate K_{ow} of naphthalene from its K_{hw} value:

$$\log K_{ow} = \frac{\log K_{hw} + 0.76}{1.28} \tag{D-4}$$

Insertion of the K_{hw} value given above into Eq. D-4 yields

$$\log K_{ow} = \frac{\log (2.1 \times 10^3) + 0.76}{1.28}$$

$$= 3.19$$

or

$$K_{ow}\text{ (naphthalene)} = 1.5 \times 10^3$$

This result is somewhat worse than the estimate using the aqueous solubility of liquid naphthalene (see above), but the value obtained is still quite reasonable.

As can be seen from *Figure 7.6*, for the more polar compound aniline, an even greater discrepancy between predicted and experimental value has to be expected and is also found. Here, the LFER including polar compounds has to be used:

$$\log K_{ow} = \frac{\log K_{hw} + 2.26}{1.66} \tag{D-5}$$

Using the K_{hw} value of aniline given above, a K_{ow} value of

$$K_{ow}\ (\text{aniline}) = 18.5$$

is obtained.

Answer (c)

Since both naphthalene and aniline are quite simple molecules with no complicated intramolecular interactions (no F-values have to be applied), reasonably good results can be expected when estimating K_{ow} from structural group contributions. For naphthalene, the fragment contributions include (*Table 7.3*)

$$\begin{aligned}\log K_{ow}\ (\text{naphthalene}) &= \sum_i f_i\ (\text{naphthalene})\\ &= 8\, f_{c,arom} + 2 f_{c,arom.between\ rings} + 8\, f_H^{\phi}\\ &= 8(0.13) + 2(0.23) + 8(0.23)\\ &= 3.34\end{aligned}$$

and, hence,

$$K_{ow}\ (\text{naphthalene}) = 2.2 \times 10^3$$

In the case of aniline, take benzene ($\log K_{ow} = 2.13$, see *Appendix*) as a starting point:

$$\begin{aligned}\log K_{ow}\ (\text{aniline}) &= \log K_{ow}\ (\text{benzene}) - f_H^{\phi} + f_{NH_2}^{\phi}\\ &= 2.13 - 0.23 - 1.00\\ &= 0.90\end{aligned}$$

or,

$$K_{ow}\ (\text{aniline}) = 7.9$$

which, in this case, exactly matches the experimental value.

Estimating Bioconcentration from K_{ow}

Problem
Estimate the maximum concentration of 1,2,4-trichlorobenzene (1,2,4-TCB) in a rainbow trout swimming in water exhibiting a 1,2,4-TCB concentration, C_w, of 10 nM.

1,2,4-TCB

mw	$181.5\ g \cdot mol^{-1}$
K_{ow} (25°C)	$10^{4.0}$

Answer
It has long been recognized (*Neely et al., 1974*) that nonpolar organic compounds accumulate in organisms to an extent that is directly related to the chemical's incompatibility with water (i.e., γ_w). Since γ_w values for chemicals are also related to organic solvent-water partition coefficients (particularly when using solvents that mimic well the lipids present in organism, such as is the case for n-octanol, see *Chiou, 1985*), it may not be too surprising to find that the extent of bioaccumulation for any particular compound is also related to the compound's K_{ow}. *Veith et al. (1979)* give a correlation that relates the bioconcentration factor (BCF), defined as the concentration in fish divided by the concentration in the water, to octanol-water partition constants based on tests on fathead minnows, rainbow trouts, and bluegills:

$$\log BCF \left(\frac{\text{mol/g wet fish}}{\text{mol/mL water}} \right) = 0.85 \log K_{ow} - 0.70 \qquad \text{(D-6)}$$

$$n = 59, R^2 = 0.90$$

Because it is thought that most of the accumulated organic contaminant is localized in fatty tissues and cellular components (e.g., membranes), *Chiou (1985)* has normalized the fish concentrations to the lipid content of the individual type of fish tested and observed:

$$\log \text{BCF}\left(\frac{\text{mol/g fish lipid}}{\text{mol/mL water}}\right) = 0.89 \log K_{ow} + 0.61 \tag{D-7}$$

$$n = 18, R^2 = 0.90$$

These correlations, and others available in the literature, are fundamentally the same but differ in how fish concentrations are normalized. Comparison of Eqs. D-6 and D-7 indicates that the lipid content of a fish is about 5% of its wet weight (the intercepts differ by about 1.3). *Note that such correlations are limited to cases in which metabolism of the contaminant is slow or nonexistent.*

Thus, to estimate the 1,2,4-TCB concentration in a trout exposed to 10 nM of this compound in water, first estimate the BCF using Eq. D-6:

$$\log \text{BCF} = 0.85 \log (10^4) - 0.7 = 2.7$$

or

$$\text{BCF} = 500 \text{ mL water/g wet fish}$$

Note that if the BCF had been calculated based on the lipid content of the fish (Eq. D-7), the BCF would have had an about ten times larger value. Hence, the concentration in the fish can be calculated by

$$\text{concentration in trout} = \text{BCF} \cdot C_w$$

$$= 500 \frac{\text{mL water}}{\text{g wet fish}} \cdot (10 \times 10^{-12}) \frac{\text{mol}}{\text{mL water}}$$

$$= 5 \times 10^{-9} \text{ mol/g wet fish}$$

or about 1 ppm

PROBLEMS

● *D-1 Extraction of Organic Pollutants from Water Samples*

For analyzing organic pollutants in water, the compounds are commonly preconcentrated either by adsorption, stripping, or extraction with an organic solvent. You have the job to determine 1,2,4-trichlorobenzene (1,2,4-TCB, see above) in a contaminated groundwater. You decide to extract a 1 L sample of this water with hexane and you wonder how much hexane you should use. Calculate the volume of hexane that you need at minimum, if you want to extract at least 95% of the total 1,2,4-TCB present in the water sample. How could you improve the extraction efficiency? If you wanted to extract 2-chlorophenol from the same water sample, would you also use hexane as a solvent? If not, what other solvent would you suggest and what precautions would you have to take?

OH

Cl

2-chlorophenol

K_{ow} (25°C) $\simeq$ 150
(other solvent water partition constants for 2-chlorophenol can be found in *Hansch and Leo, 1979*).

● *D-2 Some Additional K_{ow} Estimation Exercises Using Fragment Constants and Intramolecular Interaction Factors*

Estimate the K_{ow} values of the four compounds (I-IV) indicated below by using (a) only fragment constants and interaction factors and (b) by starting with the K_{ow} value of a structurally related compound that you choose from the *Appendix*.

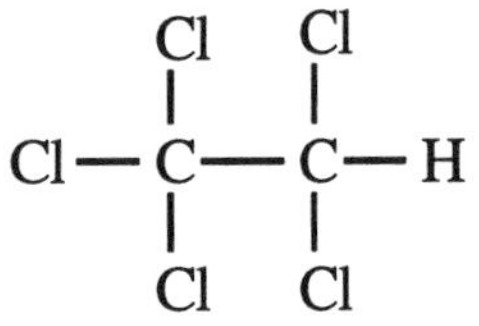

pentachloroethane
I

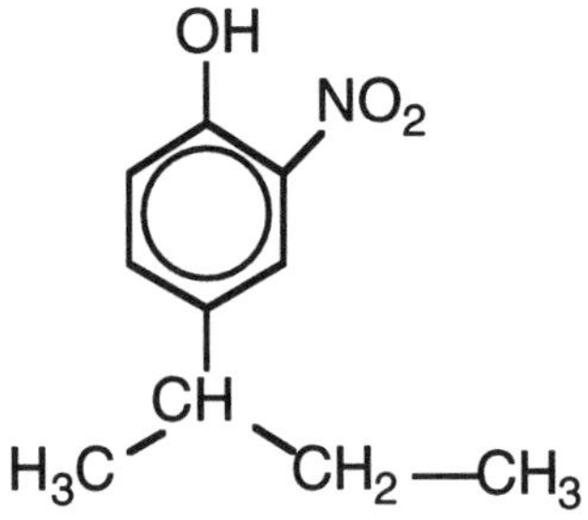

4-s-butyl-2-nitrophenol
II
(log K_{ow} (2-nitrophenol) = 1.89;
see *Schwarzenbach et al., 1988*)

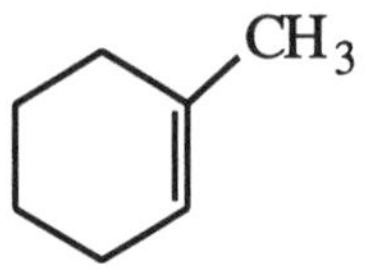

1-methyl-cyclohexene
III

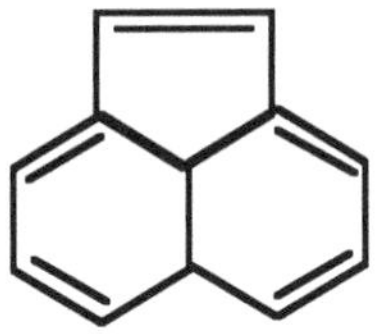

acenaphthylene
IV

• *D-3 Correlating Acute Toxicity With Octanol-Water Partitioning - How Well Does it Work?*

In a large number of studies in which fish or Daphnia were used as test organisms, it was found for various compound classes, that the acute toxicities of the compounds correlated quite well with their octanol-water partition constants (see review by *Hermens, 1989*). Using, for example, LC_{50} as a measure of acute toxicity (LC_{50} is the concentration causing 50% mortality after a certain time period), numerous relationships of the general form

$$\log \frac{1}{LC_{50}} = a \cdot \log K_{ow} + b$$

have been reported. Estimate the 14-day LC_{50} of 1,2,4-trichlorobenzene (1,2,4-TCB) and of 2,4-dinitrotoluene (2,4-DNT) for guppies using the data given below, and compare these values with the indicated experimental LC_{50}'s of the two compounds. Plot ($-\log LC_{50}$) versus log K_{ow} and discuss your findings. The LC_{50} (14 days) data for the methyl-

and chlorobenzenes are from *Könemann (1981),* the data for the nitroaromatic compounds are from *Deneer et al. (1987).*

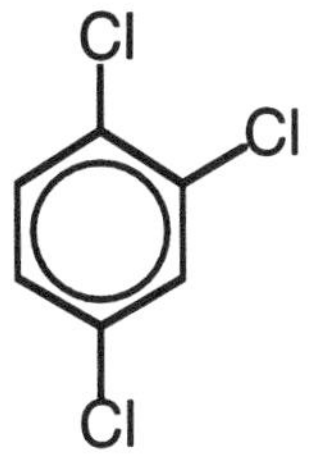

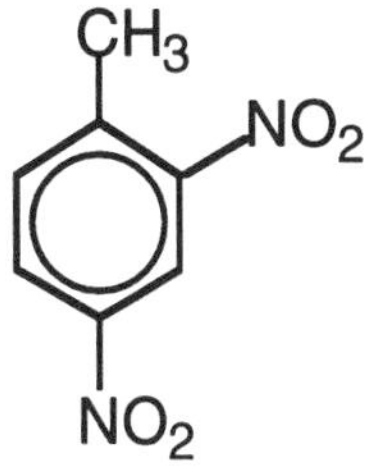

	1,2,4-TCB	2,4-DNT
log K_{ow}	4.00	1.98
LC_{50} (14 day, exp)	1.3 x 10^{-5} mol·L^{-1}	6.9 x 10^{-5} mol·L^{-1}

Compound	LC_{50} (14 days, guppies) (μmol·L^{-1})	log K_{ow}
Toluene	740	2.69
1,3-Dimethylbenzene	355	3.15
Chlorobenzene	170	2.92
1,2-Dichlorobenzene	40	3.38
1,4-Dichlorobenzene	27	3.38
1,2,3,4-Tetrachlorobenzene	3.7	4.55
1,2,3,5-Tetrachlorobenzene	3.7	4.65
Pentachlorobenzene	0.70	5.03
Nitrobenzene	500	1.83
2-Nitrotoluene	240	2.30
4-Nitrotoluene	270	2.39
2-Chloronitrobenzene	190	2.26
4-Chloronitrobenzene	38	2.35
3,5-Dichloronitrobenzene	30	3.13
1,2-Dinitrobenzene	7	1.58
1,4-Dinitrobenzene	2.3	1.47

Personal Notes

E. ORGANIC ACIDS AND BASES: ACIDITY CONSTANT AND PARTITIONING BEHAVIOR

ILLUSTRATIVE EXAMPLES

Estimating Acidity Constants of Aromatic Acids and Bases Using the Hammett Equation

Problem
Estimate the pK_a-values at 25°C of (a) 3,4,5-trichlorophenol (3,4,5-TCP), (b) pentachlorophenol (PCP), (c) 4-nitrophenol (4-NP), (d) 3,4-dimethylaniline (3,4-DMA, pK_a of conjugate acid), and (e) 2,4,5-trichlorophenoxy acetic acid (2,4,5-T).

Answer
Use the Hammett relationship *Eq. 8-23*

$$pK_a = pK_{aH} - \rho \sum_i \sigma_i \qquad (8\text{-}23)$$

to estimate the pK_a-values of compounds (a) - (e). Get the necessary σ_i, pK_{aH}, and ρ-values from *Tables 8.4* and *8.5*. The results are

(a)

3,4,5-TCP

pK_{aH} (phenol)	9.92
ρ	2.25
σ_{meta} (Cl)	0.37
σ_{para} (Cl)	0.23

$$pK_a = 9.92 - (2.25)\,[2\,(0.37) + 0.23] = 7.74$$

The reported experimental value is 7.73 *(Schellenberg et al., 1984)*.

(b)

PCP

pK_{aH} (phenol)	9.92
ρ	2.25
$\sigma_{ortho}^{phenols}$ (Cl)	0.68
σ_{meta} (Cl)	0.37
σ_{para} (Cl)	0.23

$$pK_a = 9.92 - (2.25)\,[2(0.68) + 2(0.37) + 0.23] = 4.68$$

The reported experimental values are 4.75 *(Schellenberg et al., 1984)*, and 4.83 *(Jafvert et al., 1990)*.

(c)

4-NP

pK_{aH} (phenol)	9.92
ρ	2.25
σ_{para}^{-} (NO_2)	1.25

$$pK_a = 9.92 - (2.25)(1.25) = 7.11$$

Note that because the nitro group is in resonance with the OH-group, the σ^-_{para} and not σ_{para}-value has to be used. The reported experimental values are 7.08, 7.15, and 7.18 (see *Schwarzenbach et al., 1988,* and refs. cited therein).

(d)

NH_2

CH_3

CH_3

3,4-DMA

pK_{aH} (C_6H_5–$NH_3^{\oplus}$)	4.63
ρ	2.89
σ_{meta} (CH_3)	– 0.07
σ_{para} (CH_3)	– 0.17

$$pK_a = 4.63 - (2.89)(-0.07 - 0.17) = 5.32$$

The reported experimental value is 5.28 *(Johnson and Westall, 1990).*

(e)

O-CH_2-COOH

Cl

Cl

Cl

2,4,5-T

pK_{aH} (2-CPAA)	3.05
ρ	0.30
σ_{meta} (Cl)	0.37
σ_{para} (Cl)	0.23

Since there are no σ_{ortho}-values available, use 2-chlorophenoxy acetic acid (2-CPAA, pK_a = 3.05, *CRC Handbook of Chemistry and Physics*) as the starting value.

$$pK_a = 3.05 - (0.30)(0.37 + 0.23) = 2.87$$

The reported experimental values are 2.80 and 2.83 *(Jafvert et al., 1990).*

Calculating the Speciation of Organic Acids and Bases in Natural Waters

Problem
Calculate the fraction of pentachlorophenol (PCP, $pK_a = 4.75$, see above) and of 3,4-dimethylaniline (3,4-DMA; $pK_a = 5.28$, see above) present at 25°C as neutral species in a rain drop (pH = 4.0) and in lake water (pH = 8.0).

Answer
Calculate the fraction in the acid form (α_a) using *Eq. 8-16*

$$\alpha_a = \frac{1}{1 + 10^{(pH-pKa)}} \qquad (8\text{-}16)$$

For PCP, the acid form is the neutral species (i.e., fraction = α_a), while for 3,4-DMA it is the base form (i.e., fraction = $(1 - \alpha_a)$). Insertion of the corresponding pK_a and pH-values into *Eq. 8-16* yields

$$\alpha_a\,(\text{PCP, pH4}) = 0.85 \qquad \alpha_a\,(\text{PCP, pH8}) = 0.00056$$

$$(1-\alpha_a)\,(\text{3,4-DMA, pH4}) = 0.049 \qquad (1-\alpha_a)\,(\text{3,4-DMA, pH8}) = 0.998$$

Calculating the Air-Water Equilibrium Partitioning of Organic Acids and Bases

Problem

Consider a rain drop (V_w = 0.1 mL) in a volume of air (V_a = 100 L) containing a certain amount of (a) 4-chloro-2-nitrophenol (4-Cl-2NP), and (b) aniline (AN). Calculate the fraction of total 4-Cl-2NP and AN, respectively, present in the rain drop at equilibrium at 25°C for pH=4 and pH=7. *Assumption: Only neutral species in the gas phase.*

Answer

Use *Eqs. 8-26* and *8-27* to calculate the air-water distribution ratios of 4-Cl-2NP and AN as a function of pH:

$$D_{aw}(HA,A^-) = \alpha_a \cdot K_H' \qquad (8\text{-}26)$$

for 4-Cl-2NP, and

$$D_{aw}(BH^+,B) = (1-\alpha_a)\, K_H' \qquad (8\text{-}27)$$

for AN. Calculate α_a using *Eq. 8-16.*

As derived earlier (see section on air-water partitioning), the fraction of the compound in the aqueous phase (i.e., in the rain drop) is given by

$$f_w = \frac{1}{1 + D_{aw}\dfrac{V_a}{V_w}}$$

The results of the calculations are

(a)

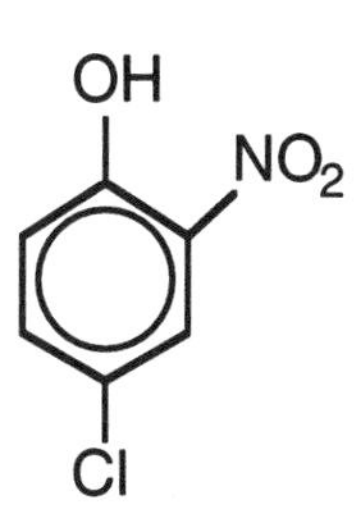

4-Cl-2NP

pK_a = 6.44

K_H' (25°C) = 5.2×10^{-4}

(data from *Schwarzenbach et al., 1988*)

$$f_w\,(\text{pH4}) = \frac{1}{1 + (5.2 \times 10^{-4})\,10^{6}} = 0.002 \;\; (0.2\,\%)$$

$$f_w\,(\text{pH7}) = \frac{1}{1 + (1.1 \times 10^{-4})\,10^{6}} = 0.009 \;\; (0.9\,\%)$$

(b) NH_2

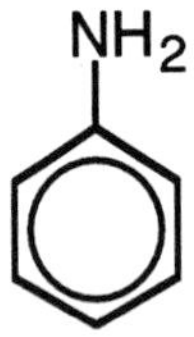

AN

pK_a = 4.63

$K'_H(25°C)$ = 1.4×10^{-4}

$$f_w\,(\text{pH4}) = \frac{1}{1 + (2.7 \times 10^{-5})\,10^{6}} = 0.036 \;\; (3.6\,\%)$$

$$f_w\,(\text{pH7}) = \frac{1}{1 + (1.4 \times 10^{-4})10^{6}} = 0.007 \;\; (0.7\,\%)$$

Calculating the Octanol-Water Distribution Ratio of Organic Acids and Bases

Problem

Calculate the octanol-water distribution ratio at 25°C of pentachlorophenol (PCP, see structure above, pK_a = 4.75, log K_{ow} = 5.24 *(Schellenberg et al., 1984)* and 3,4-dimethylaniline (3,4-DMA, see structure above, pK_a = 5.28, log K_{ow} = 1.84 (*Johnson and Westall, 1990*)) at pH=4 and at pH=7. *Assume that you can neglect the partitioning of the charged species into the octanol.* Note that for PCP at higher pH-values (pH > 8), and for 3,4-DMA at very low pH (pH < 2), the partitioning of the charged species cannot be neglected (for details see *Jafvert et al., 1990,* and *Johnson and Westall, 1990).*

Answer

Use *Eqs. 8-29* and *8-30* to calculate the octanol-water distribution ratios of PCP and 3,4-DMA as a function of pH:

$$D_{ow}\,(HA,A^-) \simeq \alpha_a \cdot K_{ow}\,(HA) \qquad (8\text{-}29)$$

for PCP, and

$$D_{ow}\,(BH^+,B) \simeq (1-\alpha_a) \cdot K_{ow}\,(B) \qquad (8\text{-}30)$$

for 3,4-DMA. Calculate α_a using *Eq. 8-16.*

The results are

D_{ow} (PCP, pH4) = 1.5×10^5 D_{ow}(PCP, pH7) = 9.7×10^2

D_{ow} (3,4-DMA, pH4) = 3.4 D_{ow} (3,4-DMA, pH7) = 68

Note that pH has a large effect on the octanol-water partitioning of organic acids and bases.

PROBLEMS

● *E-1 Speciation of Organic Acids and Bases in Natural Waters*

Represent graphically (as shown in *Fig. 8.2*) the speciation of (a) 4-methyl-2,5-dinitrophenol, (b) 3,4,5-trimethylaniline, and (c) o-phthalic acid as a function of pH (pH-range 2 to 10) at 25°C. Estimate, if necessary, the pK_a-values of the compounds.

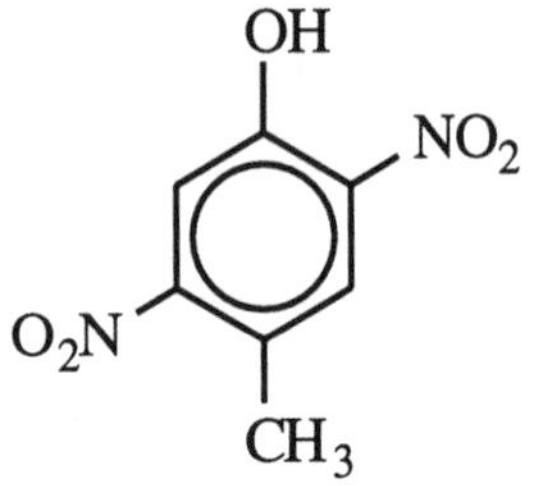

4-methyl-2,5-dinitrophenol

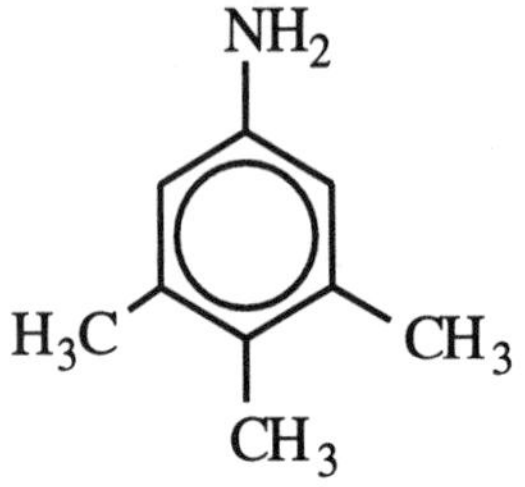

3,4,5-trimethylaniline

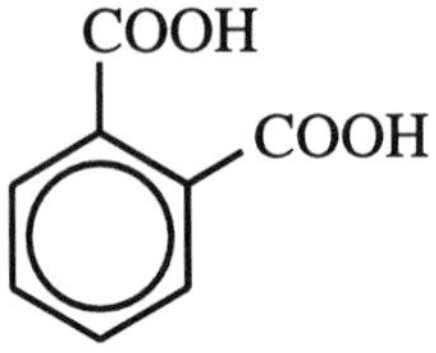

o-phthalic acid
$pK_{a,1} = 2.89$
$pK_{a,2} = 5.51$

● *E-2 Air-Water Distribution of Organic Acids and Bases in Fog*

Represent graphically the approximate fraction of (a) total 2,4-dinitrophenol and (b) total 4-chloroaniline present in fog water (air-water volume ratio $\simeq 10^6$) as a function of pH (pH-range 2 to 7) at 10°C. Assume that only the neutral species is present in the gas phase and that you may neglect the temperature dependence of K_a. Comment on all of your other assumptions.

2,4-dinitrophenol
K_H (20°C) $\simeq 3 \times 10^{-4}$ atm·L·mol^{-1}
(from *Schwarzenbach et al., 1988*)

4-chloroaniline
K_H = ?

● *E-3 A Strange Result?*

Estimate and compare the activity coefficients for nondissociated 2-nitrophenol (2-NP) and 4-nitrophenol (4-NP) at 20°C in water (γ_w) and octanol (γ_o) using the data (from *Schwarzenbach et al., 1988*) given below. Comment on all of your assumptions and discuss the result.

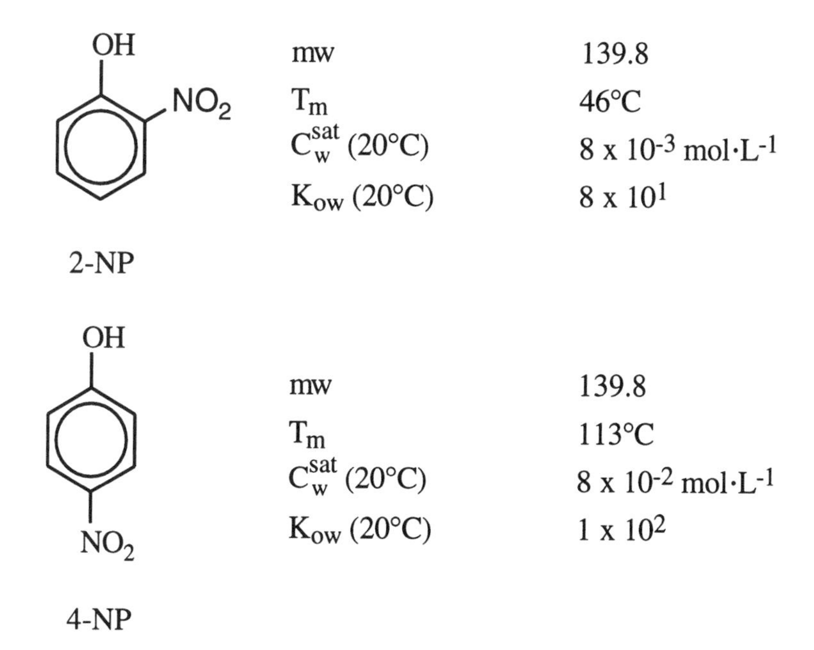

2-NP

mw	139.8
T_m	46°C
C_w^{sat} (20°C)	8×10^{-3} mol·L^{-1}
K_{ow} (20°C)	8×10^{1}

4-NP

mw	139.8
T_m	113°C
C_w^{sat} (20°C)	8×10^{-2} mol·L^{-1}
K_{ow} (20°C)	1×10^{2}

Personal Notes

F. PHYSICAL-CHEMICAL PROPERTIES AND EQUILIBRIUM PARTITIONING - ADDITIONAL PROBLEMS

● *F-1 How Much of These Compounds Can Be Dissolved in Water?*

The aqueous solubility of benzoic acid is reported to be 22 mM at 18°C and pH 2.0.

(a) What is the aqueous activity coefficient at saturation γ_w^{sat} of this compound at 18°C when pure liquid benzoic acid is chosen as the reference state? Would you expect that this activity coefficient is also valid for a dilute aqueous solution of benzoic acid?

(b) How much benzoic acid (in molar units) can be dissolved in water at pH 6.0 and 18°C.

(c) Estimate how much 4-bromo-benzoic acid could be dissolved in water at pH 6.0. Comment on your assumptions.

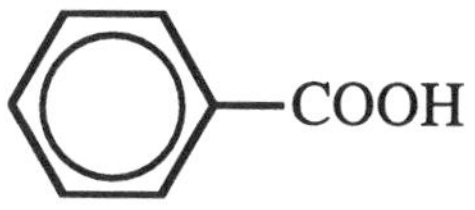

benzoic acid

mw	122.1 $g \cdot mol^{-1}$
T_m	122.1°C
density (15°C)	1.266 $g \cdot cm^{-3}$

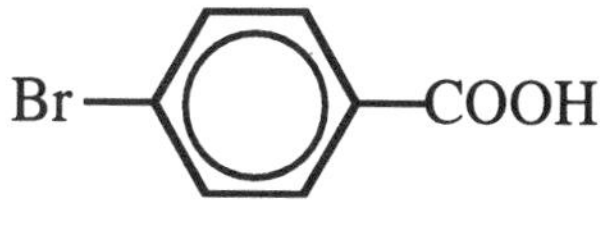

4-bromo-benzoic acid

mw	201.0 $g \cdot mol^{-1}$
T_m	254.5°C
density (20°C)	1.894 $g \cdot cm^{-3}$

● *F-2 Estimating Aqueous Solubility from K_H*

Using the structural unit contribution method of *Hine and Mookerjee (1975) (Table 6.4)*, estimate the aqueous solubility (in molar units) of hexachlorobenzene at 25°C. The only additional data available are T_m and T_b.

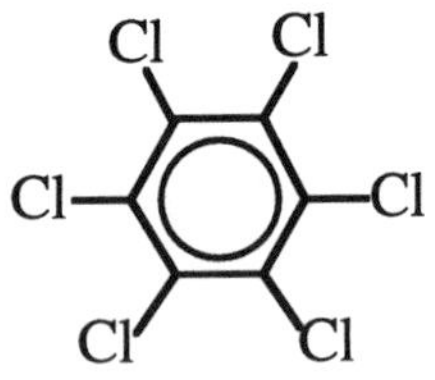

hexachlorobenzene

mw	284.8 $g \cdot mol^{-1}$
T_m	230.0°C
T_b	322.0°C

● *F-3 Estimating Solubilities in Organic Solvents and Organic Solvent-Water Mixtures*

The aqueous solubility of perylene at 25°C is 2 nM.

(a) Estimate the solubility (in molar units) at 25°C of perylene in benzene.

(b) What do you estimate the solubility of perylene to be in a 20% / 80% (v/v) mixture of methanol and water?

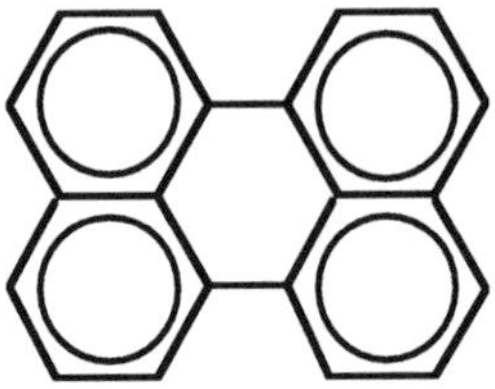

perylene

mw	252.3 $g \cdot mol^{-1}$
T_m	278°C

● *F-4 Evaluating Enthalpies, ΔH_{Henry}, of Air-Water Partitioning*

The Henry's law constants of some volatile chlorinated ethenes have been carefully studied as a function of temperature (*Gossett, 1987*), and the following relationships were found for the temperature range 10°C to 35°C:

tetrachloroethene	$\ln K_H =$	$12.45 - 4918$ K/T
1,1-dichloroethene	$\ln K_H =$	$8.85 - 3729$ K/T
cis - 1,2 -dichloroethene	$\ln K_H =$	$8.48 - 4192$ K/T
trans - 1,2-dichloroethene	$\ln K_H =$	$9.34 - 4182$ K/T

(Note that K_H is given in atm·m^3·mol^{-1}·K^{-1}.)

(a) What is the enthalpy (in kJ·mol^{-1}) of air-water partitioning of tetrachloroethene?

(b) Why is the enthalpy of air-water partitioning of 1,1-dichloroethene so different from the enthalpies of the 1,2-isomers?

● *F-5 Getting the "Right" K_H-Value for Benzyl Chloride*

In Chapter O, the rate of elimination by gas exchange of benzyl chloride (BC) in a river will be calculated. To this end, the Henry's law constant of BC must be known. In the literature (*Mackay and Shiu, 1981*), you can find only vapor pressure and water solubility data for BC (see below). Because BC hydrolyzes in water with a half-life of 15 hours at 25°C (see *Table 12.7*), you wonder whether you can trust the aqueous solubility data. Approximate the K_H-value of BC by vapor pressure and aqueous solubility, and compare it to the value obtained by applying the structural unit contributions given in *Table 6.4*. (Use the K_H-value of toluene that you can find in the *Appendix* as a starting value.) Which value do you trust more?

Hints and Help
Use other compound properties that are available or that can be estimated to perform simple plausibility tests on the experimental vapor pressure and aqueous solubility data of BC at 25°C.

CH_2Cl (benzene ring structure)

benzyl chloride
(BC)

mw	126.6
T_m	– 39°C
T_b	179.3°C
P^o (25°C)	1.7 x 10^{-3} atm
C_w^{sat} (20°C)	3.5 x 10^{-3} mol·L

● *F-6 Comparing K_{ow}-Values of Closely Related Compounds*

Which compound in each of the following closely related pairs exhibits the larger octanol-water partition constant? Estimate the ratio of these K_{ow} values and describe the molecular-scale reasoning for why one isomer is more hydophobic than the other.

(a) 1,1-dichloropentane versus 1,3-dichloropentane

(b) 2-methylpentane versus 2,2-dimethylbutane

(c) n-butylbenzene versus 1-ethyl-3,4-dimethybenzene

(d) 2-hydroxy-cyclohexanone versus 4-hydroxy-cyclohexanone

● *F-7 Equilibrium Partitioning Between Three Phases: The "Soup Bowl" Problem*

Consider a system exhibiting an aqueous phase (volume = V_w), an air phase (V_a), and an organic phase (V_o). The system also contains a total mass M_{tot} of a given organic pollutant. Derive general expressions for calculating the fraction of the total amount of organic pollutant in each phase (i.e., f_w, f_a, f_o; $f_w + f_a + f_o = 1$) at equilibrium for

(a) a neutral organic compound,
(b) an organic acid (HA), and
(c) an organic base (B).

For simplification, assume that ionic species are only present in the aqueous phase. Apply the results to the following problem:

A covered soup bowl contains 1 L of a very diluted cold soup (25°C, pH 5.5), 1 L of air, and a floating blob of fat of a volume of 1 mL. The soup bowl also contains a total of 1 mg of

(a) naphthalene,
(b) dinoseb (a herbicide), and
(c) 3,4-dimethylaniline.

Estimate the equilibrium concentration of the three compounds in each phase, and calculate the amount of each compound that you would take up if you would eat only the fat blob. Use the octanol-water partition constant as a substitute for the fat-water partition constant. How much more naphthalene would you eat with the fat blob if the soup were extremely salty (1M NaCl)? If you do not find all the data necessary to answer all these questions, try to estimate the missing data, preferably from data available for structurally related compounds.

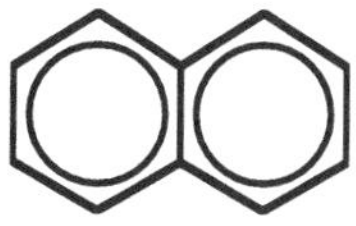

naphthalene 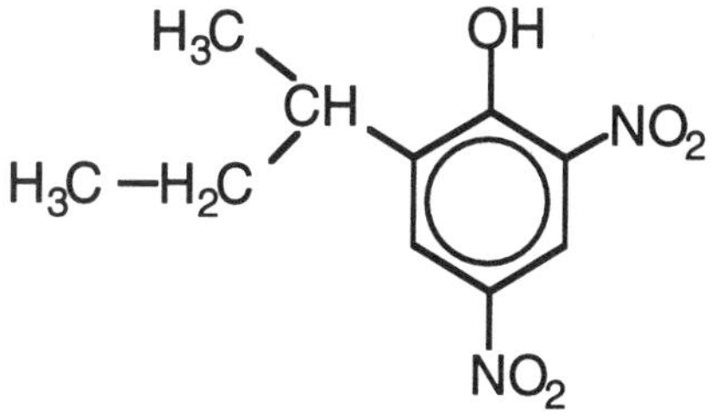 | dinoseb 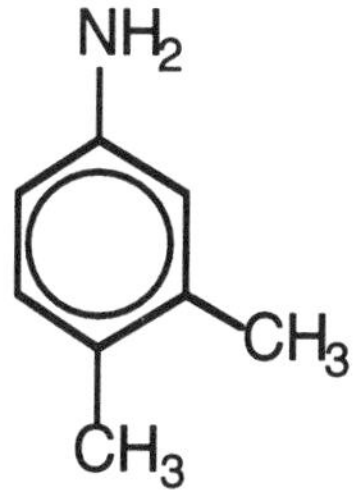 | 3,4 dimethylaniline

● *F-8 Estimating the Octanol-Air Partition Constant - A Key Parameter for Describing of Leaf-Air Exchange of Hydrophobic Organic Chemicals*

Based on measurements of a series of hydrophobic organic chemicals in Azalia leaves (*Patterson et al., 1991*) and Welsh Ray Grass (*Tolls and McLachlan, 1994*), it has been demonstrated that the octanol-air partition constant, K_{oa}, is a key parameter for a quantitative description of the equilibrium and the kinetics of leaf-air exchange of such compounds. You are involved in a study in which foliar uptake from the gas phase of a series of aromatic hydrocarbons is investigated. In this context you need to know the K_{oa} value of acenaphthene. Unfortunately, the only data you can find for this compound are its melting and boiling points (*CRC Handbook of Chemistry and Physics*). Estimate K_{oa} of acenaphthene using only these data and the structure of the compound.

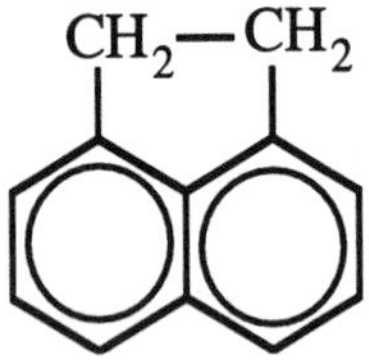

CH_2—CH_2

acenaphthene

mw	154.2
T_m	96.2°C
T_b	279°C

G. DIFFUSION

ILLUSTRATIVE EXAMPLES

Estimating Molecular Diffusion Coefficients in Air and Water

Consider the chemical dichlorodifluoromethane (also called freon-12 or CFC-12).

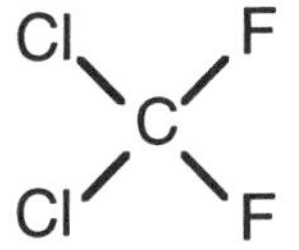

Cl, F, C, Cl, F

dichlorodifluoromethane
(freon-12 or CFC-12)

mw	120.9 g·mol^{-1}
T_m	–158°C
T_b	– 29.8°C
ρ_{liquid} (20°C)	1.328 g·cm^{-3}

In the literature we also find the following liquid densities of some related compounds:

Compound		Liquid density at 20°C ρ_{liquid} (g·cm^{-3})
Tetrachloromethane	CCl_4	1.594
Trichlorofluoromethane	CCl_3F	1.490
Dichlorofluoromethane	$CHCl_2F$	1.366
Chlorodifluoromethane	$CHClF_2$	1.213

Problem

Calculate the molar volume $\bar{V}$ of CFC-12 from (a) its liquid density and (b) with the element contribution method of *Fuller et al. (1966)* (*Table 9.3*).

Answer (a)

By definition molar volume $\bar{V}$, liquid density ρ_{liquid}, and molecular mass mw are related by (see also *Table 9.4*)

$$\bar{V} = \frac{mw}{\rho_{liquid}} = \frac{120.9 \text{ g·mol}^{-1}}{1.328 \text{ g·cm}^{-3}} = 91.0 \text{ cm}^3\text{·mol}^{-1}$$

Answer (b)

Inspection of *Table 9.4* shows that no contribution for F is available. However, by comparing the molar volumes of related compounds that contain different numbers of fluorine atoms, we get the following differences in molar volume, $\Delta\bar{V}$, if one Cl is substituted by one F:

Compound	mw	$\bar{V}$ ($cm^3·mol^{-1}$)	$\Delta\bar{V}$* ($cm^3·mol^{-1}$)
CCl_4	153.8	96.5	
CCl_3F	137.4	92.2	- 4.3
$CHCl_2F$	102.9	75.3	
$CHClF_2$	86.5	71.3	- 4.1

* Change of molar volume if one Cl is substituted by one F

These data indicate that the contribution of the fluorine atom to $\bar{V}$ is roughly 4.2 $cm^3·mol^{-1}$ less than the contribution of a chlorine atom. Since the latter is 19.5 $cm^3·mol^{-1}$ *(Table 9.3)*, the contribution of F to $\bar{V}$ is estimated to be about 15.3 $cm^3·mol^{-1}$. Thus from *Table 9.3*, completed by the value for F, it follows

$$\bar{V}(CCl_2F_2) = \bar{V}(C) + 2\bar{V}(Cl) + 2\bar{V}(F)$$

$$= (16.5 + 39.0 + 30.6) \text{ cm}^3\text{·mol}^{-1} = 86.1 \text{ cm}^3\text{·mol}^{-1}$$

Problem
Estimate the molecular diffusion coefficient in air, D_a, of CFC-12 at 25°C by the following methods:

(a) from "mean molecular velocity times mean free path" (Einstein-Smoluchowski relationship)

(b) from the molecular mass

(c) from the molar volume

(d) from the combined molecular mass and volume relationship of Fuller (*Eq. 9-22*)

(e) from the molecular diffusion coefficient of a related molecule, $D_a(CH_4) = 0.23\ cm^2{\cdot}s^{-1}$ (experimental value).

Answer (a)
According to the model of random walk (see *Fig. 9.2*), the diffusion coefficient D can be expressed as one half of the product of "jumping distance" and "jumping velocity" (*Eq. 9-8*). In the framework of the molecular theory of gases, D can be expressed as

$$D = \frac{1}{2}\,\overline{|u|} \cdot \lambda \qquad \text{(G-1)}$$

where $\overline{|u|}$ is the absolute value of the mean molecular velocity, and
λ is the mean free path, i.e., the average distance a molecule moves between collisions with other molecules.

The theory of gases also provides an expression for the mean absolute molecular velocity in the three-dimensional space (e.g., *Atkins, 1978*):

$$\overline{|u|} = \left(\frac{8RT}{\pi \cdot (mw)}\right)^{1/2} \qquad \text{(G-2)}$$

where R is the gas constant (8.315 $J{\cdot}mol^{-1}K^{-1}$), and
T is the absolute temperature (K)

Note the difference from Eq. 10-1, which expresses the mean velocity along a single fixed spatial direction.

In order to get the correct units, transform R into the cm/g/s-system. Since 1 J = 1 $kg \cdot m^2 \cdot s^{-2} = 10^7\ g \cdot cm^2 \cdot s^{-2}$, it follows that R = 8.315 x $10^7\ g \cdot cm^2 \cdot s^{-2} \cdot K^{-1} \cdot mol^{-1}$. Thus, for the CFC-12 molecules (mw = 120.9 $g \cdot mol^{-1}$) at 25°C Eq. G-2 yields

$$\overline{|u|} = \left(\frac{8 \times 8.315 \times 10^7 \times 298.2}{3.142 \times 120.9} cm^2 \cdot s^{-2} \right)^{1/2} = 2.29 \times 10^4\ cm \cdot s^{-1}$$

Calculate now the mean free path of the CFC-12 molecules, λ, by picturing the molecules as individual spheres each with a volume $\bar{v} = \bar{V}/N$ ($N = 6.02 \times 10^{23}\ mol^{-1}$) and radius $r = (3\bar{v}/4\pi)^{1/3}$. Using $\bar{V}$ calculated above from ρ_{liquid} yields $\bar{v}$(CFC-12) = 91.0 $cm^3 \cdot mol^{-1}/6.02 \times 10^{23}\ mol^{-1} = 1.51 \times 10^{-22}\ cm^3$; thus r(CFC-12) = 3.30×10^{-8} cm.

The most probable collision partners for a trace molecule in air (such as CFC-12) are molecular nitrogen (N_2) and oxygen (O_2). The CFC-12 molecule is hit whenever its center gets closer to the center of an air molecule than the critical distance r_{crit} = r(CFC-12) + r(air) (see Fig. G.1).

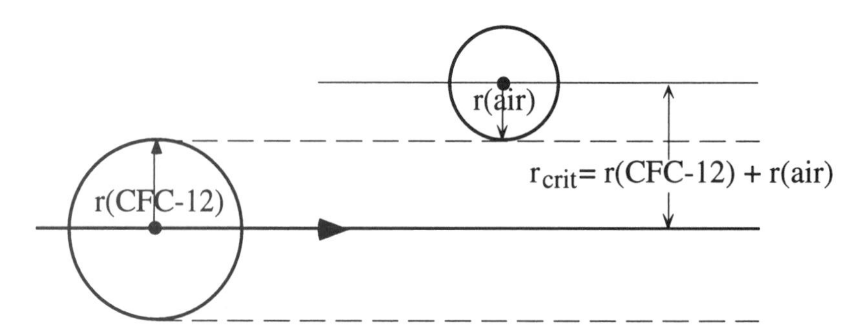

Fig. G.1 Scheme to calculate the mean free path of CFC-12 molecules in air.

The *Tables of Physical and Chemical Constants* (Longman, London, 1973) give the following collision cross-sections σ:

$$\sigma(N_2) = 0.43 \times 10^{-14}\ cm^2$$
$$\sigma(O_2) = 0.40 \times 10^{-14}\ cm^2$$

Since these values are not very different for N_2 and O_2, the following calculations are made with σ of N_2, the more abundant molecule in air.

Picturing the N_2-molecule as sphere, the molecular radius of N_2 is related to σ by: $\sigma(N_2) = \pi(r(N_2))^2$; thus

$$r(N_2) = (\sigma(N_2)/\pi)^{1/2} = \left(\frac{0.43 \text{ x } 10^{-14} \text{ cm}^2}{3.142}\right)^{1/2} = 3.70 \text{ x } 10^{-8} \text{ cm}$$

Imagine that the air molecules are fixed in space and that only one CFC-12 molecule moves around with mean velocity $|\bar{u}|$. As long as the center of the CFC-12 molecule keeps a distance larger than $r_{crit} = (3.30 + 3.70)$ x 10^{-8} cm = 7.00 x 10^{-8} cm, no collision between air and CFC-12 molecule occurs. Thus, if the CFC-12 molecule moves over the distance λ, it sweeps out a cylindrical volume $v^* = \pi r_{crit}^2 \lambda$.

How far can the CFC-12 molecule move until the corresponding volume v^* contains an air molecule?

The average volume per air molecule at normal conditions (p = 1 atm, T = 298.2 K) is

$$v_{mol} = \frac{RT}{N \cdot p} = \frac{8.206 \text{ x } 10^{-5} \text{ m}^3 \cdot \text{atm} \cdot \text{mol}^{-1} \text{ K}^{-1} \text{ x } 298.2 \text{ K}}{6.022 \text{ x } 10^{23} \text{ mol}^{-1} \text{ x } 1 \text{ atm}}$$

$$= 4.06 \text{ x } 10^{-26} \text{ m}^3 = 4.07 \text{ x } 10^{-20} \text{ cm}^3.$$

On the average a collision between CFC-12 and air molecule occurs when v^* and v_{mol} are equal. Solving for λ yields

$$\lambda = \frac{v_{mol}}{\pi r_{crit}^2} = \frac{4.07 \text{ x } 10^{-20} \text{ cm}^3}{3.141 \text{ x } (7.00 \text{ x } 10^{-8} \text{ cm})^2} = 2.64 \text{ x } 10^{-6} \text{ cm}$$

Note that in the above calculation it was assumed that the air molecules do not move. In fact, the mean free path should correctly be calculated by considering the ***relative*** *motion of the colliding molecules (e.g., Atkins, 1978). This reduces λ to*

$$\lambda^* = \left(\frac{m_{air}}{m + m_{air}}\right)^{1/2} \lambda = \left(\frac{29}{120.9 + 29}\right)^{1/2} \lambda = 0.44\,\lambda \sim 1.2 \text{ x } 10^{-6} \text{ cm}$$

According to Eq. G-1, the molecular diffusion coefficient of CFC-12 in air, D_a, is

$$D_a = \frac{1}{2} |\bar{u}| \lambda^* = \frac{1}{2} \times 2.28 \times 10^4 \text{ cm}\cdot\text{s}^{-1} \times 1.2 \times 10^{-6} \text{ cm} = 1.33 \times 10^{-2} \text{ cm}^2\cdot\text{s}^{-1}$$

A quick glance at *Fig. 9.6* shows that typical values of D_a found for other molecules are greater by a factor 5 to 8. One reason for this discrepancy may be the way in which λ was calculated. It was implicitly assumed that whenever the center of the CFC-12 molecules gets closer than r_{crit} to the center of an air molecule, the two molecules collide in such a way to completely "forget" their former momentum (velocity, direction). In fact, it is more reasonable to assume that slighter collisions would only partially change the course of the molecules. The mean free path would then be some weighted average of distances between collisions of different strength.

The diffusion volumes as calculated based on *Table 9.3* and on the average air volume $\bar{V}(air) = 20.1 \text{ cm}^3\cdot\text{mol}^{-1}$ (as it appears in *Eq. 9-22*), reflects this point of view. From $\bar{V}(air)$ one gets $r(air) = [3\bar{V}(air)/(4\pi N)]^{1/3} = 2.0 \times 10^{-8}$ cm, which is considerably smaller than the value derived from the collision cross-section σ. This means that not all collisions are "complete", or that for "complete" collisions the cross-section of the average air molecule is smaller than the σ-values given above.

The reduced value for r(air) yields $\lambda^*_{modif} = 2.0 \times 10^{-6}$ cm. Thus

$$D_{a,modified} = \frac{1}{2} \times 2.28 \times 10^4 \text{ cm}\cdot\text{s}^{-1} \times 2.0 \times 10^{-6} \text{ cm} = 2.28 \times 10^{-2} \text{ cm}^2\cdot\text{s}^{-1}$$

This is still considerably smaller than the values in *Fig. 9.6*. Compared to the following estimates of D_a, the mean-free path model, though interesting from a mechanistic point of view, seems to be less reliable with respect to the absolute size of the involved variables, i.e., the mean free path and the mean molecular velocity.

Answer (b)
Calculate D_a from the empirical relation to molar mass. From *Fig. 9.6.b* follows

$$D_a\,[\mathrm{cm^2 \cdot s^{-1}}] = \frac{1.55}{m^{0.65}} = \frac{1.55}{(120.9)^{0.65}} = 6.9 \times 10^{-2}\ \mathrm{cm^2 \cdot s^{-1}}$$

Answer (c)
Calculate D_a from the empirical relation to molar volume $\bar{V}$. From *Fig. 9.6a* follows

$$D_a\,[\mathrm{cm^2 \cdot s^{-1}}] = \frac{2.35}{(\bar{V})^{0.73}} = \frac{2.35}{(91.0)^{0.73}} = 8.7 \times 10^{-2}\ \mathrm{cm^2 \cdot s^{-1}}$$

Answer (d)
Fuller's semi-empirical relationship (*Eq. 9-22*) is

$$D_a\,[\mathrm{cm^2 \cdot s^{-1}}] = 10^{-3}\,\frac{T^{1.75}\,[(1/m_{air}) + (1/mw)]^{1/2}}{P[\bar{V}_{air}^{1/3} + \bar{V}^{1/3}]^2}$$

where

T = 298.2 K is the absolute temperature

m_{air} = 29 $\mathrm{g \cdot mol^{-1}}$ is the average molecular mass of air

mw = 120.9 $\mathrm{g \cdot mol^{-1}}$ is the molecular mass of CFC-12

P = 1 atm is the gas-phase pressure

$\bar{V}_{air}$ = 20.1 $\mathrm{cm^3 \cdot mol^{-1}}$ is the average molar volume of the molecules in air

$\bar{V}$ = 91.0 $\mathrm{cm^3 \cdot mol^{-1}}$ is the molar volume of CFC-12

Inserting these numbers into Fuller's equation yields

$$D_a = 8.5 \times 10^{-2}\ \mathrm{cm^2 \cdot s^{-1}} \qquad \textit{Fuller's Eq. 9-22}$$

Answer (e)
Take *Eq. 9-25* to calculate the diffusion coefficient of CFC-12 from the known value of methane:

$$D_a(\text{CFC-12}) = D_a(\text{CH}_4)\left(\frac{\text{mw}(\text{CH}_4)}{\text{mw}(\text{CFC-12})}\right)^{1/2}$$

$$= 0.23\ \text{cm}^2\cdot\text{s}^{-1} \times \left(\frac{16}{120.9}\right)^{1/2} = 8.4 \times 10^{-2}\ \text{cm}^2\cdot\text{s}^{-1}$$

The following table summarizes these results:

Diffusion coefficient of CFC-12 in air at 25°C, D_a

Method		D_a ($cm^2 \cdot s^{-1}$)
a)	Einstein-Smoluchowski	(1.4×10^{-2})
a*)	E-S, reduced σ(air)	(2.3×10^{-2})
b)	From molecular mass	6.9×10^{-2}
c)	From molar volume	8.7×10^{-2}
d)	From Fuller's combined expression	8.5×10^{-2}
e)	From diffusivity of CH_4	8.4×10^{-2}

Note that the Einstein-Smoluchowski model, even its corrected version, does not give a reasonable value. It is remarkable that the results from method c) through e) agree within a few percent.

Problem
Estimate the molecular diffusion coefficient of CFC-12 in water at 25°C, D_w, by the following methods:

(a) from the molecular mass
(b) from the molar volume
(c) from the diffusion coefficient of methane (CH_4) in water
(d) from Hayduk and Laudie's semiempirical expression

Use the following additional information:

Viscosity of water at 25°C, $\mu = 0.89$ centipoise ($10^{-2}\ g \cdot cm^{-1} \cdot s^{-1}$)
Molecular diffusion coefficient of CH_4 in water at 25°C, $D_w(CH_4) = 3.0 \times 10^{-5}\ cm^2 \cdot s^{-1}$

Answer (a)
The regression line shown in *Fig. 9.7b* yields for CFC-12:

$$D_w\ [cm^2 \cdot s^{-1}] = \frac{2.7 \times 10^{-4}}{(mw)^{0.71}} = \frac{2.7 \times 10^{-4}}{120.9^{\,0.71}} = 9.0 \times 10^{-6}\ cm^2 \cdot s^{-1}$$

Answer (b)
From the regression line in *Fig. 9.7a* and $\bar{V}$(CFC-12) = 91.0 $cm^3 \cdot mol^{-1}$ follows

$$D_w\ [cm^2 \cdot s^{-1}] = \frac{2.3 \times 10}{\bar{V}} = \frac{2.3 \times 10^{-4}}{91.0^{0.71}} = 9.3 \times 10^{-6}\ cm^2 \cdot s^{-1}$$

Answer (c)
From *Eq. 9-30* follows

$$D_w(\text{CFC-12}) = D_w(CH_4) \left(\frac{mw(CH_4)}{mw(\text{CFC-12})} \right)^{1/2}$$

$$= 3.0 \times 10^{-5}\ cm^2 \cdot s^{-1} \times \left(\frac{16}{120.9} \right)^{1/2} = 11 \times 10^{-6}\ cm^2 \cdot s^{-1}$$

Answer (d)

The semi-empirical relationship by *Hayduk and Laudie (1974)*, a slightly modified version of an expression given earlier by *Othmer and Thakar (1953)*, is (see *Eq. 9-26*)

$$D_w = \frac{13.26 \times 10^{-5}}{\mu^{1.14} \cdot \bar{V}^{0.589}} \quad (cm^2 \cdot s^{-1})$$

where μ, the viscosity of water in centipoise , and $\bar{V}$, the molar volume of CFC-12 were given above.

Inserting these values yields

$$D_w = \frac{13.26 \times 10^{-5}}{(0.89)^{1.14} \times (91.0)^{0.589}} \, cm^2 \cdot s^{-1} = 11 \times 10^{-6} \, cm^2 \cdot s^{-1}$$

In summary:

Diffusion coefficient of CFC-12 in water at 25°C, D_w,

	Method	D_w ($cm^2 \cdot s^{-1}$)
a)	From molar mass	9.0×10^{-6}
b)	From molar volume	9.3×10^{-6}
c)	From D_w of CH_4	11×10^{-6}
d)	From Hayduk and Laudie's relationship	11×10^{-6}

As found for D_a, the agreement between the different estimation methods is remarkably good.

Calculating Mass Fluxes due to Diffusion and Advection

Problem

In a small lake (surface area $A_o = 2 \times 10^4$ m^2, depth 10 m) the following vertical profiles of water temperature and dichloro-difluoromethane (CFC-12) concentration were measured:

Depth (m)	Temperature (°C)	CFC-12 (10^{-12} mol·L^{-1})
0	15.8	9.5
1	15.8	10
2	15.7	10
3	15.3	8.0
4	15.0	6.1
6	14.3	2.0
8	14.1	1.2
10	14.0	0.4

At the bottom of the lake, a subsurface spring adds water to the lake at a rate of Q_{sp} = 10 L·s^{-1}. The only outlet of the lake is at the surface. Calculate the size and direction of the vertical flux of CFC-12 per unit time and area through the thermocline of the lake located between 2 m and 6 m depth. The thermocline of a lake is the depth zone in which the strongest drop in the water temperature from the warm surface to the colder deep water occurs (Fig. G.2).

Note: The coefficient of thermal expansion of water, α, is defined as

$$\alpha = -\frac{1}{\rho}\frac{d\rho}{dT} \quad [K^{-1}] \qquad \text{(G-3)}$$

where

g = 9.81 m·s^{-2} is the acceleration of gravity,
ρ is the density of water, and
T is the water temperature.

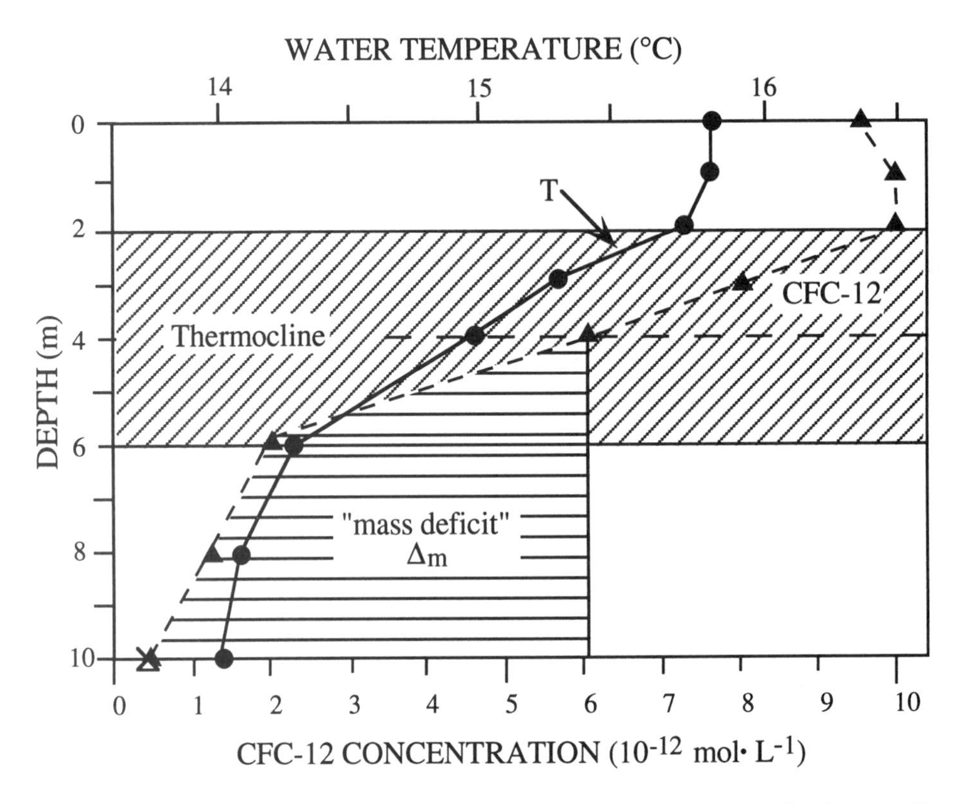

Figure G.2 Vertical distribution of temperature and CFC-12 concentration in a small lake

The *CRC Handbook of Chemistry and Physics* gives the following values for α as a function of T:

T (°C)	5	10	15	20
α (10^{-4} K^{-1})	0.16	0.88	1.51	2.07

Answer

The following transport processes have to be considered:
(a) molecular diffusion, (b) turbulent diffusion, and (c) advection.
The fluxes shall be calculated at depth z = 4 m, i.e., at the center of the thermocline.

(a) Fick's first law (*Eq. 9-2*) yields

$$F_{z,mol}(z = 4m) = - D_w \left.\frac{dC}{dz}\right|_{z=4m} \qquad \text{(G-4)}$$

where $F_{z,mol}(z = 4m)$ is the vertical flux at 4 m depth due to molecular diffusion, $D_w = 10 \times 10^{-6}\ cm^2{\cdot}s^{-1} = 10 \times 10^{-10}\ m^2{\cdot}s^{-1}$ is the molecular diffusion coefficient of CFC-12 in water (see preceding illustrative example), and C is the concentration of CFC-12 in water.

One way to estimate the vertical gradient of the CFC-12 concentration at 4 m depth, $\left.\frac{dC}{dz}\right|_{z=4m}$, consists of calculating the average slopes between 3 m and 4 m and between 4 m and 6 m, respectively, and then taking the mean of the two slopes:

Slope between 3 m and 4 m:

$$\left.\frac{dC}{dz}\right|_{3/4} = \frac{C(4m) - C(3m)}{4\ m - 3\ m} = \frac{6.1 - 8.0}{1}\ 10^{-12}\ mol{\cdot}L^{-1}{\cdot}m^{-1}$$
$$= -\ 1.9 \times 10^{-12}\ mol{\cdot}L^{-1}\ m^{-1}$$

Slope between 4 m and 6 m:

$$\left.\frac{dC}{dz}\right|_{4/6} = \frac{C(6m) - C(4m)}{6\ m - 4\ m} = \frac{2.0 - 6.1}{2}\ 10^{-12}\ mol{\cdot}L^{-1}{\cdot}m^{-1}$$
$$= -\ 2.05 \times 10^{-12}\ mol{\cdot}L^{-1}\ m^{-1}$$

Mean slope at 4 m

$$\left.\frac{dC}{dz}\right|_{4m} \cong -\ 2.0 \times 10^{-12}\ mol{\cdot}L^{-1}\ m^{-1}$$

Note that evaluation of Eq. G-4 requires that all spatial units be expressed in meters: $1L = 10^{-3}\ m^3$, $1\ cm^2 = 10^{-4}\ m^2$. Thus, $D_w = 10 \times 10^{-6}\ cm^2{\cdot}s^{-1} = 10 \times 10^{-10}\ m^2{\cdot}s^{-1}$, and $\left.\frac{dC}{dz}\right|_{4m} = -\ 2.0{\times}10^{-12}\ mol{\cdot}L^{-1}\ m^{-1} = -\ 2.0{\times}10^{-9}\ mol{\cdot}m^{-4}$. Insertion into Eq. G-4 yields

$$F_{z,mol}(z = 4\text{ m}) = (-10\text{x}10^{-10})\text{ m}^2\cdot\text{s}^{-1})\,(-2.0\text{x}10^{-9}\text{ mol}\cdot\text{m}^{-4})$$
$$= 2.0\text{x}10^{-18}\text{ mol}\cdot\text{m}^{-2}\cdot\text{s}^{-1}$$

Since the numerical value of $F_{z,mol}$ has a positive sign, the flux is directed along the positive z-axis, which, according to Fig. G.2, is downwards: *Molecular diffusion yields a downward flux of CFC-12 in the lake.*

(b) According to *Eq. 9-44*, the flux by turbulent diffusion, $F_{z,turb}$, is described by the same kind of expression as Eq. G-2, except that the molecular diffusion coefficient D_w is replaced by the vertical eddy diffusion coefficient, E_z. In case E_z has not been measured, it can be estimated from the stability frequency N^2 (*Fig. 9-14*), which, in turn, is calculated from the vertical temperature profile given in Fig. G.2 (see also *Eq. 9-49*):

$$N^2 = \frac{g}{\rho}\frac{d\rho}{dz} = \frac{g}{\rho}\frac{d\rho}{dT}\cdot\frac{dT}{dz}$$

The variation of density ρ with temperature T, $d\rho/dT$, can be expressed by the coefficient of thermal expansion, α, defined in Eq. G-3:

$$\frac{d\rho}{dT} = -\alpha\cdot\rho$$

Thus,

$$N^2 = \frac{g}{\rho}(-\alpha\cdot\rho)\frac{dT}{dz} = -g\cdot\alpha\frac{dT}{dz} \qquad \text{(G-5)}$$

The vertical temperature gradient at z = 4 m is evaluated according to the method employed for the concentration gradient of CFC-12:

$$\left.\frac{dT}{dz}\right|_{z=4m} = \frac{1}{2}\left(\frac{T(4) - T(3)}{1\text{ m}} + \frac{T(6) - T(4)}{2\text{ m}}\right)$$

$$= \frac{1}{2}(-0.3 - 0.35)\ K\cdot m^{-1} = -\ 0.325\ K\cdot m^{-1}$$

With $\alpha(15°C) = 1.51 \times 10^{-4}\ K^{-1}$ and Eq. G-5 it follows that

$$N^2 = -\ 9.81\ m\cdot s^{-2} \times 1.51 \times 10^{-4}\ K^{-1} \times (-\ 0.325\ K\cdot m^{-1}) = 4.8 \times 10^{-4}\ s^{-2}$$

Note that for the calculation of N^2, it is tacitly assumed that water temperature is the only factor which contributes to $d\rho/dz$. In fact, vertical salinity gradients could also contribute to the vertical density gradient.

Visual inspection of the right-hand plot of *Fig. 9.14* (the left-hand plot describes the extreme situation during a storm) for $N^2 = 5\text{x}10^{-4}s^{-2}$ yields roughly $E_z = 1.5 \times 10^{-2}\ cm^2\cdot s^{-1} = 1.5 \times 10^{-6}\ m^2\cdot s^{-1}$. This value may be too large, since Urnersee, for which *Fig. 9.14* has been drawn, is a much larger lake than the one considered in this example. Large lakes are generally more strongly exposed to wind and are thus more turbulent.

Nevertheless, this value will be used to calculate $F_{z,turb}$ from Eq. G-4 by replacing D_w by E_z. Note that $\frac{dC}{dz}$ is the same as for molecular diffusion. Thus,

$$F_{z,turb}(z{=}4\ m) = (-\ 1.5\text{x}10^{-6}\ m^2\cdot s^{-1})\text{x}(-2.0\text{x}10^{-9}\ mol\cdot m^{-4})$$
$$= 3\text{x}10^{-15}\ mol\cdot m^{-2}\cdot s^{-1}$$

Not surprisingly, $F_{z,turb}$ is much larger than $F_{z,mol}$. Since both fluxes "feel" the same concentration gradient, but E_z is much larger than D_w, molecular diffusion can usually be disregarded unless all currents in the water are completely suppressed (e.g., as in the pore space of the sediments).

(c) The subsurface input of water causes the water to flow upwards with an average velocity $v = -\ Q_{sp}/A_o$. The minus sign indicates that the flow occurs against the z-axis. At this point the fact that the cross-sectional area of a lake usually decreases with depth is ignored. Instead, the picture of a lake with rectangular walls is adopted in which the cross-sectional area of the lake, A, is constant from top to bottom. Thus,

$$v = -\frac{0.01\ m^3 \cdot s^{-1}}{2x10^4\ m^2} = -0.5\ x\ 10^{-6}\ m \cdot s^{-1}$$

The flux due to advection is calculated from *Eq. 9-32* :

$$F_{z,ad}(z = 4\ m) = C(4\ m) \cdot v \simeq 6.1\ x\ 10^{-9}\ mol \cdot m^{-3}\ x\ (-0.5\ x\ 10^{-6}\ m \cdot s^{-1})$$
$$\simeq -3.0\ x\ 10^{-15}\ mol \cdot m^{-2} \cdot s^{-1}$$

The two major fluxes, $F_{z,turb}$ and $F_{z,ad}$, act in opposite directions. Since they are roughly of equal absolute size, the total flux in this particular case is practically zero!

$$F_{z,tot} = F_{z,mol} + F_{z,turb} + F_{z,ad} \simeq F_{z,turb} + F_{z,ad} \sim 0$$

Note that this particular situation could be interpreted as a steady state in which the vertical profile of CFC-12 in the lake is controlled by two processes: (1) the input of water at the bottom which contains little CFC-12 (concentration less than $2x10^{-12}$ mol·L^{-1}), and (2) the addition of CFC-12 at or close to the surface which keeps the concentration at about $10x10^{-12}$ mol·L^{-1}. The CFC-12 profile then reflects the opposing influences of the downward transport by (mainly turbulent) diffusion and the upward transport by the rising water.

Problem

As hypothesized above, repeated measurements of the CFC-12 profile in the small lake indeed show that the CFC-12 concentration remains constant below 4 m depth. What can be deduced from this fact with respect to the size of the vertical turbulent diffusion coefficient, E_z, at 6 m and 8 m depth?

Answer

According to *Eq. 9-13* and the scheme shown in *Fig. 9.4*, the temporal change of the concentration at depth z is given by

$$\left.\frac{\partial C}{\partial t}\right|_{z=\text{const.}} = -\left.\frac{\partial F_z}{\partial z}\right|_{t=\text{const.}} \qquad \textit{(9-13)}$$

provided that the chemical has no *in situ* sources or sinks. F_z stands for $F_{z,tot}$ calculated in the preceding problem. Note that *Eq. 9-13* is only valid if the flux gradients along the other spatial axes, $\frac{\partial F_x}{\partial x}$ and $\frac{\partial F_y}{\partial y}$, are zero (see *Eq. 9-16a*), an assumption which is reasonable for a small lake. Since steady state means $\left.\frac{\partial C}{\partial t}\right|_{z=\text{const.}} = 0$, it follows that $\frac{\partial F_z}{\partial z}$ is zero; that is, F_z is constant at all depths below 4 m. In fact, since at 4 m $F_{z,tot}$ is zero, the total vertical flux of CFC-12 must also be zero at other depths. Thus,

$$F_{z,tot} = F_{z,turb} + F_{z,ad} = -E_z \left.\frac{dC}{dz}\right|_z + vC = 0$$

Solving for E_z yields

$$E_z = \frac{vC}{\left.\frac{dC}{dz}\right|_z} \qquad \text{(G-6)}$$

The so-called upwelling velocity $v = -0.5 \times 10^{-6}\ \text{m}\cdot\text{s}^{-1}$ is constant with depth. The concentration gradient $\left.\frac{dC}{dz}\right|_z$ is calculated as explained above. The following table summarizes all terms of Eq. G-6:

Depth z	v (10^{-6} m·s^{-1})	C (10^{-9} mol·m^{-3})	dC/dz (10^{-9} mol·m^{-4})	E_z (m^2·s^{-1})
6 m	– 0.5	2.0	– 1.23	0.82×10^{-6}
8 m	– 0.5	1.2	– 0.4	2.25×10^{-6}

Note that this particular vertical variation of E_z does not correlate with the stability frequency N as shown in Fig. 9.14.

Problem
Consider a situation in which the subsurface input of water suddenly stops. *Estimate* the approximate time needed for the CFC-12 to spread throughout the whole water column.

Answer
Do not try to derive the exact solution of the problem. You would have to solve the time-dependent partial differential *Equation 9-14* for which *Eq. 9-19* is one specific example.

Instead, try two approximative approaches. Note that once the subsurface water input ceases, the only remaining flux of importance is the flux by turbulent diffusion, $F_{z,turb}$. The first method is to compare this flux with the "mass deficit" Δm below 4 m as shown in Fig. G.2. Roughly, this deficit per unit area can be calculated from the mean concentration below 4 m, $\bar{C} = 2.1 \times 10^{-9}$ mol·m^{-3}. This value is subtracted from the concentration at 4 m, C(4m) = 6.1×10^{-9} mol·m^{-3}, and multiplied by the height of the water column from the bottom to 4 m. Thus, $\Delta m = 4.0 \times 10^{-9}$ mol·m^{-3} x 6 m = 24×10^{-9} mol·m^{-2}. Thus, the time to fill this deficit, t_1, is at least of the order

$$t_1 = \frac{\Delta m}{F_{z,turb}} = \frac{24.0 \times 10^{-9} \text{ mol} \cdot \text{m}^{-2}}{3 \times 10^{-15} \text{ mol} \cdot \text{m}^{-2}\text{s}^{-1}} = 8.0 \times 10^{6} \text{ s} \sim 3 \text{ months}$$

Note that t_1 is a lower limit. When the concentration in the deep water layer increases, the concentration gradient (and thus the diffusive flux) at 4 m depth decreases. In fact, since the concentration gradients approach zero when C becomes nearly constant with depth, strictly speaking the time needed to reach complete homogeneity is infinite. Therefore, we are interested in simple estimates rather than in exact (but useless) solutions!

As a second method, the characteristic transport time of diffusion can be used (*Eq.9-31*) with D replaced by $E_z = 1.5 \times 10^{-6}$ m^2·s^{-1} and L = 6 m. Thus,

$$t_2 = \frac{L^2}{2E_z} = \frac{(6\ m)^2}{2 \times 1.5 \times 10^{-6}\ m^2 \cdot s^{-1}} = 12 \times 10^6\ s \simeq 4.6\ months$$

Thus, both methods give results of similar magnitude.

Calculating Turbulent Diffusion Coefficients from Field Data

Problem
In a lake (maximum depth of 20 m) two vertical profiles of tetrachloroethene (PER) were measured at a time interval of one month (Table G.1). Calculate the vertical turbulent diffusivity E_z at 8, 12, and 16 m depth.

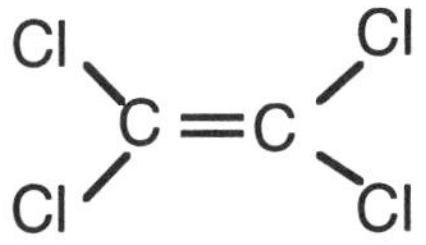

Cl Cl
C=C
Cl Cl

tetrachloroethene
(PER)

Note: This is a numerical routine problem that today, in the age of personal computers, is hardly solved by hand anymore. You can put a lot of energy and time into questions such as how to best calculate volumes and mean concentrations in individual layers of the lake. This task shall be left to the specialist. The purpose of the exercise here is to guide the reader in the simplest manner through all of the calculatorial steps and thus provide him or her with a feeling for how the different quantities enter the computations. These calculations also reveal where precision is important and where it is of less concern.

Table G.1 Cross sectional area and vertical profiles of tetrachloroethene (PER) measured in a lake

Depth z (m)	Area A (10^6 m^2)	Concentration of PER ($\mu mol \cdot m^{-3}$) Day 0	Concentration of PER ($\mu mol \cdot m^{-3}$) Day 30
0	8.0	1.0	0.5
2	7.2	5.0	2.0
4	6.4	9.5	3.5
6	5.6	5.5	5.5
8	4.8	3.9	4.2
10	3.6	3.1	3.7
12	3.2	2.6	3.4
14	2.4	2.3	3.3
16	1.6	2.1	3.25
18	0.8	2.0	3.2
20	0	1.9	3.1

Answer

PER is more or less conservative in the water column. It does not adsorb significantly to particles and is not substantially decomposed, neither by photolysis nor by biodegradation (see *Chapter 15.2*). The only two important removal processes are flushing (transport through the outlet of the lake) and air-water exchange. Both processes occur at the surface of the lake.

In order to calculate the coefficients of vertical eddy diffusion, E_z, from PER, the same scheme can be used as derived for temperature (*Fig. 9.12*). The lake is divided into layers, each 2 m thick, to calculate the total mass of PER from the lake bottom to the depths z = 16, 12, and 8 m. The volume of layer i, V_i, is taken as the product of the mean cross-section (the cross-section in the middle of the layer) and the layer thickness Δz = 2 m. The mean concentration, $\bar{C}_i$, is calculated by taking the mean of the values measured at the depths of the layer interfaces. The following scheme evolves:

Depth (m)	V_i ($10^6 m^3$)	$\bar{C}_i(t{=}0)$ ($\mu mol \cdot m^{-3}$)	$\bar{C}_i(t{=}30d)$ ($\mu mol \cdot m^{-3}$)	$\Delta\bar{C}_i$ ($\mu mol \cdot m^{-3}$)	$\Delta M_i = V_i \Delta\bar{C}_i$ (mol)	Δ_i (mol)
8-10	8.4	3.50	3.95	0.45	3.78	21.66
10-12	6.8	2.85	3.55	0.70	4.76	17.88
12-14	5.6	2.45	3.35	0.90	5.04	13.12
14-16	4.0	2.20	3.275	1.075	4.30	8.08
16-18	2.4	2.05	3.225	1.175	2.82	3.78
18-20	0.8	1.95	3.15	1.20	0.96	0.96

$\Delta\bar{C}_i$ is the concentration change in layer i from day 0 to day 30, ΔM_i is the corresponding change of total mass in this layer, and Δ_i is the cumulative sum of ΔM_i (computed from the lake bottom). Note that in *Eq. 9-52* this sum is defined as change per unit time, $\Delta = \Delta_i / \Delta t$.

In order to apply *Eq. 9-54*, the vertical concentration gradients of PER at the three depths are needed. The easiest (though not the optimal) way is to calculate them using an "averaging" scheme in the following way:

$$\left.\frac{\partial C}{\partial z}\right|_{z=8} = \frac{1}{2}\left(\frac{C(8) - C(6)}{2\text{ m}} + \frac{C(10) - C(8)}{2\text{ m}}\right) = \frac{C(10) - C(6)}{4\text{ m}}$$

and correspondingly for the other depths. This calculation is made for both PER profiles and the mean of both values taken:

Depth (m)	dC/dz ($\mu mol \cdot m^{-4}$) t = 0	t = 30 d	mean
8	–0.60	–0.45	–0.525
12	–0.20	–0.10	–0.15
16	–0.075	–0.025	–0.05

The final step follows from *Eq. 9-54* with the only difference that Δ_i has to be divided by the time interval $\Delta t = 30$ d. Thus,

$$E_z(z) = -\frac{\Delta_i}{A(z)\left.\frac{\partial C}{\partial z}\right|_z \Delta t}\quad , \qquad \Delta t = 30\text{ d}$$

The following table summarizes the relevant data and the final result. (Use $1\ m^2 \cdot d^{-1} = 8.64\ cm^2 \cdot s^{-1}$)

Depth (m)	Δ_i (mol)	A ($10^6 m^2$)	dC/dz ($\mu mol \cdot m^{-4}$)	E_z ($m^2 \cdot d^{-1}$)	E_z ($cm^2 \cdot s^{-1}$)
8	21.66	4.8	-0.525	0.29	0.033
12	13.12	3.2	-0.15	0.91	0.11
16	3.78	1.6	-0.05	1.58	0.18

Note that the value at z = 16 m is very sensitive to analytical errors and uncertainties of the lake geometry since its calculation involves division by a rather small vertical concentration gradient.

PROBLEMS

● *G-1 Assessing the Temperature and Pressure Dependence of the Molecular Diffusion Coefficient in Air*

You work in an atmospheric chemistry laboratory, and you are asked to estimate the diffusion coefficient of the freon CFC-12 (CCl_2F_2) in air at – 50°C and 0.3 atm. Since during your environmental chemistry course, you have already estimated D_a of this compound for 25°C and 1 atm ($D_a \approx 8 \times 10^{-2}\ cm^2 \cdot s^{-1}$, see Illustrative Examples), you only need to consider the effect of temperature and pressure on D_a.

Hints and Help
To get an idea of the temperature and pressure dependence of D_a, remember the random motion model of molecular diffusion, although this model does not seem to yield very accurate absolute D_a-values (see *Section 9.1* and Illustrative Examples). The temperature and pressure dependence may be expressed by

$$D_a(T) = D_a(T_{ref}) \left(\frac{T}{T_{ref}}\right)^{\gamma_T}$$

$$D_a(p) = D_a(P_{ref}) \left(\frac{P}{P_{ref}}\right)^{\gamma_p}$$

where T_{ref} and P_{ref} are reference values, e.g., 298.2 K and 1 atm, respectively. Your first task is, obviously, to determine the exponents γ_T and γ_p.

● *G-2 Estimating the Molecular Diffusion Coefficient of Two PCB Congeners in Water*

You are asked by your boss to provide her with the molecular diffusion coefficients in water (D_w) at 25°C of the two PCB congeners 2,2',4,4'-tetrachlorobiphenyl (I) and 3,3',4,4'-tetrachlorobiphenyl (II). Note that the two compounds are isomers. In order to make sure that you get a reliable result, you use three different estimation methods for D_w. Which one(s) do you favor, and what are the results of your calculations? Comment on the results taking into account the fact that one of the PCBs is planar (compound II) and the other one is not.

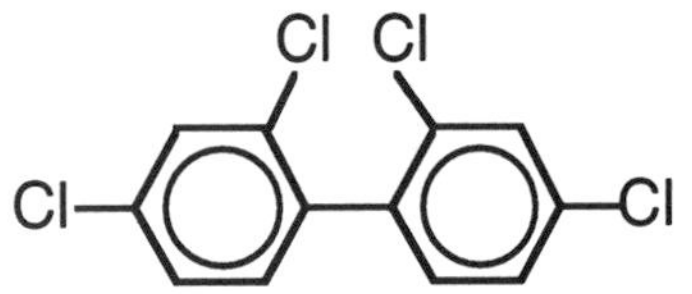

2,2'4,4'- tetrachlorobiphenyl
(I)

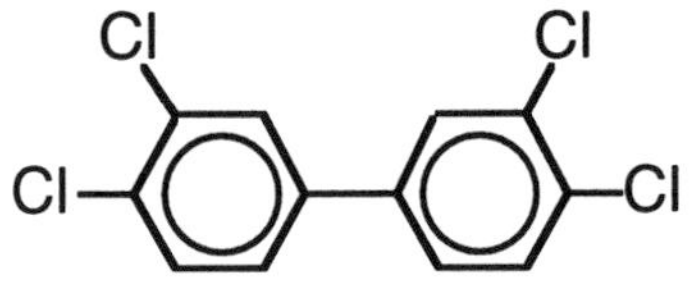

3,3', 4,4'-tetrachlorobiphenyl
(II)

● *G-3 Vertical Turbulent Diffusion Coefficient in a Lake*

(a) You are responsible for the water quality monitoring in Lake X (surface area $A_0 = 15\ km^2$, maximum depth $z_{max} = 50$ m). Among the various physical and chemical parameters you measure regularly is the vertical distribution of water temperature T. The following table gives two profiles measured during the same summer. The table also gives some information on lake topography, i.e., on the decrease of the lake cross-section with depth. What is the vertical turbulent diffusion coefficient E_z at the fixed depths of 10, 20, and 30 meters?

Depth (m)	Area A (km^2)	Water Temperature T (°C) June 15	Water Temperature T (°C) August 1
0	15.0	18.3	22.2
5	13.5	13.1	19.1
7.5	12.8	10.0	12.8
10	12.0	7.3	8.4
12.5	11.2	6.0	6.4
15	10.5	5.8	5.9
20	9.0	5.4	5.5
30	6.0	5.1	5.3
40	3.0	4.9	5.1
50 (max)	0	4.8	5.0

(b) Since you are interested in understanding the physical processes in your lake, you compare the E_z-values to the vertical stability of the water column, which is usually quantified by the square of the stability frequency, N^2. Compare your result with the $E_z = a(N^2)^{-0.5}$ relationship shown in the right graph of *Fig. 9.14.*

● *G-4 How Fast Does a Patch of Hexachlorobenzene (HCB) Disappear from the Surface of the Ocean?*

As a result of an accident a cloud of hexachlorobenzene (HCB) of approximatively circular shape floats in the surface mixed layer of the ocean. The concentration can be described by a two-dimensional Gaussian distribution of the form

$$C(x,y,t) = C_0(t) \exp\left(-\frac{r^2}{2R^2(t)}\right) = C(r,t)$$

where

$C(x,y,t) = C(r,t)$	is the concentration at distance $r = (x^2+y^2)^{1/2}$ from the center of the cloud,
x,y	are the two horizontal Cartesian coordinates
$C_0(t) = 25\ \text{nmol}\cdot\text{L}^{-1}$	is the concentration at the center of the cloud at time t,
$R(t)$	is a characteristic length scale which at time $t = 0$, when the cloud is first measured, is $R_0 = 5$ km

The cloud drifts along with the mean currents and grows in size due to horizontal diffusion, which is described by the turbulent horizontal diffusion coefficients $E_x = E_y$ (isotropic mixing). Calculate $C_0(t)$ after 24 hours (t = 24 h). Note that vertical mixing of HCB through the thermocline can be disregarded during this time period.

Hints and Help

(1) Estimate $E_x = E_y$ from *Fig. 9.10*.

(2) Assume that the total amount of HCB, M_{tot}, remains fairly constant during the period of 24 hours. M_{tot} can be calculated by

$$M = \int_{-\infty}^{\infty} dx \int_{-\infty}^{\infty} dy \int_{0}^{z_{mix}} dz \cdot C(x,y,z,t)$$

It can be assumed that $C(x,y,z,t)$ is constant over the mixed layer depth z_{mix}. Thus, the integral over z is z_{mix}.

$$M = z_{mix} \int_{-\infty}^{\infty} dx \int_{-\infty}^{\infty} dy \; C(x,y,t) = 2 \cdot \pi \cdot z_{mix} \cdot C_0(t) \int_0^{\infty} r \cdot \exp\left(-\frac{r^2}{2R^2(t)}\right) dr$$

$$= 2 \cdot \pi \cdot z_{mix} \cdot C_0(t) \cdot R^2 = \text{const.}$$

● *G-5 Vertical Transport of Nitrilotriacetic Acid (NTA) in a Lake*

Recently, your country has banned the use of polyphosphates in detergents. One of the possible substitutes for such polyphosphates is nitrilotriacetic acid (NTA). Therefore, you decide to monitor the NTA-concentration in your lake (the same as described in Problem G-3) to learn more about the dynamic behavior of this compound in the water column. The NTA profile that was measured on August 1 is shown in the following table (the temperature data are given in Problem G-3).

Depth (m)	NTA-concentration (10^{-9} mol·L^{-1})
0	7.5
5	7.5
7.5	7.2
10	5.2
11.5	3.4
15	2.5
20	2.4
30	2.3
40	2.3
50 (max)	2.3

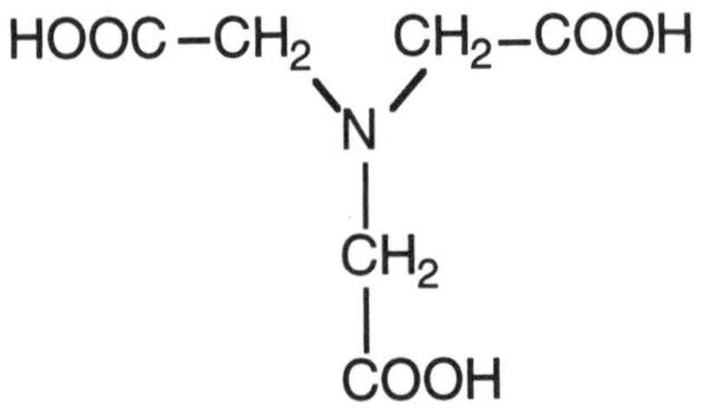

HOOC–CH_2 CH_2–COOH
N
CH_2
COOH

Nitrilotriacetic acid (NTA)

Calculate the size and direction of the vertical flux of NTA per unit area and time across the thermocline (at 10 m depth). The results from Problem G-3 should help you to solve this problem.

H. THE GAS-LIQUID INTERFACE: AIR-WATER EXCHANGE

ILLUSTRATIVE EXAMPLES

Evaluating the Direction of Air-Water Gas Exchange

C_1-and C_2-halocarbons of natural and anthropogenic origin are omnipresent in the atmosphere and in seawater. For example, for 1,1,1-trichloroethane (also called methyl chloroform, MCF) and tribromomethane (bromoform, BF), typical concentrations in the northern hemisphere air and in Arctic seawater are given in Table H.1 together with their air-seawater equilibrium partition constants, $K_{H,sw}$ (note the differences from the Henry's law constants given below). Note that for the following calculations, we neglect the (slow) hydrolysis of MCF (see *Table 12.8*).

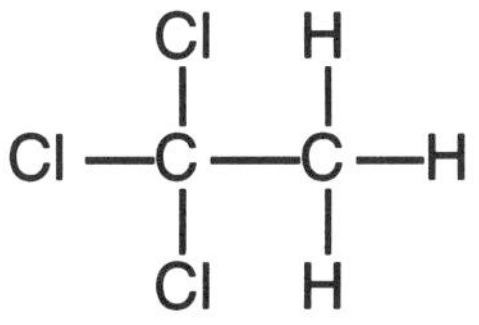

1,1,1-trichloroethane
(methyl chloroform, MCF)

mw	133.4
T_m	- 30.4°C
T_b	74.1°C
K_H(25°C)	19.5 atm·L·mol^{-1}

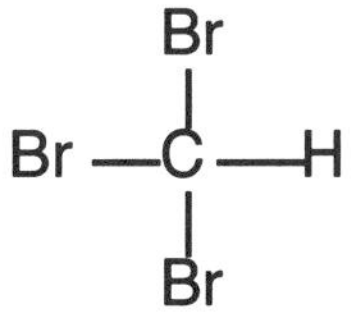

tribromomethane
(BF)

mw	252.8
T_m	8.3°C
T_b	149.5°C
K_H(25°C)	0.60 atm·L·mol^{-1}

Table H.1 Concentrations of MCF and BF in Air and Seawater; $K_{H,sw}$ Values of MCF and BF at 0°C and 25°C (all data from *Fogelqvist, 1985*).

Parameter	MCF	BF
Concentration in air (data from 1980)		
C_a (ng·L^{-1})	0.93	0.05
Concentration in surface water of the Arctic Ocean (0-10m)		
C_w^o (ng·L^{-1})	2.5	9.8
Concentration at 200 m depth in the Arctic Ocean		
C_w^{200} (ng·L^{-1})	1.6	3.0
$K_{H,sw}$ (0°C) (atm·L·mol^{-1})	6.5	0.20
$K_{H,sw}$ (25°C) (atm·L· mol^{-1})	23.8	0.86

Problem

Using the concentrations of MCF and BF given in Table H.1, evaluate whether there is a net flux of these compounds between the air and the surface waters of the Arctic Ocean assuming a temperature of (a) 0°C, and (b) 10°C. If there is a net flux, indicate its direction (i.e., sea-to-air or air-to-sea).

Answer

Independent of the model that is used to describe gas exchange at the air-seawater interface, the flux, F, can be expressed by (see *Eqs. 10-10, 10-21*, and *10-22*):

$$F = v_{tot} \cdot (C_w^o - \frac{C_a}{K'_{H.sw}}) \qquad \text{(H-1)}$$

Because the total transfer velocity (or the overall mass transfer coefficient) v_{tot} is always greater than zero, the second term in Eq. H-1 determines the direction of the flux. Note that the way Eq. H-1 is defined, a positive flux means a flux from water to air (the actual concentration in the surface water, C_w^o, is greater than the concentration that would be established in the water at equilibrium with the actual concentration of the compound in air, i.e., $C_a / K'_{H.sw}$).

(a) Divide the $K_{H,sw}$ values given in Table H-1 by RT to obtain the corresponding $K'_{H.sw}$ values (*Eq. 6-5*).

The resulting $K'_{H.sw}$ values are

$K'_{H.sw}$ (MCF, 0°C) = 0.29

$K'_{H.sw}$ (BF, 0°C) = 0.0089

Determine the sign of the second term in Eq. H-1 by using these $K'_{H.sw}$ values together with the C_w^o and C_a values given in Table H.1. The results are

Compound	C_w^o (ng·L^{-1})	$C_a/K'_{H.sw}$ (ng·L^{-1})	$C_w^o - C_a K'_{H.sw}$ (ng·L^{-1})
MCF	2.5	3.2	− 0.7
BF	9.8	5.6	+ 4.2

Thus, at 0°C, there is a net flux of MCF from the atmosphere to the sea while for BF, a net transfer occurs from the water to the atmosphere.

(b) Estimate first the $K_{H,sw}$ at 10°C from the $K_{H,sw}$ values given in Table H.1 for 0°C and 25°C using a temperature dependence of the form (see *Eq. 6-11*):

$$\ln K_{H,sw}(T) = -\frac{B}{T} + A \qquad \text{(H-2)}$$

where T is the temperature in K. Since

$$\ln\left(\frac{K_{H.sw}(T_1)}{K_{H.sw}(T_2)}\right) = \ln K_{H,sw}(T_1) - \ln K_{H,sw}(T_2) \qquad \text{(H-3)}$$
$$= B\left(\frac{1}{T_2} - \frac{1}{T_1}\right),$$

the fitting parameter B can, in this case, be calculated by

$$B = \ln \frac{K_{H.sw}(298.2K)}{K_{H.sw}(273.2K)} \cdot \left(\frac{1}{273.2K} - \frac{1}{298.2K}\right)^{-1} \qquad \text{(H-4)}$$

Inserting the $K_{H,sw}$ values given in Table H-1 into Eq. H-3 yields

$$B(\text{MCF}) = 4230\ \text{K}$$
$$B(\text{BF}) = 4750\ \text{K}$$

Use these B values and Eq. H-3 with, for example, the known $K_{H,sw}$'s at 25°C to calculate the $K_{H,sw}$ values of the two compounds at 10°C:

$$\ln K_{H,sw}\ (\text{MCF, }10°\text{C}) = \ln(23.8) + 4230\left(\frac{1}{298.2} - \frac{1}{283.2}\right)$$
$$= 2.42$$

Hence,

$$K_{H,sw}\ (\text{MCF, }10°\text{C}) = 11.2\ \text{atm·L·mol}^{-1}$$

and, therefore,

$$K'_{H.sw}(\text{MCF, }10°\text{C}) = 0.48$$

Similarly,

$$\ln K_{H,sw}\ (\text{BF,}10°\text{C}) = \ln(0.86) + 4750\left(\frac{1}{298.2} - \frac{1}{283.2}\right)$$
$$= -0.99$$

or

$$K_{H,sw}\ (\text{BF,}10°\text{C}) = 0.37\ \text{atm·L·mol}^{-1}$$

and

$$K'_{H.sw}\ (\text{BF,}10°\text{C}) = 0.016$$

Inserting these $K'_{H.sw}$ values in the second term in Eq. H-1 and using the same air and seawater concentrations as above, one obtains for 10°C:

Compound	C_w^o ($ng \cdot L^{-1}$)	$C_a/K'_{H.sw}$ ($ng \cdot L^{-1}$)	$C_w^o - C_a/K'_{H.sw}$ ($ng \cdot L^{-1}$)
MCF	2.5	1.9	+0.6
BF	9.8	3.1	+6.7

Comparison with the results obtained above for 0°C shows that, with the same air and seawater concentrations, the net flux of MCF is now directed from the water to the atmosphere.

Problem

Given the concentrations of MCF in the air and in the surface water of the Arctic Ocean (Table H.1), determine the temperature at which there is no net air-water exchange of this compound.

Answer

There is no net flux if the C_a and C_w^o of MCF are in equilibrium; that is,

$$K'_{H.sw}\ (\text{MCF}, T_x) = \frac{C_a}{C_w^o} = \frac{0.93}{2.5} = 0.37$$

Therefore,

$$K_{H,sw}(\text{MCF}, T_x) = (0.37) \cdot R \cdot T_x = (0.0304) \cdot T_x$$

Solve Eq. H-3 for T_x (= T_2) using the known $K_{H,sw}$ value for 25°C (298.2K = T_1)

$$T_x = \frac{B(\text{MCF})}{\ln\left(\dfrac{K_{H.sw}\,(298.2\text{K})}{0.0304 \cdot T_x}\right)} + \frac{B(\text{MCF})}{298.2\text{K}} \qquad \text{(H-5)}$$

Guess that T_x is on the order of 278.2°K (5°C) and get an approximate $K_{H,sw}\ (T_x) = R \cdot T_x \cdot K'_{H.sw} \simeq 8.4\ \text{atm} \cdot \text{L} \cdot \text{mol}^{-1}$. Insert this value into Eq. H-5 and you obtain:

$$T_x \simeq \frac{4230\text{K}}{\ln\left(\frac{23.8}{8.4}\right) + \frac{4230\text{K}}{298.2\text{K}}} \simeq 277.8\text{K} \ (4.6°\text{C})$$

Convince yourself that using $T_x = 4.6°C$ to recalculate $K_{H,sw}$ and then T_x from Eq. H-5 does not alter T_x anymore.

Estimating the Total Transfer Velocity (v_{tot}) from Wind Speed

Problem

Calculate the total air-water transfer velocity, v_{tot}, of MCF and BF (a) in a laboratory tank for a wind speed of 0.5 $m \cdot s^{-1}$ measured 10 cm above the water surface and (b) at the surface of a lake for a wind speed of 15 $m \cdot s^{-1}$ measured 3 m above the water. Assume an ambient (air and water) temperature of 25°C.

Answer

Use *Eq. 10-34* to express the total transfer resistance (the inverse of the total transfer velocity) by the sum of the water and air resistances:

$$\frac{1}{v_{tot}} = \frac{1}{v_w} + \frac{1}{v_a'} \qquad (10\text{-}34)$$

where $v_a' = v_a K_H'$ (see *Eq. 10-35*). Recall that the resistance ratio

$$R_{a/w} = \frac{(1/v_a')}{(1/v_w)} = \frac{v_w}{v_a'} \qquad (10\text{-}36)$$

is a measure of the relative importance of the air film resistance compared to the water film resistance. If $R_{a/w} < 0.1$, $v_{tot} \simeq v_w$; if $R_{a/w} > 10$, $v_{tot} \simeq v_a' = v_a K_H'$. Use *Eqs. 10-29* and *10-33* to approximate v_a and v_w of the compounds, by assuming an α value of 0.67 and a β value of 0.5.

(Note that when applying the two-film model, you would set $\alpha = \beta$ = 1.0.)

$$v_a(\text{compound}) = v_a(H_2O) \left(\frac{D_a(\text{compound})}{D_a(H_2O)} \right)^{0.67} \quad \text{(H-6)}$$

$$v_w(\text{compound}) = v_w(O_2) \left(\frac{D_w(\text{compound})}{D_w(O_2)} \right)^{0.5} \quad \text{(H-7)}$$

Estimate $v_a(H_2O)$ and $v_w(O_2)$ using the empirical relationships *Eqs. 10-28* and *10-32*, respectively:

$$v_a(H_2O) \sim 0.2\, u_{10} + 0.3 \quad (\text{cm}\cdot\text{s}^{-1}) \quad \textit{(10-28)}$$

$$v_w(O_2) \sim 4 \times 10^{-4} + 4 \times 10^{-5}\, u_{10}^2 \quad (\text{cm}\cdot\text{s}^{-1}) \quad \textit{(10-32)}$$

where u_{10} is the wind speed in $\text{m}\cdot\text{s}^{-1}$ measured 10 m above the surface. Hence, convert the wind speed given above for 0.1 and 3 m above the surface to the reference height of 10 m, using *Eq. 10-24*:

$$u_{10} = \left(\frac{10.4}{\ln z + 8.1} \right) u_z \quad \textit{(10-24)}$$

The result is $u_{10} = 0.9\ \text{m}\cdot\text{s}^{-1}$ for the laboratory tank (case (a)), and $u_{10} = 17\ \text{m}\cdot\text{s}^{-1}$ for the lake situation (case (b)). Insertion of these wind speeds into *Eqs. 10-28* and *10-32* yields

case (a): $v_a(H_2O) = 0.48\ \text{cm}\cdot\text{s}^{-1}$; $v_w(O_2) = 4.3 \times 10^{-4}\ \text{cm}\cdot\text{s}^{-1}$

case (b): $v_a(H_2O) = 3.7\ \text{cm}\cdot\text{s}^{-1}$; $v_w(O_2) = 1.2 \times 10^{-2}\ \text{cm}\cdot\text{s}^{-1}$

Insert these values now into Eqs. H-6 and H-7, respectively, together with the ratios of the molecular diffusion coefficients of the compounds and the reference species H_2O and O_2 (see footnote in *Table 10.3*). Approximate these ratios by the square root of the inverse ratio of the molecular weights (see *Eqs. 9-25* and *9-30*):

$$\frac{D_a(\text{compound})}{D_a(H_2O)} \simeq \left(\frac{18}{\text{mw(compound)}}\right)^{1/2} \quad \text{(H-8)}$$

and

$$\frac{D_w(\text{compound})}{D_w(O_2)} \simeq \left(\frac{32}{\text{mw (compound)}}\right)^{1/2} \quad \text{(H-9)}$$

The resulting v_a and v_w values for MCF and BF are

	case (a)		case (b)	
Compound	v_a ($cm \cdot s^{-1}$)	v_w($cm \cdot s^{-1}$)	v_a($cm \cdot s^{-1}$)	v_w($cm \cdot s^{-1}$)
MCF	0.25	3.0×10^{-4}	1.9	8.4×10^{-3}
BF	0.20	2.6×10^{-4}	1.5	7.2×10^{-3}

Finally, using the K_H' values of MCF and BF at 25°C (0.80 and 0.025, respectively) calculate the v_{tot} *(Eq. 10-34)* and $R_{a/w}$ *(Eq. 10-36)* values for the two compounds for both cases:

		case (a)		case (b)	
Compound	K_H'(25°C)	$R_{a/w}$	v_{tot}($cm \cdot s^{-1}$)	$R_{a/w}$	v_{tot}($cm \cdot s^{-1}$)
MCF	0.80	0.0015	3.0×10^{-4}	0.0056	8.4×10^{-3}
BF	0.025	0.052	2.5×10^{-4}	0.19	6.0×10^{-3}

Note that except for BF at high wind speed (case (b)), air-water exchange of the two compounds is primarily controlled by the liquid film resistance.

Calculating Air-Water Exchange Fluxes

Problem

Calculate the fluxes of MCF and BF between the air and the surface water of the Arctic Ocean at high latitudes at a temperature of 0°C and an average wind speed u_{10} =10 $m \cdot s^{-1}$, using the concentrations given in Table H.1. Neglect the temperature dependence of v_a and v_w.

Answer

Estimate $v_a(H_2O)$ and $v_w(O_2)$ using *Eqs. 10-28* and *10-32* with u_{10} = 10 $m \cdot s^{-1}$:

$$v_a(H_2O) = 2.3 \text{ cm} \cdot s^{-1}; \quad v_w(O_2) = 4.4 \times 10^{-3} \text{ cm} \cdot s^{-1}$$

Insert these values into Eqs. H-6 and H-7 together with the ratios of the molecular diffusion coefficients of the compounds and the reference species (Eqs. H-8 and H-9). Calculate v_a and v_w for MCF and BF, respectively, as well as $R_{a/w}$ *(Eq. 10-36)* using the $K'_{H.sw}$ (0°C) values given above.

Compound	$K'_{H.sw}$	$v_a(cm \cdot s^{-1})$	$v_w(cm \cdot s^{-1})$	$R_{a/w}$
MCF	0.29	1.2	3.1×10^{-3}	0.0089
BF	0.0089	0.95	2.6×10^{-3}	0.31

Insertion of these results into *Eqs. 10-34* and H-1 yields v_{tot}. Note that when using Eq. H-1 and the concentrations given in Table H-1, the units have to be converted from $cm^2 \cdot s^{-1}$ and $ng \cdot L^{-1}$ to $m \cdot s^{-1}$ and $\mu g \cdot m^{-3}$, respectively.

Compound	v_{tot} $(m \cdot s^{-1})$	$C_w^o - C_a/K'_{H.sw}$ $(\mu g \cdot m^{-3})$	flux F $(\mu g \cdot m^{-2} \cdot s^{-1})$
MCF	3.1×10^{-5}	− 0.7	-2.2×10^{-5}
BF	2.0×10^{-5}	+ 4.2	$+8.4 \times 10^{-5}$

Note again, that for MCF, the flux is negative, indicating that the source of methyl chloroform is the atmosphere. This is confirmed by the observation that the MFC concentration decreases with depth in the water column of the Arctic Ocean. In contrast, the major source of BF is thought to be biological production in the water column, always resulting in a net transport of this compound from the water to the atmosphere. For more details, see Fogelqvist (1985).

Estimating Mean Residence Times (with Respect to Air-Water Exchange)

Problem

Based on vertical concentration profiles of MCF and BF at high latitudes in the Arctic Ocean by *Fogelqvist (1985)*, the average total mass per area in the upper 1000 m, M, is estimated to be 1×10^{-3} $g \cdot m^{-2}$ for MCF and 2.6×10^{-3} $g \cdot m^{-2}$ for BF. Calculate the mean residence times of the two compounds in the top 1000 m with respect to air-water exchange, assuming the flux estimated above for 0°C.

Answer

The mean residence time, τ, is given by

$$\tau \text{ (compound)} = \frac{M}{|F|}$$

Hence, for MCF, one obtains

$$\tau \text{ (MCF)} \simeq \frac{1.0 \times 10^{-3}\ g \cdot m^{-2}}{2.2 \times 10^{-11}\ g \cdot m^{-2} \cdot s^{-1}} = 4.5 \times 10^{7} s$$

$$\simeq 530 \text{ days}$$

For BF, the result is

$$\tau \text{(BF)} \simeq \frac{2.6 \times 10^{-3}\ g \cdot m^{-2}}{8.4 \times 10^{-11}\ g \cdot m^{-2} \cdot s^{-1}} = 3.1 \times 10^{7} s$$

$$\simeq 360 \text{ days}$$

Note that in order to sustain a steady-state in the water column, a sink for MCF and a (biological) source for BF have to be postulated. One possible MCF sink mechanism could be transport and loss to the atmosphere at lower latitudes where the temperature is higher (recall from above, you have calculated that with the given air and water concentrations of MCF, the direction of the air-water flux reverses at about 5°C).

Determining v_{tot} from Field Measurements

Problem
Due to an accidental spill, a significant amount of MCF has been introduced into a small, well-mixed pond (volume $V = 1 \times 10^4 \text{ m}^3$, total surface area $A = 5 \times 10^3 \text{ m}^2$, $T = 15°C$). Measurements carried out after the spill during a period of one week showed that MCF was eliminated from the pond by a first-order process with a half-life of 40 h. Because export of MCF by the outflow of the pond can be neglected, and because it can be assumed that neither sedimentation nor transformations are important processes for MCF, the observed elimination has to be attributed to exchange to the atmosphere. Calculate the average v_{tot} of MCF during the time period considered by assuming that (i) the concentration of MCF in the air above the pond is very small, i.e., $C_w \gg C_a / K_H'$ and (ii) that practically all MCF is present in the water in dissolved form. What is the average wind speed, u_{10}, that corresponds to this v_{tot} value?

Answer
Because $C_w \gg C_a / K_H'$, the average flux, F, of MCF from the water to the atmosphere is given by

$$F = v_{tot} \cdot C_w \qquad \text{(H-10)}$$

The total mass exported by gas exchange ($A \cdot F = A \cdot v_{tot} \cdot C_w$) is equivalent to the change in total mass of MCF in the pond:

$$A \cdot v_{tot} \cdot C_w = -V \frac{dC_w}{dt} \qquad \text{(H-11)}$$

Because the measurements show a first-order elimination process, it follows that

$$-\frac{dC_w}{dt} = k_g \cdot C_w \qquad \text{(H-12)}$$

with

$$k_g = \frac{A}{V} v_{tot} = \frac{v_{tot}}{h_{mix}} \qquad \text{(H-13)}$$

where $h_{mix} = V/A = 2$ m is the average depth of the well-mixed water body. Since the observed $t_{1/2}$ is 40 h, find (see *Eq. 12-11*)

$$k_g = \frac{\ln 2}{t_{1/2}} = \frac{0.693}{40 \text{ h}} = 0.017 \text{ h}^{-1}$$

Using this result you obtain:

$$v_{tot}(\text{MCF}) = k_{mix} \cdot k_g = (0.2 \text{ m})(0.017 \text{ h}^{-1}) = 0.034 \text{ m} \cdot \text{h}^{-1} \simeq 10^{-3} \text{ cm} \cdot \text{s}^{-1}$$

Because air-water exchange of MCF is primarily liquid-film controlled (i.e., $v_{tot} \simeq v_w$), u_{10} can be calculated from *Eq. 10-32*

$$u_{10} \simeq \left(\frac{v_w(O_2) - 4 \times 10^{-4}}{4 \times 10^{-5}}\right)^{1/2} \quad (\text{m} \cdot \text{s}^{-1}) \qquad \text{(H-14)}$$

with (see Eqs. H-7 and H-9):

$$v_w(O_2) = v_{tot}(\text{MCF}) \left(\frac{\text{mw(MCF)}}{32}\right)^{1/4} \qquad \text{(H-15)}$$

$$= 1.4 \times 10^{-3} \text{ cm} \cdot \text{s}^{-1}$$

Insert this value into Eq. H-14 to obtain

$$u_{10} \simeq 5.0 \text{ m} \cdot \text{s}^{-1}$$

Estimating Evaporation Rates of Pure Organic Liquids

Problem

Consider a spill of 2 kg of pure liquid MCF forming a puddle on the ground of about 0.3 m^2 surface area. Calculate how long it takes for 50% of the liquid MCF to evaporate at 0°C and at 25°C. Assume an average MCF concentration in the air above the liquid (C_a) of 1 ng·L^{-1}, and an average wind speed 10 m above the ground of 1 m·s^{-1}. Neglect the temperature dependence of v_a and v_w.

Answer

Use *Eq. 10-28* to estimate $v_a(H_2O)$ for a puddle of water on the ground. With $u_{10} = 1$ m·s^{-1}, a $v_a(H_2O)$ of 0.5 cm·s^{-1} is obtained. As assumed in *Section 10.4* for the transfer velocity for evaporation of water from a dilute aqueous solution, consider the transfer from the pure organic liquid to the atmosphere to be completely air-film controlled. Hence, estimate v_{tot} of MCF by combining Eqs. H-6 and H-8:

$$v_{tot}(\mathrm{MCF}) = v_a(\mathrm{H_2O})\left(\frac{18}{\mathrm{mw(MCF)}}\right)^{0.335} = 0.26\ \mathrm{cm{\cdot}s^{-1}} = 2.6 \times 10^{-3}\mathrm{m{\cdot}s^{-1}}$$

The concentration of MCF in the air right at the interface above the liquid MCF, $C_{a/l}$ is determined by the vapor pressure of MCF. Hence, when assuming ideal gas behavior, one obtains

$$C_{a/l} = \frac{P^o}{RT}$$

Estimate $P^o(0°C)$ and $P^o(25°C)$ from the boiling point temperature T_b given above (*Eq. 4-18 and Table 4.4*) with $K_F = 1.0$. The resulting $C_{a/l}$ values are

$$C_{a/l}(0°\mathrm{C}) = 2.1 \times 10^{-3}\ \mathrm{mol{\cdot}L^{-1}} = 2.1\ \mathrm{mol{\cdot}m^{-3}}$$

and

$$C_{a/l}(25°\mathrm{C}) = 6.8 \times 10^{-3}\ \mathrm{mol{\cdot}L^{-1}} = 6.8\ \mathrm{mol{\cdot}m^{-3}}$$

Because in both cases, $C_{a/l} \gg C_a$, the total flux, F_{tot}, of MCF is given by (assume that the total surface area of the puddle does not change significantly over the time period considered):

$$F_{tot}(0°\mathrm{C}) = A \cdot v_{tot} \cdot C_{a/l}(0°\mathrm{C}) = 1.6 \times 10^{-3}\ \mathrm{mol{\cdot}s^{-1}}$$

and

$$F_{tot}(25°\mathrm{C}) = A \cdot v_{tot} \cdot C_{a/l}(25°\mathrm{C}) = 5.3 \times 10^{-3}\ \mathrm{mol{\cdot}s^{-1}}$$

Finally, the time for evaporation of 50% of the liquid MCF (= 1 kg = 7.5 mol) can be calculated by

$$t_{50\%}\,(0°C) = \frac{7.5\ \text{mol}}{1.6 \times 10^{-3}\ \text{mol} \cdot \text{s}^{-1}} = 4700\ \text{s} = 78\ \text{min}$$

$$t_{50\%}\,(25°C) = \frac{7.5\ \text{mol}}{5.3 \times 10^{-3}\ \text{mol} \cdot \text{s}^{-1}} = 1400\ \text{s} = 24\ \text{min}$$

PROBLEMS

● *H-1 What Is the Source of Benzene in the Water of a Pond?*

Part of your job as a consultant to the State Water Authority is to survey the water quality of several ponds located in a recreation area. Among the volatile organic compounds, your laboratory monitors the concentration of benzene in the water of various ponds. When inspecting the results, you realize that on certain weekends during the summer, the benzene concentration in the surface pond water is up to ten times higher than in the same pond in the middle of the week or during the winter time. For example, in one of the ponds, you measure a peak concentration of 1 $\mu g \cdot L^{-1}$ on a sunny Sunday. You wonder whether this elevated benzene concentration is due to air pollution by the heavy car traffic in the area during the summer weekends, or whether the input occurs primarily by leakage of gasoline and oil from the numerous water boats cruising on the ponds. You realize that for assessing the direction of gas exchange of benzene between the air and the pond water you need to know something about the benzene concentration in the air. Since you have no measurements from the area, you search the literature for typical values. In a review by *Field et al. (1992)* you find the following typical benzene concentrations reported by various groups for urban and remote air.

Location	Mean benzene concentration in air (ppbv)[a]
Remote areas:	
Brasil	0.5
Pacific	0.5
Urban areas:	
Hamburg	3.2 (peak: 20)[b]
London	8.8

[a]Parts per billion on a volume base; use the ideal gas law to convert the numbers to molar concentrations. [b]*Bruckmann et al., 1988.*

Considering these data, answer the following questions:

(a) Is the atmosphere a likely source for the elevated benzene concentrations (i.e., 1 $\mu g \cdot L^{-1}$) found in the pond water during the summer weekends? (Assume a temperature of 25°C).

(b) Estimate the direction and rate of air-water exchange of benzene under the following conditions:

Temperature (water and air)	T	25°C
Average wind speed 10 m above the water	u_{10}	2 $m \cdot s^{-1}$
Benzene concentrations		
in the air	C_a	10 ppbv
in the water	C_w	1 $\mu g \cdot L^{-1}$

(c) What is the direction and the size of the air-water exchange of benzene for a well-mixed shallow pond located in the center of a big city for a typical winter situation (no motor boats) assuming the following conditions?

Temperature (water and air)	T	5°C
Average wind speed 10 m above the water	u_{10}	2 $m \cdot s^{-1}$
Benzene concentrations:		
in the air	C_a	10 ppbv
in the water	C_w	0.1 $\mu g \cdot L^{-1}$

You will find all relevant compound properties in the textbook, particularly in the *Appendix* and in *Table 6.1*.

● *H-2 A Lindane Accident in a Drinking Water Reservoir*

Due to an accident, an unknown amount of the insecticide, lindane (γ-HCB, see Chapter B, illustrative examples) is introduced into a well-mixed pond that is used as the drinking water reservoir for a small town. The water inflow and outflow (intake by the water works) are immediately stopped, and the water is analyzed for lindane. In water samples taken at various locations and depths, an average lindane concentration of 5.0 ± 0.2 $\mu g \cdot L^{-1}$ is determined. As a resident of the area, you ask the person in charge of the water works what they intend to do about this problem, since 5 $\mu g \cdot L^{-1}$ is far above the drinking water limit for this compound. "Oh, no problem!", the person tells you, "within a few days, all the lindane will have escaped into the atmosphere!" Being well-trained in environmental organic chemistry, and having dealt with some of the physical-chemical properties of lindane in Chapter B, you are very suspicious about this answer. In order to convince yourself, answer the following questions. Comment on all assumptions that you make.

(a) Calculate the initial flux of lindane per day from the reservoir surface (reservoir surface area $A = 5 \times 10^5\ m^2$, volume $V = 2.5 \times 10^6\ m^3$, water temperature $T = 20°C$), using the following additional information. The people from the water works tell you that the evaporation rate of water from the pond is 5 mm per day at the prevailing wind conditions and relative humidity of 70%, which, at 20°C, corresponds to a water vapor pressure of 0.0162 atm in the air above the pond. (Note that from this latter information you can calculate the K_H-value of water at 20°C.)

(b) Calculate the relative loss of lindane from the reservoir to the atmosphere and estimate the half-life of this compound in the reservoir, assuming that the reservoir volume remains constant, and that no other elimination processes occur (see Eqs. H-10 to H-13).

(c) Compare the relative loss of lindane from the reservoir to the relative water loss by evaporation. Would the lindane concentration increase or decrease as a result of the simultaneous water-to-air fluxes of H_2O and lindane if the reservoir inlets and outlets were to remain closed?

● H-3 *Experimental Determination of the Total Exchange Mass Transfer Coefficient v_{tot} for Two Chlorinated Hydrocarbons in a River*

In a field study in the River Glatt in Switzerland, the concentrations of tetrachloroethene (PER) and 1,4-dichlorobenzene (1,4-DCB) were measured at four locations along a river section of 2.4 km length (see data given below). In this river section, it could be assumed that no input of water nor of these compounds occurred. The measurements were made in a way that a specific water parcel was followed downstream and sampled at the appropriate distances. Furthermore, the study was conducted in a part of the river where the average depth ($h_{mix} = 0.4$ m) and the mean velocity ($\bar{u} = 0.67\ m \cdot s^{-1}$) of the water did not vary significantly. Calculate the average air-water transfer velocities, v_{tot}, of the two compounds for the river section by assuming that exchange to the atmosphere is the only elimination process from the water and that the concentrations of the two compounds in the air can be neglected. What are the half-lives of the two compounds in the river under these conditions (see also Eqs. H-12 and H-13). Comment on the results!

Concentrations determined for PER and 1,4-DCB in the River Glatt (data from *Schwarzenbach, 1983*):

Distance (m)	[PER] ($ng \cdot L^{-1}$)	[1,4-DCB] ($ng \cdot L^{-1}$)
0	690 ± 40	234 ± 5
600	585 ± 30	201 ± 5
1200	505 ± 6	180 ± 8
2400	365 ± 10	130 ± 5

● *H-4 An Inadvertent Air-Water Exchange Experiment*

You have worked hard to get through the theory of turbulent diffusion and you (hopefully) spent a lot of time with the illustrative example on pages G-19 - G-21, in which vertical profiles of tetrachloroethene (PER) in a lake have been used to calculate turbulent diffusion coefficients. A friend of yours is more interested in the air-water exchange process. She is very happy with your Table G.1 and immediately calculates the mean air-water exchange velocity for PER for the period day 0 to day 30. How does she do that? At least she tells you that she needs to assume that the input and output of PER by rivers and input by other sources such as treatment plants are negligible for the time period of the PER measurement. What other assumptions does she make and what is her result?

J. SORPTION: SOLID-AQUEOUS SOLUTION EXCHANGE

ILLUSTRATIVE EXAMPLES

Estimating the Solid-Water Distribution Ratio K_d of Organic Compounds

Depending on the structural moieties of the compound, on the types and relative abundance of solids present in a particular environment, and on the solution conditions (e.g., pH, major ion composition), a given organic chemical may sorb by different sorption mechanisms. Therefore, when deriving overall solid-water distribution ratios, K_d's, one must first evaluate which sorption mechanisms have to be taken into account. Then, a general expression for K_d (see example given by *Eq. 11-9*) can be formulated in order to identify the compound-specific and system-specific properties that have to be known for quantification of this sorption parameter. Note that the system-specific properties may not always be easy to quantify. However, in many cases, assumptions may be made that allow one to assess the relative importance of a given sorption mechanism.

Problem

Estimate the K_d-values for the four substituted benzenes (a) toluene, (b) trinitrotoluene (TNT), (c) aniline, and (d) pentachlorophenol (PCP) between a particular solid matrix and groundwater. Assume that the compounds are present at low concentrations (i.e., assume linear sorption isotherms), and assume the following solid matrix and groundwater composition.

solid matrix composition (% by weight)		groundwater composition (relevant species)	$(mol \cdot L^{-1})$
quartz	85	H^+	10^{-6}
kaolinite	10	O_2	$10^{-3.6}$
iron oxide	4	Na^+	10^{-3}
organic matter	1	K^+	10^{-4}
		Ca^{2+}	10^{-4}
		Cl^-	10^{-3}
		HCO_3^-	$10^{-3.5}$

Answer (a)

CH_3

toluene

mw	$92.1\ g \cdot mol^{-1}$
T_m	$-95°C$
T_b	$110.6°C$
C_w^{sat} (25°C)	$5.6 \times 10^{-3}\ mol \cdot L^{-1}$
K_{ow} (25°C)	4.9×10^2

Considering that toluene is a neutral, nonpolar chemical exhibiting no reactive moieties, and considering the relatively high organic matter content of the solid matrix, the association of this compound with the organic matter will be the dominant sorption mechanism. Hence, the general distribution ratio equation (*Eq. 11-9*) can be simplified:

$$K_d = (C_{om} \cdot f_{om} + C_{min} \cdot A + C_{ie} \cdot \sigma_{ie} \cdot A + C_{rxn} \cdot \sigma_{rxn} \cdot A) / (C_{w,neut} + C_{w,ion})$$
$$\simeq C_{om} \cdot f_{om} / C_{w,neut}$$

or, as indicated by *Eq. 11-16*:

$$K_d(\text{toluene}) = f_{om} \cdot K_{om}(\text{toluene}) \qquad \text{(J-1)}$$

For estimating K_{om}(toluene), use either *Eq. 11-21a* or *Eq. 11-22a*:

$$\log K_{om} = -0.93 \cdot \log C_w^{sat}(l,L) - 0.17 \qquad (11\text{-}21a)$$

or

$$\log K_{om} = 1.01 \cdot \log K_{ow} - 0.72 \qquad (11\text{-}22a)$$

Given that toluene is a liquid at room temperature (see T_m and T_b data above), you may use the logarithm of this compound's solubility directly in *Eq. 11-21a* to get:

$$\log K_{om}(\text{toluene}) = -0.93 \cdot (-2.25) - 0.17 = 1.92$$

or

$$K_{om}(\text{toluene}) = 83 \text{ L} \cdot \text{kg}_{om}^{-1}$$

Based on K_{ow} one obtains

$$\log K_{om}(\text{toluene}) = 1.01 \cdot (2.69) - 0.72 = 2.00$$

or

$$K_{om}(\text{toluene}) = 100 \text{ L} \cdot \text{kg}_{om}^{-1}$$

Note that these two estimates differ by about 20%. There is no *a priori* reason to prefer one estimate over the other; thus one possibility is to take a one-significant-figure mean value of K_{om}, i.e., 90 $\text{L} \cdot \text{kg}_{om}^{-1}$.

Next, you need to know the organic matter content of the solid matrix, which is indicated as 1% by weight. Thus, $f_{om} = 0.01 \text{ kg}_{om} \cdot \text{kg}_s^{-1}$. (Note: this corresponds to an f_{oc} of about 0.005.)

Insertion of f_{om} and K_{om} into Eq. J-1 then yields

$$K_d \simeq 0.01 \text{ kg}_{om} \cdot \text{kg}_s^{-1} \cdot 90 \text{ L} \cdot \text{kg}_{om}^{-1} \simeq 0.9 \text{ L} \cdot \text{kg}_s^{-1}$$

Answer (b)

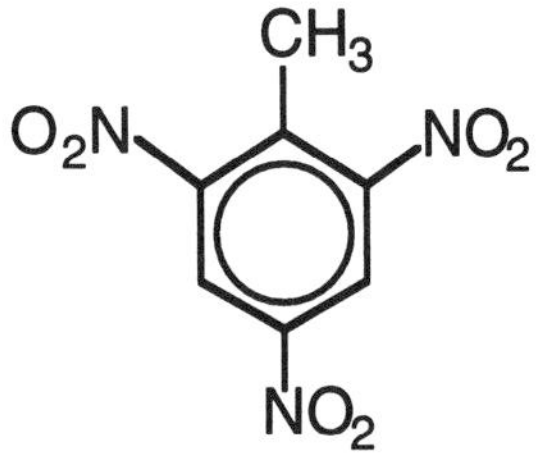

2,4,6-trinitrotoluene (TNT)		
	mw	227.1 $g \cdot mol^{-1}$
	T_m	82.0 °C
	T_b	240°C (explode)
	K_{om}	1 x 10^2 $L \cdot kg_{om}^{-1}$
	K_{EDA} (K^+-kaolinite)	2 x 10^3 $L \cdot kg_{K^+\text{-kao}}^{-1}$
	(estimated from data of *Haderlein et al., 1995*)	

In the case of nitroaromatic compounds (NACs), particularly of polynitroaromatic chemicals such as TNT, two types of sorption mechanisms have to be considered. In addition to hydrophobic partitioning into organic matter, adsorption due to specific interactions with clay mineral surfaces bearing exchangeable NH_4^+ or K^+ cations may contribute significantly to the overall K_d-value. This type of interaction has been postulated to involve an electron donor acceptor (EDA) complex where oxygen atoms at the siloxane surfaces of clay minerals act as electron donors and NACs act as electron acceptors (for details see *Haderlein and Schwarzenbach, 1993*; *Haderlein et al., 1995*). It has been shown that weakly hydrated cations, e.g., NH_4^+ or K^+, adsorbed to the negatively charged clay surfaces allow strong EDA complex formation. Only very weak interactions of the NACs with the clay surface are found if strongly hydrated cations (e.g., Na^+, Ca^{2+}, Mg^{2+}, Al^{3+}) are present at the cation exchange sites.

Haderlein et al. (1995) have determined K_{EDA}-values for a variety of environmentally relevant NACs for several homoionic clay minerals, including K^+-kaolinite, K^+-illite, and K^+-montmorillonite. The K_{EDA}-values reported represent slopes of the linear part of the adsorption isotherms; the concentration in the solid phase is expressed as the mass of compound per unit weight of the corresponding clay mineral:

$$K_{EDA} = \frac{C_{s.EDA}\ (mol \cdot kg_s^{-1})}{C_{w.neut}\ (mol \cdot L_{water}^{-1})} \tag{J-2}$$

For TNT adsorption on clay minerals, K_{EDA} is valid for aqueous TNT concentrations up to about 0.1 - 0.5 μM (0.02 - 0.1 $mg \cdot L^{-1}$). Finally, it should be noted that the K_{EDA}-values of TNT and other planar NACs for adsorption on other clay minerals (K^+-illite and K^+-montmorillonite) are about 5 and 10 times larger, respectively, as compared to K^+-kaolinite.

Using the K_{EDA}(K^+-kaolinite) given above, the K_d-value for TNT can then be expressed as

$$K_d = f_{om} \cdot K_{om} + f_{K^+\text{-kaolinite}} \cdot K_{EDA}(K^+\text{-kaolinite}) \qquad \text{(J-3)}$$

where $f_{K^+\text{-kaolinite}}$ is the weight fraction of K^+-kaolinite present in the soil. Inserting f_{om}, K_{om}, and K_{EDA} into Eq. J-3 yields

$$K_d = 1 + (2 \times 10^3)\, f_{K^+\text{-kaolinite}} \; (L \cdot kg_s^{-1}) \qquad \text{(J-4)}$$

Inspection of Eq. J-4 shows that, in the case of TNT, only a very small fraction of K^+-kaolinite is required (i.e., $f_{K^+\text{-kaolinite}} = 5 \times 10^{-4}$) to make EDA-complex formation as important as hydrophobic partitioning. Considering that 10% of the soil consists of kaolinite, it is very likely that EDA complex formation contributes significantly to the overall sorption. Hence, $f_{K^+\text{-kaolinite}}$ needs to be estimated.

The major cations present in the system are Na^+ (10^{-3} M), K^+ (10^{-4} M), and Ca^{2+}(10^{-4} M). In order to get an estimate of the fraction of cation exchange sites that are occupied by K^+ ions under these conditions, average binary exchange coefficients (or selectivity coefficients) relative to a given ion (e.g., Na^+) can be used (for details see *Appelo and Postma, 1993*; *Chapter 5*):

$$Na^+ + {}^1\!/_i \cdot I^{i+} - \text{Surf} \rightleftharpoons Na^+ - \text{Surf} + {}^1\!/_i\, I^{i+}$$

with (J-5)

$$K_{Na/I} = \frac{\beta_{Na} \cdot [I^{i+}]^{1/i}}{\beta_I^{1/i} \cdot [Na^+]}$$

β_{Na} and β_I are the fraction of cation exchange sites occupied by Na^+ and I^{i+} (e.g., K^+, Ca^{2+}), respectively, and $[Na^+]$ and $[I^{i+}]$ are the molar concentrations of the corresponding ions in the aqueous solution. The values given by *Appelo and Postma (1993)* for $K_{Na/K}$ and $K_{Na/Ca}$ are 0.2 and 0.4, respectively. Note that Eq. J-5 conforms to the Gaines-Thomas convention for describing the thermodynamic standard state of exchangeable cations. β_K and β_{Ca} can thus be expressed in terms of β_{Na}:

$$\beta_K = \frac{\beta_{Na} \cdot [K^+]}{K_{Na/K} \cdot [Na^+]}$$

and

$$\beta_{Ca} = \frac{\beta_{Na}^2 \cdot [Ca^{2+}]}{K_{Na/Ca}^2 \cdot [Na^+]^2}$$

Since $\beta_{Ca} + \beta_K + \beta_{Na} = 1$ (no other cations assumed to be present), one can derive a quadratic equation for β_{Na}:

$$\beta_{Na}^2 \left(\frac{[Ca^{2+}]}{K_{Na/Ca}^2 \cdot [Na^+]^2} \right) + \beta_{Na} \left(\frac{(K^+)}{K_{Na/K} \cdot [Na^+]} + 1 \right) -1 = 0$$

Substitution of the constants and concentrations given above yields

$$625\, \beta_{Na}^2 + 1.5\, \beta_{Na} - 1 = 0$$

which gives $\beta_{Na} = 0.04$. Subsequently, from back substitution $\beta_{Ca} = 0.94$ and $\beta_K = 0.02$. Thus, the fraction of K^+-kaolinite present in the soil considered here is estimated as

$$f_{K^+\text{-kaolinite}} = 0.1 \cdot 0.02 = 0.002$$

Insertion of this value into Eq. J-4 then yields

$$K_d \simeq 1 + (2 \times 10^3)(0.002) = 1 + 4 = 5\ L \cdot kg_s^{-1}$$

This calculation indicates that 20% of sorbed TNT molecules will be associated with the organic matter, while 80% will be bound to the kaolinite surfaces.

Note that in contrast to hydrophobic partitioning, EDA complex formation and thus the magnitude of K_d of a given NAC is not only strongly dependent on the major ion composition of the system, but may also be influenced by other competing water constituents including other NACs.

Answer (c)

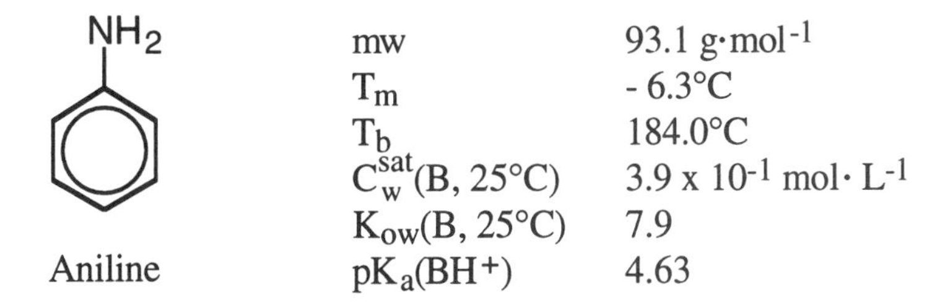

mw	$93.1\ g \cdot mol^{-1}$
T_m	$-6.3°C$
T_b	$184.0°C$
$C_w^{sat}(B, 25°C)$	$3.9 \times 10^{-1}\ mol \cdot L^{-1}$
$K_{ow}(B, 25°C)$	7.9
$pK_a(BH^+)$	4.63

Aniline is a weak base. At pH 6, the fraction of aniline present as protonated cationic species, α_a, is (*Eq. 8-16*)

$$\alpha_a = \frac{1}{1 + 10^{(pH - pK_a)}} = 0.04$$

or about 4%. Thus, you have to consider both sorption of the neutral aniline molecules into the soil organic matter and cation exchange of the charged aniline species. When neglecting adsorption of aniline to mineral oxide surfaces, as well as (irreversible) reactions with organic soil constituents (e.g., *Larson and Weber, 1994*), the general solid-water distribution ratio *Eq. 11-9* simplifies to

$$K_d(\text{aniline, soil}) \simeq (C_{om} \cdot f_{om} + C_{ie} \cdot \sigma_{ie} \cdot A) / (C_{w,neut} + C_{w,ion})$$

By an algebraic manipulation (chiefly by dividing both numerator and denominator by $C_{w,neut}$), you can group sorbed and dissolved species that are directly in equilibrium with one another:

$$K_d = (C_{om}/C_{w,neut}) f_{om} / (1 + C_{w,ion}/C_{w,neut}) \quad \text{(J-6)}$$

$$+ (C_{ie}/C_{w,ion}) (C_{w,ion}/C_{w,neut}) (\sigma_i \cdot A) / (1 + C_{w,ion}/C_{w,neut})$$

Calculate the various concentration ratios by using the appropriate relationships:

(i) Sorption to organic matter (*Eq. 11-15*):

$$C_{om} / C_{w,neut} = K_{om} \quad \text{(J-7)}$$

(ii) Sorption by ion exchange: Assume that the ion exchange sites are occupied primarily by Ca^{2+} (see example (b) above). Hence, the monovalent aniline cation has to compete with the bivalent Ca^{2+} cation. Further assume that the aniline concentration is much smaller than the calcium concentration. Then, $C_{ie}/C_{w,ion}$ can be described using *Eq. 11-111* (note that C_{ie} = [X-surf] / ($\sigma_{ie} \cdot A$)):

$$C_{ie} / C_{w,ion} = (\sigma_{ie} \cdot A)^{-1} \left(\frac{K_{ie} \cdot \sigma_{ie} \cdot A}{2[Ca^{+}]} \right)^{1/2} \quad \text{(J-8)}$$

(iii) Speciation of aniline in aqueous solution (*Eq. 8-11*):

$$C_{w,ion} / C_{w,neut} = \{H^+\} / K_a = 10^{(pK_a - pH)} \quad \text{(J-9)}$$

With pH = 6.0 and pK_a = 4.63, you obtain (Eq. J-9):

$$C_{w,ion} / C_{w,neut} = 10^{(4.63-6.0)} = 0.043$$

and

$$(1 + C_{w,ion}/C_{w,neut})^{-1} = 0.96$$

Insertion of this value and Eqs. J-7 and J-8 into Eq. J-6 then yields

$$K_d = (0.96)\, K_{om} \cdot f_{om} + (0.04) \left(\frac{K_{ie} \cdot \sigma_{ie} \cdot A}{2[Ca^{2+}]} \right)^{1/2} \quad \text{(J-10)}$$

Since there is no LFER for aromatic amines available in *Table 11.2*, use as a first approximation either one of the averaged expressions:

$$\log K_{om} \simeq -0.75 \log C_w^{sat}(1,L) + 0.44 \quad \textit{(11-23)}$$

or

$$\log K_{om} \simeq 0.82 \log K_{ow} + 0.14 \quad \textit{(11-24)}$$

Given aniline's solubility of $10^{-0.41}$ mol·L^{-1}, *Eq. 11-23* yields a K_{om} value of 5.6 L·kg_{om}^{-1}; from the compound's K_{ow} ($10^{0.9}$), a K_{om} value of 7.6 is estimated using *Eq. 11-24*. Again, there is no reason to favor one estimate over the other; thus, one can take a one-significant-figure mean value of $K_{om} \simeq 6$ L · kg_{om}. In any case, insertion of this value into Eq. J-10 with f_{om} = 0.01 $kg_{om} \cdot kg_s^{-1}$ shows that hydrophobic partitioning does not add a very large value to the overall K_d:

$$K_d \text{ (hydrophobic partitioning)} = (0.96)\,(6)\,(0.01) \simeq 0.06\ \mathrm{L \cdot kg_s^{-1}}$$

Next, consider aniline's K_{ie}. This type of selectivity constant has not been well studied for organic compounds, but some data suggest that its magnitude is proportional to the incompatibility with water of the hydrophobic part of the molecule (see *Eqs. 11-88 - 11-96*). Using benzene's liquid solubility at 25°C ($C_w^{sat} = 10^{-1.64}$ mol·L^{-1}) as a surrogate for the aqueous incompatibility of the hydrophobic part of aniline, K_{ie} is estimated as (see *Eqs. 11-96* and *11-116*):

$$K_{ie} \simeq [1.1\,(C_w^{sat}\,(\text{benzene}))^{-0.19}]^2 \simeq 5\ \mathrm{L \cdot kg_s^{-1}}$$

Note that the empirical expression given by *Eq. 11-96* has to be squared, because two aniline cations are required to exchange one calcium ion. Furthermore, for the same reason, K_{ie} is not dimensionless as would have been the case for a one-to-one exchange stoichiometry (see the example given by *Eq. 11-79*), but has the units of L·kg$_s^{-1}$ (see the example given by *Eq. 11-107*). Finally, this simple approach to estimate K_{ie} of an organic compound does not distinguish between the various monovalent (e.g., Na^+, K^+) and bivalent cations (e.g., Mg^{2+}, Ca^{2+}), respectively. Nevertheless, for an order of magnitude estimate of the relative importance of cation exchange for the overall sorption of an organic compound, it can be used as a reasonable starting point.

To complete the ion exchange term in Eq. J-10, the cation exchange capacity (CEC = $\sigma_{ie} \cdot A$) of the soil must be estimated. In the soil considered, only the CEC of quartz, kaolinite and the organic matter have to be taken into account because the iron oxide surface is positively charged at pH 6 (the pH_{zpc} is between 7 and 8, see *Table 11.5*). From *Table 11.5*, typical A and σ_{ie} values can be obtained:

solid	weight fraction	typical A ($m^2 \cdot kg_s^{-1}$)	typical σ_{ie} [a] (mol charges· m^{-2})
quartz	0.85	~ 10^2	10^{-7}
kaolinite	0.10	~ 10^4	$(0.2\text{-}1) \times 10^{-5}$
organic matter (humus)	0.01	~ 10^3	$(1\text{-}10) \times 10^{-3}$

[a] Note that in *Table 11.5* σ_{ie} is denoted as CEC!

Multiplication of $A \cdot \sigma_{ie}$ with the corresponding weight fractions demonstrates that the natural organic matter is probably the biggest contributor to the total CEC of the soil, which comes out to be on the order of 0.01 to 0.1 mol exchange sites $\cdot$ kg_s^{-1}. Inserting these values into Eq. J-10, and with $K_{ie} = 5\ L \cdot kg_s^{-1}$, $[Ca^{2+}] = 10^{-4}\ mol \cdot L^{-1}$, and (0.96) $K_{om} \cdot f_{om} = 0.06\ L \cdot kg_s^{-1}$, one then obtains

$$K_d \simeq 0.06 + (0.04) \left(\frac{5 \cdot (0.01 \text{ to } 0.1)}{2 \cdot 10^{-4}} \right)^{1/2} = 0.7 \text{ to } 2\ L \cdot kg_s^{-1}$$

This calculation suggests that, although only about 4% of the aniline is present as anilinium cation at the pH of the groundwater, cation exchange (presumably at organic matter carboxylate groups) is important enough to dominate the overall solid-water distribution ratio.

Note once again that the present knowledge on the magnitude of K_{ie} of organic cations is rather poor and that, therefore, this result should be considered with caution.

Answer (d)

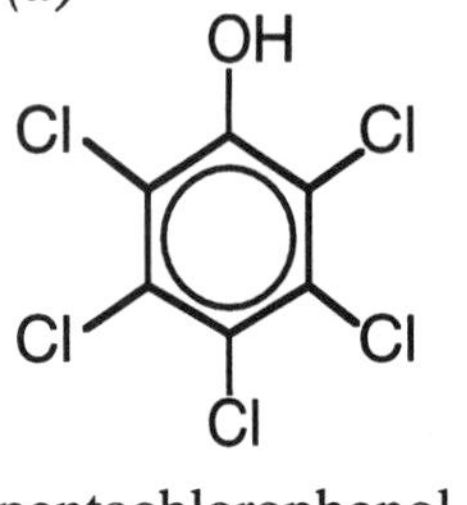

OH, Cl, Cl, Cl, Cl, Cl

pentachlorophenol
(PCP)

mw	266.3 $g \cdot mol^{-1}$
T_m	174 °C
T_b	310 °C
K_{ow}(HA,25°C)	1.7×10^5
pK_a(HA)	4.8 (see Section E)

The fraction of PCP present as nondissociated species at pH 6 is

$$\alpha_a = \frac{1}{1 + 10^{(6.0-4.8)}} \simeq 0.06$$

or about 6%. As pointed out when discussing the sorption of 2,4,5-trichlorophenol in *Section 11.6*, in cases where the solution pH is less than about two units above the acid's pK_a, one may, in general, neglect the partitioning of the anionic species (here the PCP^-) into particulate organic matter, at least at low salt concentrations (see also *Jafvert, 1990*). Thus, when postulating that no sorption mechanism other than hydrophobic partitioning is important, K_d can be expressed simply by (see also *Eq. 11-52*):

$$K_d \simeq \alpha_a \cdot K_d(PCP) = \alpha_a \cdot K_{om}(PCP) \cdot f_{om} \qquad \text{(J-11)}$$

$K_{om}(PCP)$ can be estimated from $K_{ow}(PCP)$ using *Eq. 11-22e*, which has been determined for chlorinated phenols:

$$\log K_{om} = 0.81 \log K_{ow} - 0.25 \qquad \textit{(11-22e)}$$

The result is $K_{om}(PCP) \simeq 10^4 \ L \cdot kg_s^{-1}$. Insertion of this value into Eq. J-11 with $f_{om} = 0.01$ yields a first (lower limit) estimate of K_d:

$$K_d \text{ (hydrophobic partitioning)} \simeq (0.06)\,(10^4)\,(0.01) \simeq 6 \ L \cdot kg_s^{-1}$$

It should be pointed out, however, that K_d values have been reported for PCP that are substantially larger than one would have predicted from Eq. J-11 (e.g., *Schellenberg et al., 1984*). One additional sorption mechanism is, of course, sorption of the anionic species (i.e., PCP^-) by anion exchange. As can be seen from *Table 11.5*, various mineral oxides including iron oxides and clay minerals may, depending on the pH (see Table *11.4)*, exhibit a significant anion exchange capacity (AEC). With the present knowledge on anion exchange processes, particularly of hydrophobic organic anions, it is, however, very difficult to give a reasonable quantitative estimate of the contribution of this process to the overall K_d of PCP. Such an estimate is difficult to make because natural organic matter tends to adsorb strongly to anion exchange sites (see, e.g., *Tipping, 1981*), thus making such sites unavailable to other anionic species such as PCP^-.

Determining K_d-Values from Experimental Data

A common way to determine K_d values experimentally is to measure sorption isotherms in batch experiments. To this end, the equilibrium concentrations of a given compound in the solid phase (C_s) and in the aqueous phase (C_w) are determined at various compound concentrations and/or solid-water ratios. A plot of C_s versus C_w is then fitted using an algebraic sorption equation, in many cases a Freundlich isotherm (see *Section 11.2*):

$$C_s = K \cdot C_w^n \qquad (11\text{-}1)$$

At low concentrations, *Eq. 11-1* can, in general, be simplified to the linear expression:

$$C_s = K \cdot C_w$$

Note that in this case, K is equal to K_d.

Consider now the sorption of 1,4-dinitrobenzene (1,4-DNB) to K^+-illite at pH 7.0 and 20°C. Recall that 1,4-DNB forms EDA-complexes with clay minerals (see case (b) in the preceding illustrative examples). In a series of batch experiments, *Haderlein et al. (1995)* have measured the following data:

O_2N–C_6H_4–NO_2 1,4-dinitrobenzene (1,4-DNB)

C_w ($\mu mol \cdot L^{-1}$)	C_s ($\mu mol \cdot kg_s^{-1}$)
0.06	97
0.17	241
0.24	363
0.34	483
0.51	633
0.85	915

C_w ($\mu mol \cdot L^{-1}$)	C_s ($\mu mol \cdot kg_s^{-1}$)
1.8	1640
2.8	2160
3.6	2850
7.6	4240
19.5	6100
26.5	7060

Problem

Using the experimental data given above, estimate the K_d-values for 1,4-DNB in a K^+-illite/water suspension (pH 7.0, 20°C) for equilibrium concentrations of 1,4-DNB in the aqueous phase of (a) 0.2 μM and (b) 15 μM, respectively.

Answer

Plot C_w versus C_s to get an idea of the shape of the sorption isotherm:

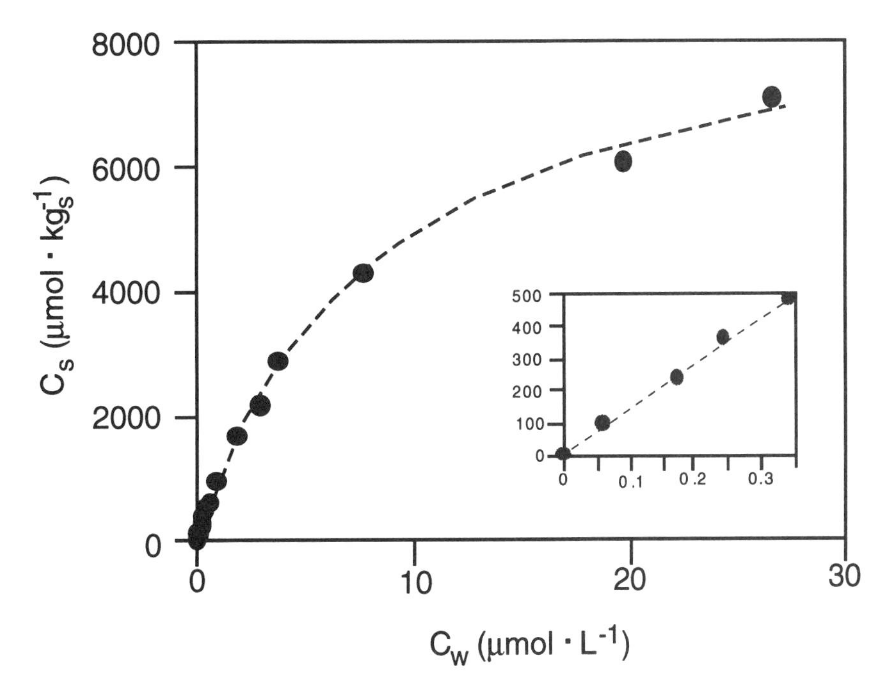

Figure J.1 Plot of C_s versus C_w. The dotted line represents the fitted Langmuir equation (Eq. J-14 see below).

(a) For estimating K_d(1,4-DNB) at C_w = 0.2 μM, assume a linear isotherm for the concentration range 0-0.5 μM. Perform a least square fit of C_s versus C_w using the origin and the first 4 data points only (see the insert in Fig. J.1). The resulting regression equation is

$$C_s = 1430 \cdot C_w \quad (R = 0.99) \tag{J-12}$$

Hence,

$$K_d = 1430 \ \mathrm{L \cdot kg_s^{-1}}$$

(b) For deriving K_d at C_w = 15 µM, fit the experimental data with the Freundlich equation (*Eq. 11-1*). To determine the K and n values, use the logarithmic form of *Eq. 11-1*:

$$\log C_s = n \cdot \log C_w + \log K$$

Perform a least square fit of log C_s versus log C_w using all data points given above:

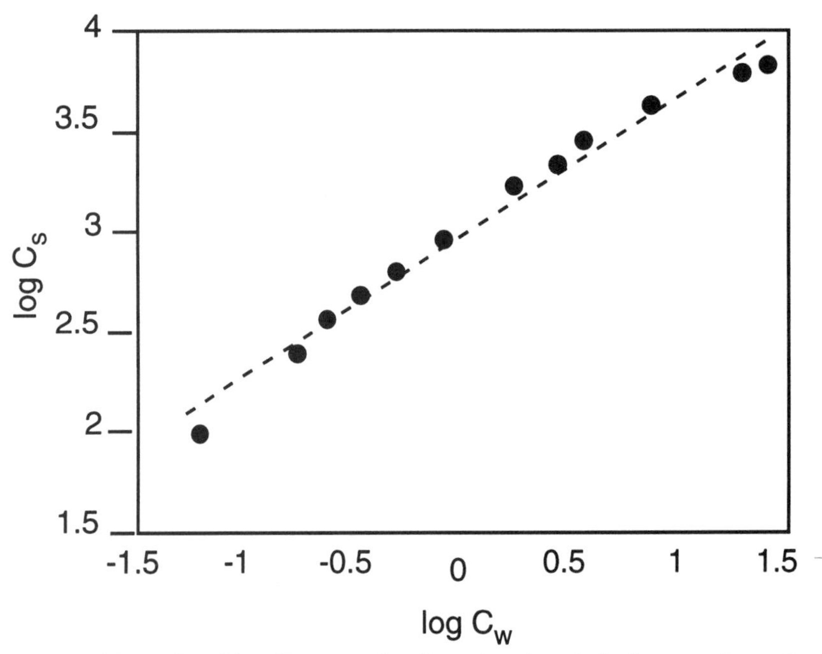

Figure J.2 Plot of log C_s versus log C_w using the whole data set given above.

The resulting regression equation is

$$\log C_s = 0.70 \log C_w + 2.97 \ (R = 0.99)$$

Hence, the Freundlich isotherm for describing the sorption of 1,4-DNB to K^+-illite at aqueous 1,4-DNB concentrations between 0.06 and 26.5 $\mu mol \cdot L^{-1}$ is given by

$$C_s = 10^{2.97} \cdot C_w^{0.7} \simeq 1000 \cdot C_w^{0.7} \qquad \text{(J-13)}$$

Insertion of C_w = 15 $\mu mol \cdot L^{-1}$ yields a C_s value of about 6700 $\mu mol \cdot kg_s^{-1}$ and thus a K_d value of

$$K_d \simeq \frac{6700\ \mu\text{mol}\cdot\text{kg}_s^{-1}}{15\ \mu\text{mol}\cdot\text{L}^{-1}} \simeq 450\ \text{L}\cdot\text{kg}_s^{-1}$$

Note that this K_d value is significantly smaller than the K_d obtained in the linear part of the isotherm (i.e., at low 1,4-DNB concentrations). Furthermore, as can be seen from Figure J.2, Eq. J-13 overestimates C_s (and thus K_d) at both the low and the high end of the concentration range considered. In fact, inspection of Figure J.1 reveals that at very high concentrations, the K^+-illite surface seems to become saturated with 1,4-DNB, which is not surprising considering that only limited adsorption sites are available. In such a case, the sorption isotherm can also be approximated by a Langmuir equation (see *Eq. 11-85* and *Fig. 11.21)*:

$$C_s = \frac{K_{Langmuir} \cdot C_w}{1 + K_{Langmuir} \cdot C_w} \cdot C_{s,max} \qquad \text{(J-14)}$$

where $K_{Langmuir}$ is the Langmuir constant (e.g., in $L \cdot mol^{-1}$) and $C_{s,max}$ is the maximum surface concentration of the compound. The dotted line in Figure J.1 represents the fitted Langmuir equation J-14. The fitting coefficients are $K_{Langmuir} = 0.122\ L \cdot \mu mol^{-1}$ and $C_{s,max} = 8980\ \mu mol \cdot kg_s^{-1}$, respectively. At very low concentrations, i.e., $K_{Langmuir} \cdot C_w << 1$, Eq. J-14 simplifies to the linear relationship

$$Cs = K_{Langmuir} \cdot C_w \cdot C_{s,max}$$

Hence, for this part of the isotherm, one obtains

$$K_d \simeq K_{Langmuir} \cdot C_{s,max} \qquad \text{(J-15)}$$

Insertion of the corresponding values for 1,4-DNB into Eq. J-15 gives a K_d value of about 1100 $L \cdot kg_s^{-1}$, which is somewhat smaller than the K_d value determined from the linear regression analysis using only the first four data points (i.e., $K_d = 1430\ L \cdot kg_s^{-1}$, see above). This again is not too surprising, when considering that the Langmuir model assumes that all surface sites exhibit exactly the same affinities for the sorbate, which is, of course, not necessarily the case. It is very likely that sites with higher affinities are occupied first. Therefore, a linear fit of data points determined at low concentrations can be expected to yield a higher apparent sorption constant as compared to the constant calculated from nonlinear extrapolation of data covering a wide concentration range (see also *Haderlein and Schwarzenbach, 1993*). Consequently, when estimating K_d values from experimental data, depending on the concentration range of interest, one has to make an optimal choice with respect to the selection of the experimental data points as well as with respect to the type of isotherm used to fit the data.

Assessing the Relative Importance of Sedimentation in a Well-Mixed Water Body

Problem

Consider a well-mixed eutrophic water body with a volume $V = 10^4$ m^3, a surface area $A = 2 \times 10^3$ m^2, and an average water throughflow $Q = 10^2$ $m^3 \cdot d^{-1}$. Assume a particle steady-state concentration, $r_{sw} \simeq [S]$, of 5×10^{-3} $kg_s \cdot m^{-3}$, of which 40% consists of organic material (i.e., $f_{om} = 0.4$ $kg_{om} \cdot kg_s^{-1}$). The average flux of particles to the sediments, F_s, is 5×10^{-3} $kg_s \cdot m^{-2} \cdot d^{-1}$. Calculate how large the K_d-value of an organic compound has to be, in order to make sedimentation an equally important elimination process for the compound as compared to export by flushing. What is the corresponding K_{om}-value of the compound if hydrophobic partitioning is the dominant sorption mechanism?

Answer

Start with a simple mass balance equation (see *Section 15.2*) by assuming that only flushing and sedimentation are responsible for the export of the compound from the water body:

$$\frac{dM}{dt} = \underset{\text{input}}{I_{in}} - \underset{\substack{\text{export}\\\text{by}\\\text{flushing}}}{Q \cdot C_t} - \underset{\substack{\text{elimination}\\\text{by}\\\text{sedimentation}}}{S}$$

M is the total mass (mol) and C_t is the total concentration ($mol \cdot L^{-1}$) of the compound in the water body. Express elimination via sedimentation by

$$S = A \cdot F_s \cdot C_s$$

where C_s = concentration ($mol \cdot L^{-1}$) of the compounds on the solids ($mol \cdot kg_s^{-1}$). Assume sorption equilibrium and a linear sorption isotherm (*Eq. 11-2*):

$$C_s = K_d \cdot C_w$$

with (*Eqs .11-3* to *11-5):*

$$C_w = f_w \cdot C_t = \frac{1}{1 + r_{sw} \cdot K_d} C_t$$

Since $r_{sw} \cong [S]$ (the total volume is about equal to the water volume), you get

$$S = A \cdot F_s \cdot \frac{K_d}{1 + [S] \cdot K_d} \cdot C_t$$

By setting $Q \cdot C_t = S$ and by rearranging the equation you obtain

$$K_d = \frac{Q}{A \cdot F_s - Q \cdot [S]} \simeq 10 \ m^3 \cdot kg_s^{-1} \simeq 10^4 \ L \cdot kg_s^{-1}$$

If only hydrophobic sorption is important (i.e., $K_d = f_{om} \cdot K_{om}$; see *Eq. 11-16)*, sedimentation is equally important to flushing if

$$K_{om} = K_d \cdot f_{om}^{-1} = 2.5 \times 10^4 \ L \cdot kg_{om}^{-1}$$

Note that by assuming a linear sorption isotherm, you have expressed sedimentation by a first-order process. Dividing the mass balance equation given above by the total volume yields ($M = V \cdot C_t$)

$$\frac{dC_t}{dt} = \frac{I_{in}}{V} - \frac{Q}{V} C_t - \frac{S}{V}$$

As discussed in Chapter N (and *Chapter 15*), $Q/V = k_w = 0.01\ d^{-1}$ is the characteristic first-order rate constant for elimination by flushing (Eq. N-5). For sedimentation, one obtains

$$\frac{S}{V} = \frac{A \cdot F_s}{V} \cdot \frac{K_d}{1 + [S] \cdot K_d} \cdot C_t$$

Hence, a characteristic rate constant, k_{sed}, can be derived:

$$k_{sed} = \frac{A \cdot F_s}{V} \cdot \frac{K_d}{1 + [S] \cdot K_d}$$

Setting $V/A = h_{mix}$ (average depth) and $F_s/[S] = v_s$ (average settling velocity of particles), k_{sed} can be expressed as (Eqs. N-9 and N-10)

$$k_{sed} = \frac{v_s}{h_{mix}} \cdot \frac{K_d \cdot [s]}{1 + [S] \cdot K_d} = \frac{v_s}{h_{mix}} (1 - f_w)$$

In the case considered, $v_s = 1\ m{\cdot}d^{-1}$, $h_{mix} = 5$ m, and, therefore

$$k_{sed} = (0.2)(1 - f_w)\ d^{-1}$$

Thus, about 5 % of the compound has to be present in particulate form to yield $k_{sed} = k_w = 0.01\ d^{-1}$.

Assessing the Effect of Organic Colloids on Air-Water Partitioning of Organic Compounds

Associations of organic compounds with settleable solids and/or colloids (i.e., particles that are too small to settle) may have a significant effect on those processes that involve only "truly dissolved" molecules. Such processes include air-water exchange and passive uptake by organisms.

One way to measure the affinity of hydrophobic organic compounds for colloidal organic matter (COM) is to determine the effect such organic matter has on the compound's apparent solubility. Colloidal organic matter has traditionally been included in the pool of organic matter that passes through a filter (e.g., 0.4 μm pore openings) and is often incorrectly referred to as "dissolved". In the case of a particular polychlorinated biphenyl, 2,4,4'-trichlorobiphenyl (TCBP), *Chiou et al. (1986)* have found that additions of colloidal organic matter to water caused this particular compound's apparent solubility to change:

$$C_w^{sat} \text{ (apparent) in } \mu\text{mol}\cdot\text{L}^{-1} = (7.1 \times 10^3)\,[\text{COM}] + 0.42 \qquad \text{(J-16)}$$

where C_w^{sat} (apparent, μM) is the total TCBP in the dissolved plus colloid-associated phases and [COM] has units of $kg_{om}\cdot L^{-1}$. Note that this measurement reflects saturated conditions; thus, the similar assumptions have to be made as in the case in which Henry's law constants were approximated using aqueous solubility data (see *Chapter 6.2*).

Problem

(a) Calculate the sorption coefficient, K_c (see *Eq. 11-36*), for the association of TCBP with the colloidal organic matter used by *Chiou et al. (1986)*, and compare it with the K_{om} you would predict from TCBP's K_{ow}.

(b) Consider a well-sealed flask with 50 mL of water (V_w) and 5 L of air (V_a) at 25°C. Suppose the water also contains 50 $mg\cdot L^{-1}$ of colloidal organic matter ([COM] = 5×10^{-5} $kg_{om}\cdot L^{-1}$). If 10^{-8} moles of TCBP were added to this system, what concentration of TCBP would you calculate for the water-plus-colloidal "phase"?

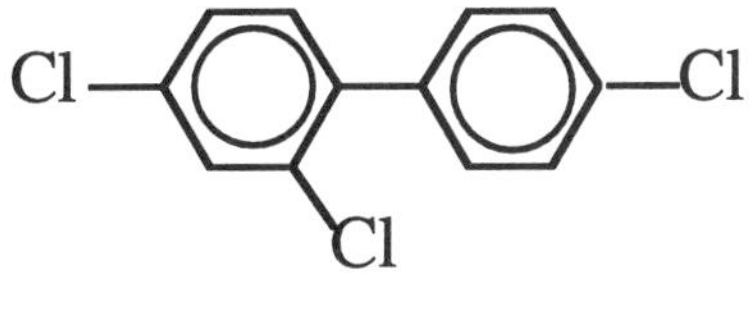

2,4,4'-trichlorobiphenyl
(2,4,4'-TCBP)

Answer

(a) The coefficients in Eq. J-16 reveal both the true aqueous solubility of TCBP and the sorption coefficient involved (see *Fig. 11.11)* since

$$C_T = C_w^{sat} \cdot K_c \cdot r_{cw} + C_w^{sat}$$

When [COM] = 0, C_w^{sat} = 0.42 μM. Also, if [COM] is taken as a measure of r_{cw}, then the slope in the *Chiou et al. (1986)* regression must be

$$C_w^{sat} \cdot K_c = 7.1 \times 10^3 \ \mu mol \cdot kg_{om}^{-1}$$

Thus,

$$K_c = 7.1 \times 10^3 \ \mu mol \cdot kg_{om}^{-1} \ / \ 0.42 \ \mu mol \cdot L^{-1}$$

$$= 1.7 \times 10^4 \ L \cdot kg_{om}^{-1}$$

To estimate a value of K_{om} for TCBP, use regression *Eq. 11-22b*:

$$\log K_{om} = 0.88 \log K_{ow} - 0.27$$

and a value of K_{ow} for this particular trichlorobiphenyl ($\log K_{ow} \simeq 5.7$ (*Chiou et al., 1986*)). Thus, one finds

$$\log K_{om} \simeq 0.88 \times 5.7 - 0.27$$

and

$$\simeq 4.7$$

$$K_{om} \simeq 5.0 \times 10^4 \ L \cdot kg_{om}^{-1}$$

This estimated value is somewhat larger (factor of 3) than the value measured by *Chiou et al. (1983)*; generally it is felt that colloidal organic matter in aqueous suspension is likely to be more polar than the organic matter condensed on sediment or soil particles. Hence, one may not be too surprised to see that such a K_c has a somewhat lower value than the comparable K_{om}.

(b) The "concentration" requested is a sum of dissolved plus colloid-bound concentrations, $C_w + C_c \cdot r_{cw}$ (where C_w is in units of $mol \cdot L^{-1}$, C_c is in units of $mol \cdot kg_{om}^{-1}$, and r_{cw} is in $kg_{om} \cdot L^{-1}$). Since C_c is related to C_w via the expression

$$K_c = \frac{C_c}{C_w}$$

the total water-borne "concentration" is also equal to $C_w \cdot (1 + r_{cw} \cdot K_c)$. Hence, one needs to solve for C_w.

The total mass (M) of TCBP added to the closed system distributes itself between the air, water, and colloidal phases

$$M_{total} \text{ (in moles)} = C_a \cdot V_a + C_w \cdot V_w + C_c \cdot r_{cw} \cdot V_w$$

and the fraction, f_w, which is truly dissolved in the water is

$$f_w = \frac{C_w \cdot V_w}{M_{total}} \qquad \text{(J-17)}$$

Dividing both the numerator and denominator of the right hand side of Eq. J-17 by the product, $C_w \cdot V_w$, and substituting equilibrium constants for the resulting concentration ratios gives

$$f_w = \frac{1}{(K_H' \cdot r_{aw} + 1 + K_c \cdot r_{cw})} \qquad \text{(J-18)}$$

where r_{aw} is the ratio of the air volume in the system to the water volume.

For 2,4,4'-TCBP, K_H' is 0.0082 $L_a \cdot L_w^{-1}$ *(Brunner et al., 1990)*, and as you have already considered above, K_c is 1.7×10^4 $L \cdot kg_{om}^{-1}$. Insertion of these values together with the corresponding r-values yields

$$f_w = 1 / \{(0.0082 \, L_a \cdot L_w^{-1} \times 5 \, L_a / 0.05 \, L_w) + 1$$

$$+ 1.7 \times 10^4 \, L \cdot kg_{om}^{-1} \times 5 \times 10^{-5} \, kg_{om} \cdot L^{-1})\}$$

$$= 0.37$$

Hence, of the total 10^{-8} mol TCBP added, 37% is dissolved in the water giving a water concentration of

$$C_w = 0.37 \times 10^{-8} \text{ mol} / 0.05 \text{ L}$$

$$= 7.4 \times 10^{-8} \text{ mol} \cdot L^{-1}$$

Finally, to deduce the total water-borne "concentration":

$$\text{total waterborne} = C_w (1 + r_{cw} \cdot K_c)$$

$$= 7.4 \times 10^{-8}\ \text{mol} \cdot \text{L}^{-1} \cdot (1 + 5 \times 10^{-5} \times 1.7 \times 10^{4})$$

$$= 1.4 \times 10^{-7}\ \text{mol} \cdot \text{L}^{-1}$$

In other words, the presence of colloids causes nearly a doubling of the total TCBP ultimately distributed into water plus colloid-bound phase. Thus, due to colloid-associated species, the air-water distribution ratio (see *Eq. 11-35*) differs substantially from the Henry's law constant for this compound under these conditions:

$$D_{aw} = C_a / (C_w + C_c \cdot r_{cw}) = \{(10^{-8}\ \text{mol} - 1.4 \times 10^{-7}\ \text{mol} \cdot \text{L}^{-1} \times 0.05\ \text{L}) / 5\ \text{L}_a\} / (1.4 \times 10^{-7}\ \text{mol} \cdot \text{L}_w^{-1})$$

$$= 0.0043\ \text{L}_w \cdot \text{L}_a^{-1}$$

which is about half of K_H' ($0.0082\ L_w \cdot L_a^{-1}$).

PROBLEMS

● *J-1 Estimating K_d and the Mobility of Organic Pollutants in the Clay Liner of a Landfill*

As an employee of a company that is running a hazardous waste site, you are asked to assess whether certain organic pollutants present in the site are able to diffuse through the clay liner built around the site to protect the groundwater in the area. In this context, you are interested in calculating the effective molecular diffusion coefficient, D_{eff}, of the compounds of interest, assuming that the clay is saturated with water (see also *Eq. 11.155*):

$$D_{eff} = \frac{D_w}{R_f}$$

where D_w is the molecular diffusion coefficient in water and R_f is the retardation factor and is given by (see *Section 11.2*):

$$R_f = 1 + r_{sw} \cdot K_d$$

Estimate D_{eff} for 1,3-dichlorobenzene, 4-chloronitrobenzene, 2,4,6-trichlorophenol, 4-chloro-2-nitrophenol assuming that the clay liner consists primarily of kaolinite ($f_{om} = 0.001\ kg_{om} \cdot kg_s^{-1}$, density $\rho_s = 2.65\ g \cdot cm_s^{-1}$, porosity $\Phi = 0.4$), and that the major cations present in the aqueous phase (pH = 7.2, T = 10°C) are Na^+ ($5 \times 10^{-3}\ mol \cdot L^{-1}$) and K^+ ($10^{-4}\ mol \cdot L^{-1}$). Also assume linear sorption isotherms. What is the average distance that each of the compounds travels by molecular diffusion in 1 year? For calculating the average travel distance, use a tortuosity $\chi = 1.5$ (see Chapter P). χ accounts for the extra average diffusion distance for the molecules going around the particles. Hence, the travel distance is given by $x = (2 \cdot D_{eff} \cdot t / \chi)^{1/2}$.

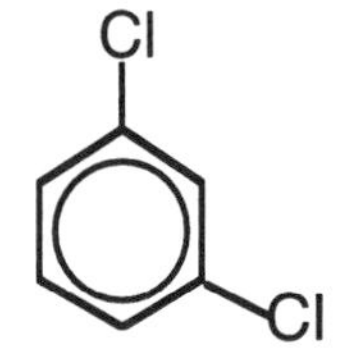

1,3-dichlorobenzene
(log $K_{ow} = 3.30$)

4-chloronitrobenzene
(log $K_{ow} = 2.40$;
log K_{EDA}(K^+-kaolinite) =
0.65 (K_{EDA} in $L \cdot kg_s^{-1}$)
estimated from *Haderlein and Schwarzenbach, 1993*)

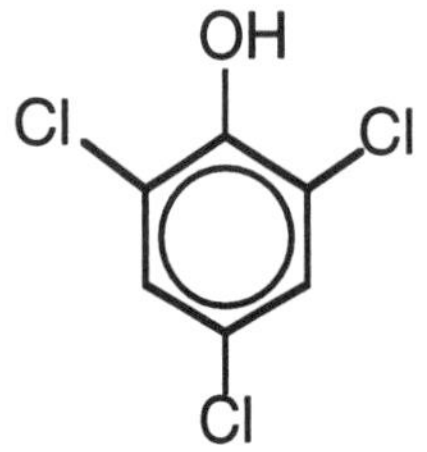

2,4,6-trichlorophenol
(log $K_{ow} = 3.72$; $pK_a = 6.13$;
see *Schellenberg et al., 1984*)

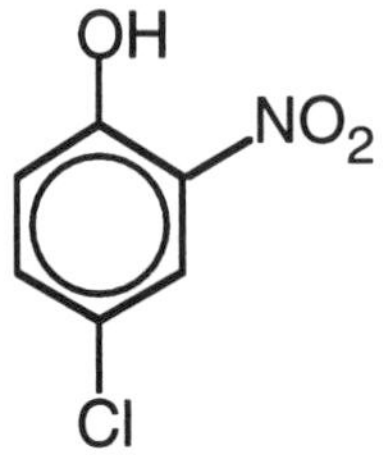

4-chloro-2-nitrophenol
(log K_{ow} (HA) = 2.46;
$pK_a = 6.44$; log K_{EDA} (HA, K^+-kaolinite) = 1.1 (K_{EDA} in $L \cdot kg_s^{-1}$)
estimated from *Haderlein and Schwarzenbach, 1993*)

● *J-2 Evaluating Experimental Sorption Data*

As part of your PhD thesis dealing with the mobility of organic pollutants in clay liners of landfills, you investigate the sorption of a series of nitrophenols to a variety of clay minerals. In this context, an undergraduate student (whom you supervise) measures the equilibrium partitioning of 2,6-dinitro-4-methylphenol (2,6-DNC) between K^+-montmorillonite and water in batch reactors at pH 2.8. He generates the data set given below. You ask him to plot the K_d value of 2,6-DNC as a function of pH (pH-range 4 to 9) by assuming a total 2,6-DNC concentration in the aqueous phase of (a) 0.05 μM, and (b) 2.0 μM. The student tells you that he does not know how to do this, so you have to do it yourself. When inspecting the data, you realize that case (b) is somewhat more tricky than case (a)! You also realize that for different concentration ranges, you may have to use different fitting equations. What does your plot (K_d versus pH) look like?

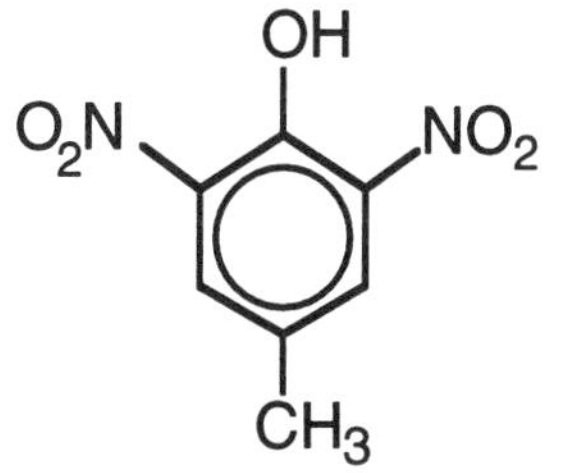

OH

O_2N NO_2

CH_3

2,6-dinitro-4-methylphenol
(2,6-DNC)

$pK_a = 4.06$

Results from the batch experiments (pH 2.8, T = 21°C):

C_w ($\mu mol \cdot L^{-1}$)	C_s ($\mu mol \cdot kg_s^{-1}$)	C_w ($\mu mol \cdot L^{-1}$)	C_s ($\mu mol \cdot kg_s^{-1}$)
0.062	924	0.314	4970
0.075	1490	0.784	9920
0.095	1990	1.57	14840
0.129	2490	2.30	19770
0.158	2980		

● *J-3 Estimating the Mobility of Organic Compounds during River Water-Groundwater Infiltration Using Data from Batch Sorption Experiments*

A river in your area is occasionally loaded with the six organic compounds (I-VI) listed below that are present in the wastewater of a chemical factory. As a consequence, the concentrations of these compounds in the river water vary significantly with time. Being a consultant of the water works that operates several groundwater pumping stations near the river downstream of the wastewater input, you are interested in the mobility of the compounds in the ground between the river and the groundwater wells. Since you have also studied Chapter P of this book, you are aware that knowledge of the retardation of the compounds in the ground is not only important for assessing their average travel times, but also for estimating the attenuation of concentration fluctuations during the ground passage. You assume that most of the retardation of the compounds takes place in the first few meters of infiltration (primarily through old river sediment). Therefore, you ask one of your coworkers to measure sorption isotherms in batch systems containing old river sediment and water from the infiltration site. Since your budget only allows you to investigate two compounds, you have to make a selection. Among the six compounds of interest you choose 1,2,4-trichlorobenzene (1,2,4-TCB, I) and 1,4-dinitrobenzene (1,4-DNB, IV). Is this the best choice? Give some reasons for your selections.

From your laboratory you get the data sets given below together with the information that the measurements were carried out at 20°C, that the pH of the water was 7.5, and that the average organic matter content of the sediment used was 1% (i.e., $f_{om} = 0.01\ kg_{om} \cdot kg_s^{-1}$).

Estimate the K_d-values and retardation factors ($R_f = f_w^{-1} = 1 + (\rho_s(1-\Phi)/\Phi)K_d$, see *Section 11.2*) of the six compounds (I-VI), by assuming a sediment density $\rho_s = 2.5\ g \cdot cm_s^{-3}$ and a porosity $\Phi = 0.2$. Assume also that sorption of the phenolate species (A^-) can be neglected in all cases, and that relative K_{EDA}-values are independent of the type of clay minerals present (if you do not know what K_{EDA} means, read Answer (b) of the first illustrative example of this chapter).

Compounds of interest:

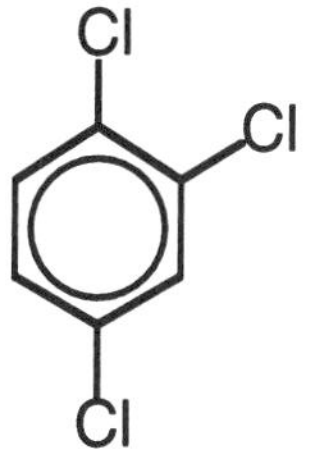

I
1,2,4-trichlorobenzene
(1,2,4-TCB; log K_{ow}=4.00;
see *Appendix*)

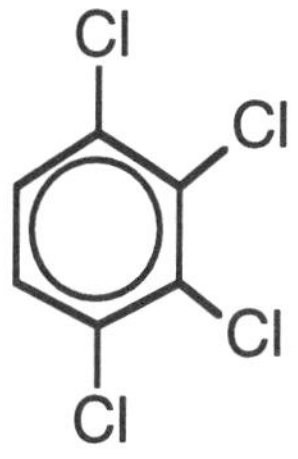

II
1,2,3,4-tetrachlorobenzene
(1,2,3,4-TeCB; log K_{ow}=4.55;
see *Appendix*)

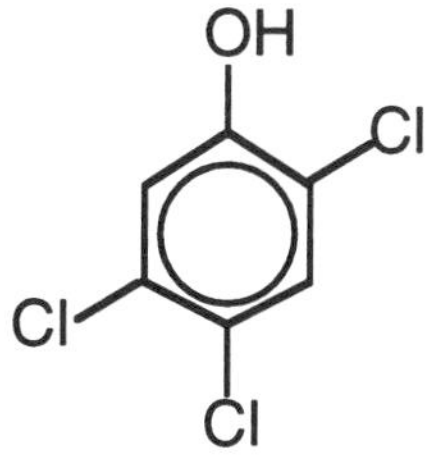

III
2,4,5-trichlorophenol
(2,4,5-TCP; log K_{ow} (HA) = 4.19;
pK_a =6.94;
see *Schellenberg et al., 1984*)

IV
1,4-dinitrobenzene
(1,4-DNB; log K_{ow}=1.48;
log K_{EDA}(K^+-montmorillonite)=
3.49 (K_{EDA} in $L \cdot kg_s^{-1}$)
see *Haderlein et al., 1995*)

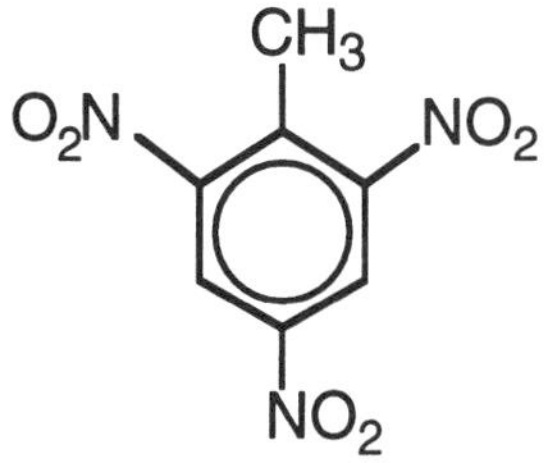

V
2,4,6-trinitrotoluene
(TNT, log K_{ow}=1.74; log K_{EDA}
(K^+-montomorillonite)=
4.33 (K_{EDA} in $L \cdot kg_s^{-1}$));
see *Haderlein et al., 1995*)

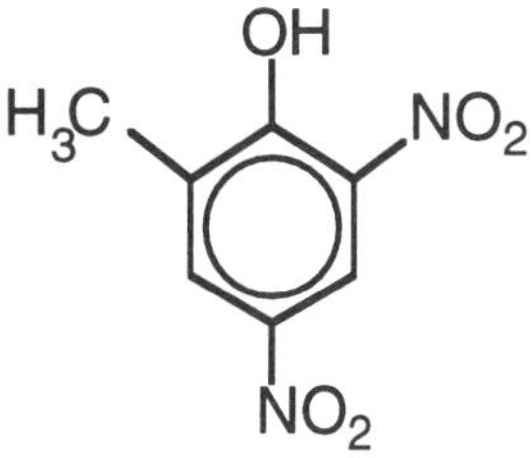

VI
2,4-dinitro-ortho-cresol
(DNOC, log K_{ow} (HA)=2.12; log
K_{EDA}(HA, K^+-montmorillonite)=
4.57 (K_{EDA} in $L \cdot kg_s^{-1}$); pK_a=4.31;
see *Haderlein et al., 1995*)

Results from the batch experiments:

1,2,4-TCB (I)		1,4-DNB(IV)	
C_w ($\mu mol \cdot L^{-1}$)	C_s ($\mu mol \cdot kg_s^{-1}$)	C_w ($\mu mol \cdot L^{-1}$)	C_s ($\mu mol \cdot kg_s^{-1}$)
0.2	6.0	0.2	0.8
0.5	17.2	0.45	1.9
1.0	32.4	0.95	4.3
3.2	100.0	2.8	12.0
6.2	201.5	5.8	26.5

● *J-4 How Much Phenanthrene Is Removed by Sedimentation?*

With a wastewater effluent, 0.1 kg of phenanthrene is introduced each day into a small lake (volume $V = 10^5\ m^3$, surface area $A = 2 \times 10^4\ m^2$, average throughflow $Q = 10^3\ m^3 \cdot d^{-1}$). The steady-state particle concentration in the lake water, [S], is $3 \times 10^{-3}\ kg_s \cdot m^{-3}$, of which about 40% consists of organic material. The average flux of particles to the sediment, F_s, is $5 \times 10^{-3}\ kg_s \cdot m^{-2} \cdot d^{-1}$. Calculate what fraction of the total input of phenanthrene is transported from the water column to the sediments, assuming that loss by sedimentation is half as important as loss by water-air exchange and that no transformation reactions take place.

mw	$178.2\ g \cdot mol^{-1}$
K_{ow} (25°C)	3.7×10^4

phenanthrene

● *J-5 What Fraction of These PCB Congeners Is Truly Dissolved in the Water Column of Lake Michigan?*

Achman et al. (1993) determined the distribution of PCBs between filterable and unfilterable phases in water samples from Green Bay, Lake Michigan. They passed the water samples through 0.8 μm glass fiber filters, and they measured the concentrations of the individual PCB congeners both in the filtered water and on the particles trapped on the filters. They also measured the organic carbon content (POC) of the particles. By normalizing the PCB concentrations to the particulate organic carbon on the filters and then by dividing these values by the corresponding PCB concentrations in the filtrate, they calculated "field" K_{oc} values for each PCB congener. For each site and date, they then plotted the thus obtained log K_{oc}'s versus the log K_{ow}'s of the compounds, and fitted the data with a linear regression line. An example is given below. In most cases, the slopes of the regression lines were in the order of 0.2-0.3.

Try to explain these observed trends (in a quantitative way!). What fraction of a particular PCB congener with $K_{ow} = 10^{5.5}$ and $K_{ow} = 10^{7}$, respectively, would you estimate to have been truly dissolved in the water column at site 4 on June 10, 1989 (see graph below).

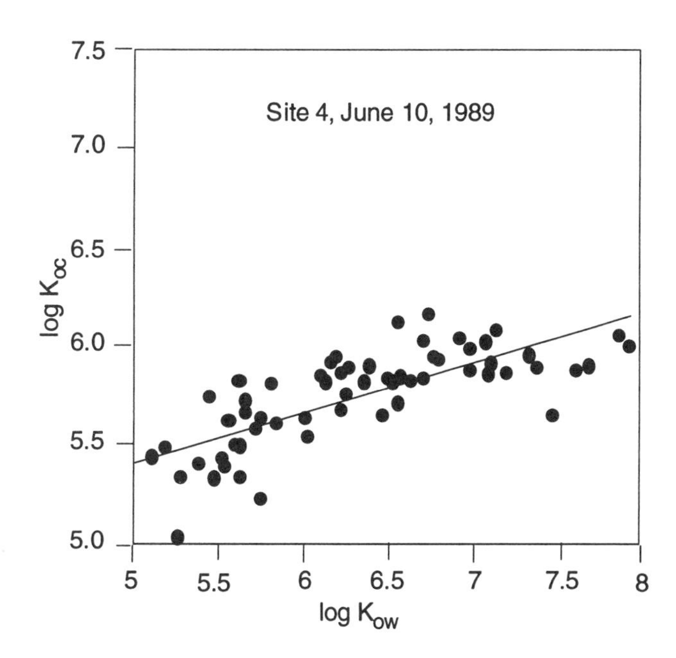

Personal Notes

K. CHEMICAL TRANSFORMATION REACTIONS

ILLUSTRATIVE EXAMPLES

Evaluating the Thermodynamics of Hydrolysis and Other Nucleophilic Substitution Reactions

Problem

Consider the hydrolysis of methyl bromide (CH_3Br) at 25°C in an aqueous solution (pH 7.0 (buffered), T = 25°C) containing 1 mM Br^-:

$$CH_3Br + H_2O \rightleftharpoons CH_3OH + Br^- + H^+ \qquad \text{(K-1)}$$

To what extent is CH_3Br ultimately converted to CH_3OH when assuming that CH_3Br is initially present at a low concentration (i.e., $[CH_3Br] < 1\ \mu M$)? In the literature, you find the following data (all for 25°C):

Species	ΔG_f^0 (kJ·mol^{-1})	P^o (atm)	C_w^{sat}(1atm) (mol·L^{-1})
CH_3Br (g)	- 28.2	2.1	1.6×10^{-1}
CH_3OH (aq)	- 175.4		
H_2O (l)	- 237.2		
H^+ (aq)	0		
Br^- (aq)	- 104.0		

Answer
Using the infinite dilution state as the reference state (except for H_2O), and assuming activity coefficients of 1 for all species involved in the reaction, the free-energy change of the reaction Eq. K-1 is given by (see also *Eq. 3-32)*

$$\Delta G_{reaction} = \Delta G^o + RT \ln \frac{[CH_3OH]\,[Br^-]\,[H^+]}{[CH_3Br] \cdot 1} \qquad \text{(K-2)}$$

Note that the concentration (mole fraction) of H_2O is more or less equal to its concentration in the standard state, and, therefore, a "1" appears in the denominator on the right hand side of Eq. K-2. Set $\Delta G_{reaction} = 0$ and rearrange Eq. K-2 to get an expression for the ratio of $[CH_3OH]/[CH_3Br]$ at equilibrium:

$$\ln \frac{[CH_3OH]}{[CH_3Br]} = -\frac{\Delta G^o}{RT} - \ln [Br^-]\,[H^+] \qquad \text{(K-3)}$$

The standard free energy change of the reaction Eq. K-2, ΔG^o, is given by

$$\begin{aligned}\Delta G^o = & -\Delta G_f^0(CH_3Br(aq)) - \Delta G_f^0\,(H_2O(l)) \\ & + \Delta G_f^0(CH_3OH(aq)) + \Delta G_f^0(Br^-(aq)) + \Delta G_f^0(H^+(aq))\end{aligned} \qquad \text{(K-4)}$$

For CH_3Br the $\Delta G_f^0(aq)$ value can be calculated from the $\Delta G_f^0\,(g)$ value and from the Henry's law constant of the compound:

$$\Delta G_f^0(aq) = \Delta G_f^0(g) + RT \ln K_H \qquad \textit{(12-101)}$$

K_H may be approximated from the water solubility at 1 atm partial pressure

$$K_H = \frac{1\ \text{atm}}{1.6 \times 10^{-1}\ \text{mol·L}^{-1}} \simeq 6.2\ \text{atm·L·mol}^{-1}$$

Insertion of K_H into *Eq. 12-101* gives

$$\Delta G_f^0(CH_3Br(aq)) = -28.2\ kJ \cdot mol^{-1} + (8.315)\,(298.2)\ \ln(6.2)\ kJ \cdot mol^{-1}$$
$$= -23.7\ kJ \cdot mol^{-1}$$

Using this and the other ΔG_f^0 -values given above yields a ΔG^o of

$$\Delta G^o = -(-23.7) - (-237.2) + (-175.4) + (-104.0) + 0$$
$$= -18.5\ kJ \cdot mol^{-1}$$

Insertion of this value into Eq. K-3 and setting $[Br^-] = 10^{-3}$ and $[H^+] = 10^{-7}$ gives

$$\ln \frac{[CH_3OH]}{[CH_3Br]} \simeq 30.5 \quad \text{or} \quad \frac{[CH_3OH]}{[CH_3Br]} = 1.7 \times 10^{13}$$

Hence, virtually all CH_3Br will ultimately hydrolyse to CH_3OH.

Problem
Consider the reversible reaction of methyl bromide (CH_3Br) to methyl chloride (CH_3Cl):

$$CH_3Br + Cl^- \rightleftharpoons CH_3Cl + Br^- \qquad \text{(K-5)}$$

In which direction does this reaction occur at 25°C in a leachate containing 50 mM Cl^-, 1mM Br^-, and 100 times more CH_3Cl than CH_3Br? The ΔG_f^0(aq)-values for CH_3Cl and Cl^- are (– 53.9) $kJ \cdot mol^{-1}$ and (– 131.3) $kJ \cdot mol^{-1}$, respectively.

Answer

The $\Delta G_{reaction}$ is given by

$$\Delta G_{reaction} = \Delta G^o + RT \ln \frac{[CH_3Cl][Br^-]}{[CH_3Br][Cl^-]}$$

with

$$\Delta G^o = -(-23.7) - (-131.3) + (-53.9) + (-104.0) = -2.9 \text{ kJ·mol}^{-1}$$

Thus,

$$\Delta G_{reaction} = -2.9 + 2.48 \ln \frac{(100)(10^{-3})}{(1)(5 \times 10^{-2})} = -1.2 \text{ kJ·mol}^{-1}$$

The reaction occurs from left to right. Note that $\Delta G_{reaction} = 0$ (equilibrium) at $[CH_3Cl] / [CH_3Br] \simeq 160$.

Deriving Kinetic Parameters for Hydrolysis Reactions from Experimental Data

Consider the hydrolysis of 2,4-dinitrophenyl acetate (2,4-DNPA), a compound for which the acid-catalyzed reaction is unimportant at pH > 2 (see *Fig. 12.8)*:

$$\text{2,4-DNPA} + H_2O/OH^- \xrightarrow{k_h} \text{2,4-dinitrophenol (O(H))} + (H)O{-}C(=O){-}CH_3 \qquad \text{(K-6)}$$

2,4-DNPA

In a laboratory class, the time course of the change in concentration of 2,4-DNPA in homogeneous aqueous solution has been followed at various conditions of pH and temperature using an HPLC method.

Problem
Determine the (pseudo) first-order reaction rate constants, k_h, for reaction Eq. K-6 at pH 5.0 and pH 8.5 at 22.5°C using the following data sets:

pH 5.0[a], T = 22.5°C		pH 8.5, T = 22.5°C	
Time (min)	C (μM)	Time (min)	C (μM)
0	100.0	0	100.0
11.0	97.1	4.9	88.1
21.5	95.2	10.1	74.3
33.1	90.6	15.4	63.6
42.6	90.1	25.2	47.7
51.4	88.5	30.2	41.2
60.4	85.0	35.1	33.8
68.9	83.6	44.0	26.6
75.5	81.5	57.6	17.3

[a] Note that very similar results were also found at pH 4.0 and 22.5°C.

Answer
Assuming a (pseudo) first-order rate law *Eq. 12-10*, k_h can be determined from a least square fit of ln [C(t) / C(t = O)] versus time (see also *Fig. 12.1*):

$$\ln [C(t) / C(t = 0)] = -k_h \cdot t \qquad \text{(K-7)}$$

The resulting k_h-values are (see also Fig. K.1):

$$k_h(\text{pH } 5.0, 22.5°\text{C}) = 2.6 \times 10^{-3}\ \text{min}^{-1} = 4.4 \times 10^{-5}\ \text{s}^{-1}$$

$$k_h(\text{pH } 8.5, 22.5°\text{C}) = 3.1 \times 10^{-2}\ \text{min}^{-1} = 5.1 \times 10^{-4}\ \text{s}^{-1}$$

Note that k_h increases with increasing pH indicating that the base-catalyzed reaction is important at higher pH-values.

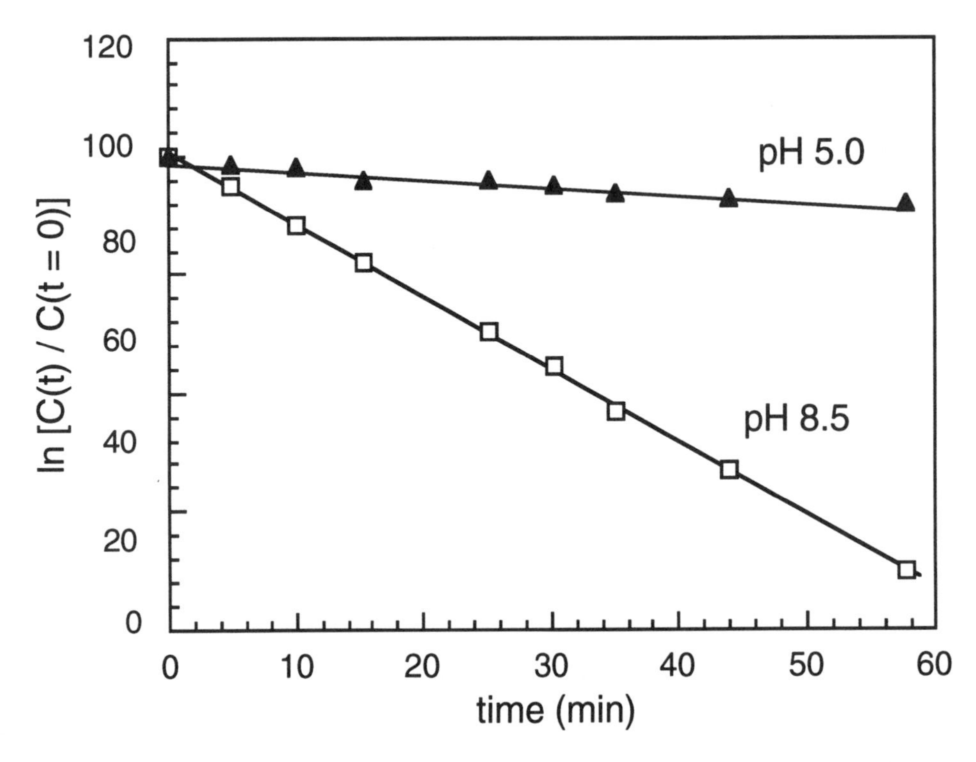

Figure K-1 Plot of ln [C(t) / C(t = 0)] versus time for 2,4-DNPA in aqueous solution at 22.5°C for two different pH-values.

Problem

Using the data given above, derive the rate constants for the neutral (k_N) and base-catalyzed (k_B) hydrolysis of 2.4-DNPA at 22.5°C. At what pH are the two reactions equally important?

Answer

When assuming that the acid-catalyzed reaction is not important in the pH-range considered, *Eq. 12-47* simplifies to

$$k_h = k_N + k_B \cdot [OH^-] \tag{K-8}$$

The fact that very similar k_h-values have been found at pH 4.0 and pH 5.0 indicates that at pH 5.0, the base-catalyzed reaction can be neglected, and, therefore,

$$k_N\,(22.5°C) = k_h\,(\text{pH } 5.0,\, 22.5°C) = 4.4 \times 10^{-5}\ s^{-1}$$

Using this k_N-value, k_B can be determined by rearranging Eq. K-7:

$$k_B\,(22.5°C) = \frac{k_h(\text{pH } 8.5,\, 22.5°C) - k_N\,(22.5°C)}{[OH^-]}$$

with (see *Eq. 8-12*)

$$[OH^-] = \frac{K_w}{[H^+]}$$

Note that the ionization constant of water, K_w, is strongly temperature dependent. At 22.5°C, $K_w = 10^{-14.08}$ (*CRC Handbook of Chemistry and Physics*). Hence, at pH 8.5 (i.e., $[H^+] = 10^{-8.5}$), $[OH^-] = 10^{-5.58}$, and

$$k_B\,(22.5°C) = \frac{4.7 \times 10^{-4}}{10^{-5.58}} = 180\ M^{-1} \cdot s^{-1}$$

The pH-value, I_{NB}, at which the neutral and the base-catalyzed reactions are of equal importance is given by (see *Fig. 12.9*)

$$I_{NB} = \log \frac{k_N}{k_B \cdot K_w} = \log \frac{4.4 \times 10^{-5}}{179 \cdot 10^{-14.08}} = 7.47$$

Thus, at pH 8.5, the hydrolysis of 2,4-DNPA is dominated by the base-catalyzed reaction.

Problem

Derive the Arrhenius activation energy, E_a, for the neutral hydrolysis of 2,4-DNPA using the following data:

Temperature (°C)	k_N (s^{-1})
17.7	3.1×10^{-5}
22.5	4.4×10^{-5}
25.0	5.2×10^{-5}
30.0	7.5×10^{-5}

Answer

According to *Eq. 12-17b*, the temperature dependence of a rate constant can be described by

$$\ln k = -\frac{E_a}{R} \cdot \frac{1}{T} + \text{const.}$$

Note that for the temperature range considered, E_a is assumed to be constant. Convert temperatures in °C to K and calculate 1/T-values. Also take the natural logarithms of the k_N-values:

1/T (K^{-1})	ln k_N (in s^{-1})
0.00344	- 10.38
0.00338	- 10.03
0.00335	- 9.86
0.00330	- 9.50

Perform a least square fit of ln k_N versus 1/T. The resulting slope is

$$\text{slope} = -\frac{E_a}{R} = -6318 \text{ K}$$

and, therefore,

$$E_a = -R \cdot \text{slope} = 8.315\ (6318) = 52.5 \text{ kJ} \cdot \text{mol}^{-1}$$

The E_a-value determined for the base-catalyzed reaction is 60.0 kJ·mol^{-1}.

Calculating Hydrolysis Reaction Times as a Function of Temperature and pH

Problem
Calculate the time required to decrease the concentration of 2,4-DNPA by hydrolysis to 50% (half-life) and to 5% of its initial concentration (a) in the epilimnion of a lake ($T = 22.5°C$, $pH = 8.5$), and (b) in the hypolimnion of the same lake ($T = 5°C$, $pH = 7.5$).

Answer
The hydrolysis half-life is calculated by

$$t_{1/2} = \frac{\ln 2}{k_h} = \frac{0.693}{k_h} \qquad \textit{(12-45)}$$

By analogy, the time required to reduce the concentration to 5% (i.e., $C(t)/C(t=0) = 0.05$) is given by (see Eq. K-7)

$$t_{0.05} = \frac{\ln(1/0.05)}{k_h} = \frac{3}{k_h} \qquad \text{(K-9)}$$

(a) Calculate k_h (Eq. K-8) for 22.5°C and pH 8.5 using the above derived k_N- and k_B-values and $[OH^-] = 10^{-5.58}$

$$k_h\,(22.5°C) = (180)\,(10^{-5.58}) + 4.4 \times 10^{-5} = 5.1 \times 10^{-4}\,s^{-1}$$

Note that at pH 8.5 and 22.5°C, hydrolysis is dominated by the base-catalyzed reaction. Insertion of k_h into *Eqs. 12.45* and K-9 then yields

$$t_{1/2}(22.5°C) = \frac{0.693}{5.1 \times 10^{-4}\,s^{-1}} = 1360\,s = 22.7\,min$$

$$t_{0.05}\,(22.5°C) = \frac{3}{5.1 \times 10^{-4}\,s^{-1}} = 5880\,s = 1.63\,h$$

(b) Calculate the k_N- and k_B-values for 5°C (278.2 K) from the corresponding rate constants derived above for 22.5°C (295.7 K) using (see *Table 12.2*)

$$k(T_1) = k(T_2) \cdot e^{(E_a/R)(1/T_2 - 1/T_1)} \qquad \text{(K-10)}$$

where $T_2 = 295.7$ K and $T_1 = 278.2$ K, and E_a is the activation energy calculated as given above. The result obtained is

$$k_N\,(5°C) = 1.1 \times 10^{-5} \cdot s^{-1}$$

$$k_B\,(5°C) = 38.6\ M^{-1} \cdot s^{-1}$$

Since $K_w = 10^{-14.73}$ at 5°C, the OH^- concentration at pH 7.5 is $10^{-7.23}$, resulting in a k_h-value of

$$k_h\,(5°C) = (38.6)\,(10^{-7.23}) + 1.1 \times 10^{-5} = 1.3 \times 10^{-5}$$

Note that in contrast to the epilimnion, in the hypolimnion, the hydrolysis of 2,4-DNAP is dominated by the neutral reaction. The corresponding reaction times are

$$t_{1/2}\,(5°C) = \frac{0.693}{1.3 \times 10^{-5}\ s^{-1}} = 53300\ s = 14.8\ h$$

$$t_{0.05}\,(5°C) = \frac{3}{1.3 \times 10^{-5}\ s^{-1}} = 230000\ s = 62.9\ h$$

Hence, under the assumed conditions, 2,4-DNPA hydrolyzes about 40 times faster in the epilimnion of the lake as compared to the hypolimnion.

Estimating Rates of Nucleophilic Substitution Reactions Using the Swain-Scott Relationship

Problem
Estimate the half-life (with respect to chemical transformation) of methyl bromide (CH_3Br) present at low concentration (i.e., < 1 µM) in a homogeneous aqueous solution (pH = 7.0, T = 25°C) containing 100 mM Cl^-, 2 mM NO_3^-, 1 mM HCO_3^-, and 0.1 mM CN^-. In pure water at pH 7.0 and 25°C, the half-life of CH_3Br is 20 days (see *Table 12.7*).

Answer
Methyl bromide reacts with nucleophiles by an S_N2 mechanism. Since all nucleophiles are present in excess concentrations, the reaction of CH_3Br can be expressed by a pseudo-first-order rate law with a pseudo-first-order rate constant, k_{obs}, that is given by (see *Eq. 12-43*):

$$k_{obs} = k_N + \sum_i k_{nucl,i} \text{ [nucl i]} \tag{K-11}$$

where $k_N = k_{H_2O}[H_2O]$. Inspection of *Table 12.6* shows that the reaction with NO_3^- (as well as with OH^-) can be neglected. For estimation of the rate constants for the reactions with the other nucleophiles, use the Swain-Scott relationship (*Eq. 12-34*) with s = 1:

$$\log \frac{k_{nucl.i}}{k_{H_2O}} = n \tag{K-12}$$

or

$$k_{nucl,i} = k_{H_2O} \cdot 10^n \tag{K-13}$$

Insert n-values from *Table 12.5* into Eq. K-13, and substitute $k_{nucl,i}$ into Eq. K-11:

$$k_{obs} = k_{H_2O}[H_2O] + k_{H_2O} \cdot 10^3 \cdot Cl^- + k_{H_2O} \cdot 10^{3.8} \cdot [HCO_3^-] + k_{H_2O} \cdot 10^{5.1} \cdot [CN^-]$$

Insertion of the concentrations of the various nucleophiles then yields

$$k_{obs} = k_{H_2O}\{(55.5) + 10^3(10^{-1}) + 10^{3.8}(10^{-3}) + 10^{5.1}(10^{-4})\}$$

$$= k_{H_2O}\ (174.4) = k_N \frac{(174.4)}{(55.5)} = 3.14\ k_N$$

Hence, the observed pseudo-first-order rate constant is 3.14 times larger than the (neutral) hydrolysis rate constant, and, therefore

$$t_{1/2} = \frac{t_{1/2}(\text{pure water})}{3.14} = \frac{20\text{d}}{3.14} = 6.4\ \text{d}$$

Estimating Hydrolysis Rate Constants Using the Hammett Equation

Consider the base-catalyzed hydrolysis of 3,4,5-trichlorophenyl N-phenyl carbamate:

3,4,5-trichlorophenyl N-phenyl carbamate $+ OH^- \xrightarrow{k_B}$ aniline (NH_2) $+ CO_2 +$ 3,4,5-trichlorophenolate (O^-) (K-14)

> ***Problem***
> Estimate the second-order rate constant, k_B, at 25°C, for reaction Eq. K-14 using the k_B-values given below (Table K.1) for other substituted phenyl N-phenyl carbamates.

Table K.1 Second-Order Rate Constants k_B at 25°C for the Hydrolysis of Some Substituted Phenyl N-Phenyl Carbamates[a]

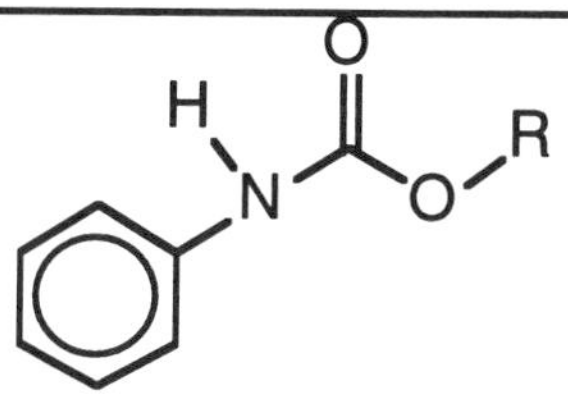

(Structure: phenyl–NH–C(=O)–O–R)

R	k_B ($M^{-1} \cdot s^{-1}$)	R	k_B ($M^{-1} \cdot s^{-1}$)
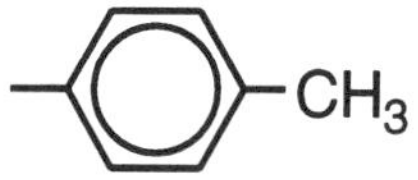 4-CH_3-phenyl	3.0×10^1	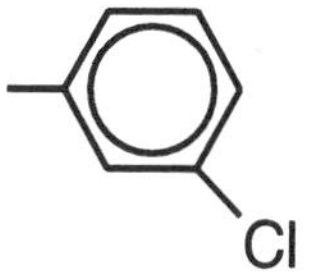 3-Cl-phenyl	1.8×10^3
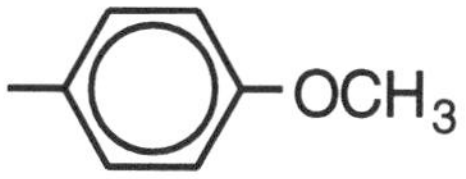 4-OCH_3-phenyl	2.5×10^1	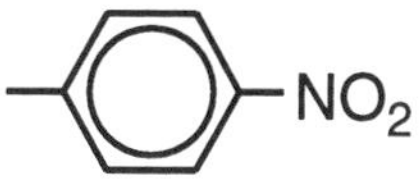 4-NO_2-phenyl	2.7×10^5
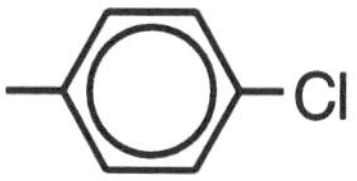 4-Cl-phenyl	4.2×10^2	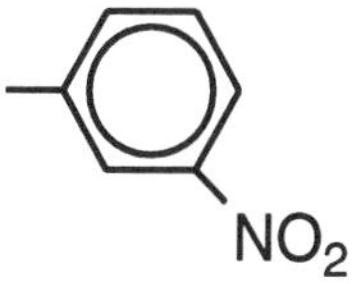 3-NO_2-phenyl	1.3×10^4

[a]Data from references given in *Table 12.13.*

Answer

The base-catalyzed hydrolysis of substituted phenyl N-phenyl carbamates occurs by an elimination mechanism with the dissociation of the leaving group (i.e., the $^-$OR-group) as rate-determining step (see *Fig. 12.15).* Use the Hammett equation to relate the k_B-values of this group of carbamates (see also *Eq. 12-28)*:

$$\log k_B = \rho \cdot \sum_i \sigma_i + C \qquad \text{(K-15)}$$

where ρ is the susceptibility factor and C is a constant corresponding to the log k_B of the unsubstituted compound (i.e., of phenyl N-phenyl carbamate). Determine ρ and C from a least square fit of log k_B versus σ_i for the monosubstituted compounds given in Table K.1 using the σ_i-values from *Table 8.4* (see also Fig. K.2):

$$\log k_B \text{ (in } M^{-1}s^{-1}) = 2.82 \sum_i \sigma_i + 2.02 \qquad \text{(K-16)}$$

Note that for substituents in the para-position that may be in resonance with the phenol group (i.e., OCH_3, NO_2), σ^-_{para}-values instead of σ_{para}-values have to be used.

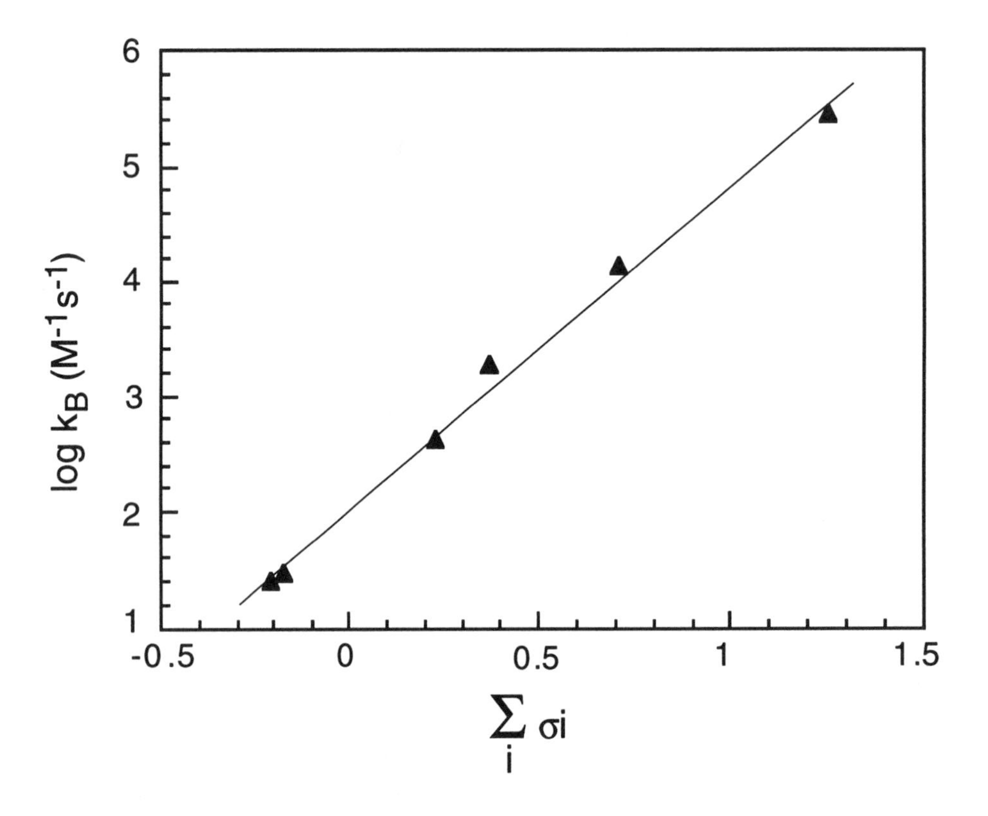

Figure K.2 Plot of log k_B versus σ_i for the monosubstituted phenyl N-phenyl carbamates given in Table K.1.

Insertion of $\Sigma\sigma_i = 2 \cdot (0.37) + (0.23) = 0.97$ for 3,4,5-trichlorophenyl N-phenyl carbamate into Eq. K-16 yields

$$\log k_B \text{ (in } M^{-1} \cdot s^{-1}) = 4.76 \quad \text{or } k_B = 5.7 \times 10^4 \, M^{-1} \cdot s^{-1} \text{ (at 25°C)}$$

As is illustrated in *Fig. 12.14*, the log k_B-value of an N-phenyl carbamate can also be related to the pK_a of the leaving group (i.e., of the alcohol group):

$$\log k_B \text{ (in } M^{-1} \cdot s^{-1}) = -1.15 \, pK_a + 13.6$$

The pK_a of 3,4,5-trichlorophenol is 7.73 *(Schellenberg et al., 1984)*, thus yielding a log k_B-value of 4.7 or $k_B = 5.1 \times 10^4 \, M^{-1}s^{-1}$ at 25°C.

Calculating Reduction Potentials of Half-Reactions at Various Conditions of pH and Solution Composition

Problem

Calculate the standard reduction potentials E_H^0 and $E_H^0(w)$ at 25°C for the reduction of azobenzene (AzB) to aniline (An) using the ΔG_f^0(aq)-values given below.

$$C_6H_5{-}N{=}N{-}C_6H_5 + 4H^+ + 4e^- = 2\, C_6H_5{-}NH_2 \qquad \text{(K-17)}$$

AzB ($\Delta G_f^0 = 426.0 \text{ kJmol}^{-1}$)

An ($\Delta G_f^0 \text{ (aq)} = 152.6 \text{ kJmol}^{-1}$)

Answer

Since by convention, $\Delta G_f^0(H^+(aq)) = 0$, ΔG^o of reaction Eq. K-17 is given by

$$\Delta G^o = -(\Delta G_f^0 \text{ (AzB (aq))} + 2\, \Delta G_f^0 \text{ (An(aq))}$$
$$= (-426.0) + 2 \cdot (152.6) = -120.8 \text{ kJ mol}^{-1}$$

The standard reduction potential E_H^o is related to ΔG^o by (see also *Eq. 12-89):*

$$E_H^o = -\frac{\Delta G^o}{nF} \tag{K-18}$$

where $F = 96{,}485\ C \cdot mol^{-1} = 96{,}485\ J \cdot V^{-1} \cdot mol^{-1}$, and, in this case, $n = 4$. Hence,

$$E_H^o = \frac{(-120.8\ kJ \cdot mol^{-1})}{4 \cdot (96.485)\ kJ \cdot V^{-1} \cdot mol^{-1}} = +0.31 V$$

For calculating $E_H^o(w)$, use the Nernst Equation (see examples in *Section 12.4*):

$$E_H = E_H^o + \frac{2.3RT}{4F} \log \frac{[H^+]^4 [AzB]}{[An]^2} \tag{K-19}$$

Setting $[An] = [AzB] = 1M$ and with $[H^+] = 10^{-7}$ yields

$$E_H^o(w) = E_H^o + \frac{2.3RT}{4F} \log (10^{-7})^4$$
$$= (0.31) + (0.0148)(-28) = -0.10 V$$

Note that at pH 7.0, the reduction potential of the half-reaction Eq. K-17 is significantly more negative than the value at pH 0.

Problem

Calculate the reduction potential of reaction Eq. K-17 at pH 7.5 and $[AzB] = [An] = 10^{-6}$ M using (a) E_H^o and (b) $E_H^o(w)$ as the starting points.

Answer (a)

Use Eq. K-19 with $E_H^o = 0.31$ V:

$$E_H = (0.31) + (0.0148) \log \frac{(10^{-7.5})^4 (10^{-6})}{(10^{-6})^2}$$

$$= (0.313) + (0.0148)(-24) = -0.04 \text{ V}$$

Answer (b)

Recall from *Section 3.3* that concentrations of chemical species in thermodynamic expressions must be divided by their concentrations at standard conditions (see *Eq. 3-28*). Thus, the Nernst equation (Eq. K-19) using natural conditions (i.e., pH = 7.0) as standard conditions needs to be written as

$$E_H = E_H^o(w) + \frac{2.3RT}{nF} \log \frac{([H^+] / (10^{-7}M))^4 ([AzB] / (1M))}{([An] / (1M))^2}$$

With $E_H^o(w) = -0.10$V (see above), one then obtains

$$E_H = (-0.10) + (0.0148) \log \frac{(10^{-0.5})^4 (10^{-6})}{(10^{-6})^2}$$

$$= -(0.10) + (0.0148)(4) = -0.04 \text{ V}$$

Problem

Calculate the half-reaction reduction potential for the reduction of manganese(IV) oxide to manganese(II) carbonate (reaction Eq. K-20) under the following conditions: pH = 7.5, $[HCO_3^-] = 10^{-3}$M.

$$MnO_2(s) + HCO_3^- + 3H^+ + 2e^- = MnCO_3(s) + 2H_2O \quad \text{(K-20)}$$

The $E_H^o(w)$-value for this reaction is + 0.52 V (see *Table 12-16*).

Answer

Because, for $MnO_2(s)$ and $MnCO_3(s)$, the solid phase is chosen as the reference state, the Nernst equation for reaction Eq. K-20 is given by

$$E_H = E_H^o(w) + \frac{2.3RT}{2F} \log \frac{(1)\ ([HCO_3]/(10^{-3}M))\ ([H^+]/(10^{-7}M))^3}{1 \cdot 1^2}$$

Setting $[H^+] = 10^{-7.5}$ and $[HCO_3^-] = 10^{-3}$ yields

$$E_H = (+0.52) + (0.0295) \log (10^{-0.5})^3 = 0.48 \text{ V}$$

Calculating Free-Energy Changes of Redox Reactions from Half-Reaction Reduction Potentials

Problem

Determine which of the following reactions may occur spontaneously in aqueous media at standard environmental conditions ("w"-conditions):

(a) the reduction of nitrobenzene to aniline by $FeCO_3(s)$ assuming that $FeOOH(s)$ is formed

(b) the reduction of trichloromethane to dichloromethane by HS^- assuming that S(s) is formed

(c) the oxidation of dimethylsulfide to dimethylsulfoxide by $FeOOH(s)$ assuming that $FeCO_3(s)$ is formed

The $E_H^o(w)$-value for the reaction $S(s) + H^+ + 2e^- = HS^-$ is -0.27 V.

Answer

The ΔG of a reaction is related to the difference, ΔE, of the reduction potentials of the corresponding half reactions by

$$\Delta G = -nF\Delta E \qquad (12\text{-}87)$$

where $\Delta E = E_H(\text{oxidant}) - E_H(\text{reductant})$. Under standard environmental conditions:

$$\Delta E = E_H^o(w)(\text{oxidant}) - E_H^o(w)(\text{reductant}) \qquad \text{(K-21)}$$

Thus, if ΔE is positive, then ΔG is negative, and the reaction may occur spontaneously. As is illustrated by Figure K.3, the sign of ΔG of a given redox reaction under standard environmental conditions can be immediately seen from a graphical representation of the $E_H^o(w)$-values of the half reactions involving the oxidants and reductants considered. Inspection of Figure K.3 shows that reactions (a) and (b) should occur spontaneously under these conditions (positive ΔE, note that nitrobenzene and trichloromethane are the oxidants in these reactions), while reaction (c) would not occur (here, dimethyl sulfide is the reductant). For calculating ΔG for the various reactions, take the $E_H^o(w)$-values given above, and in Tables *12.16* and *12.17*.

(a) The reduction of nitrobenzene to aniline requires 6 electrons to be transferred. Thus, the ΔG-value for the overall reaction

$$C_6H_5\text{-}NO_2 + 6FeCO_3 + 10H_2O =$$

$$C_6H_5\text{-}NH_2 + 6FeOOH + 6HCO_3^- + 6H^+$$

under standard environmental conditions is

$$\Delta G = -(6)(96.485)(+0.42 - (-0.05)) = -272.1 \text{ kJ}\cdot\text{mol}^{-1}$$

(b) The overall reaction in which 2 electrons are transferred is given by

$$CHCl_3 + HS^- = CH_2Cl_2 + S(s) + Cl^-$$

Under standard environmental conditions

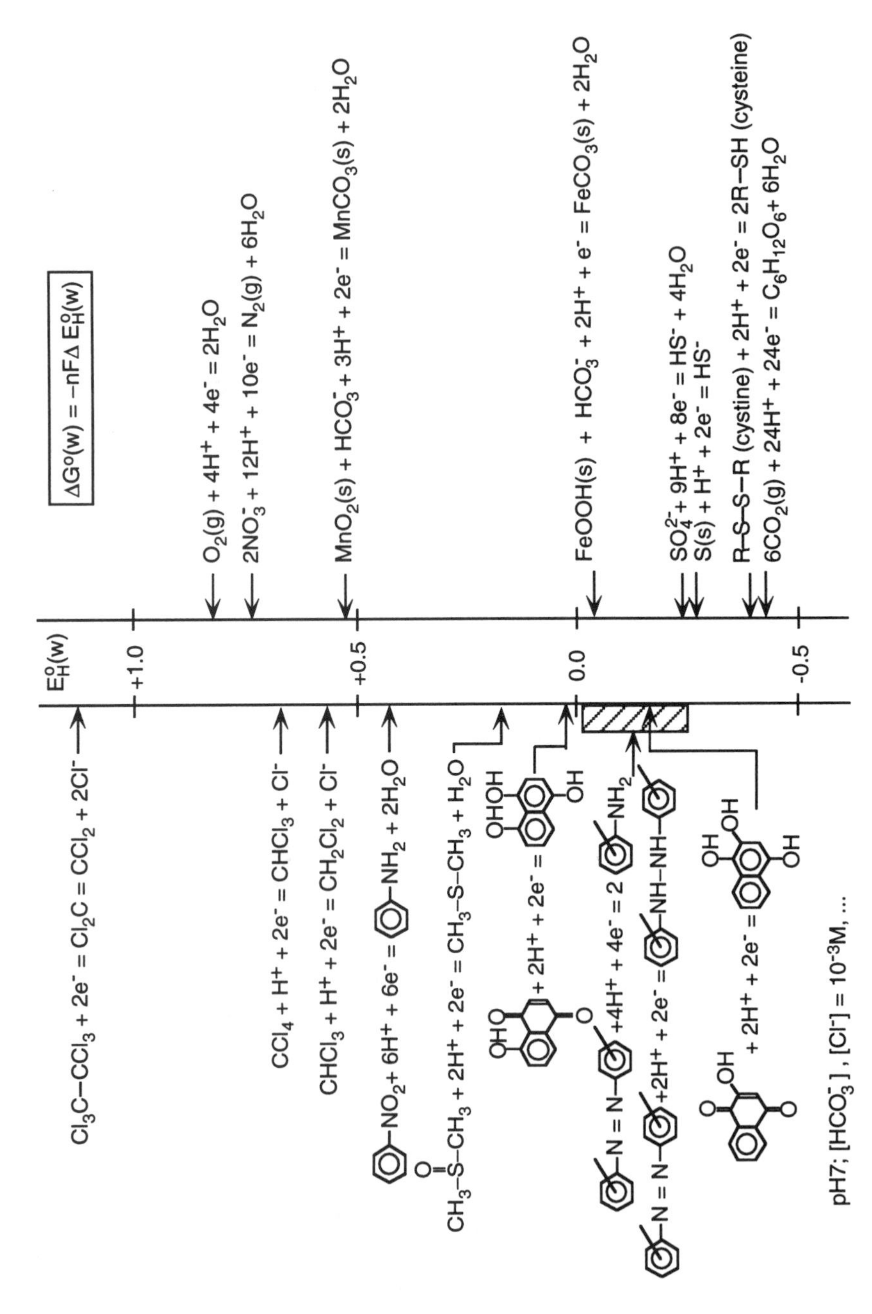

Figure K.3 Energetic considerations of redox reactions of organic pollutants. Graphical representation of the $E_H^o(w)$ values of some half reactions involving some organic pollutants (left side) as well as some environmentally important oxidants and reductants (right side). The $E_H^o(w)$ values are from *Tables 12.6* and *12.7*.

$$\Delta G = -(2)(96.485)(+0.56-(-0.27)) = -160.2 \text{ kJ}\cdot\text{mol}^{-1}$$

(c) The overall reaction in which 2 electrons are transferred from dimethyl sulfide to FeOOH(s) is given by

$$CH_3-S-CH_3 + 2FeOOH(s) + 2\,HCO_3^- + 2H^+ =$$

$$CH_3-\overset{\overset{\displaystyle O}{\|}}{S}-CH_3 + 2FeCO_3(s) + 3H_2O$$

with a Δ G-value that is positive under standard environmental conditions:

$$\Delta G = (-2)(96.485)(-0.05-0.16) = +40.5 \text{ kJ}\cdot\text{mol}^{-1},$$

Problem

Laha and Luthy (1990) have demonstrated that, in heterogeneous aqueous solutions, aniline (An) is oxidized to azobenzene (AzB) (reverse reaction Eq. K-17) by $MnO_2(s)$. Calculate the ΔG of this reaction at pH 7.5, $[HCO_3^-] = 10^{-3}$ M, $[An] = [AzB] = 10^{-6}$ M, and assuming that $MnO_2(s)$ is converted to $MnCO_3(s)$.

Answer

The ΔG for the reaction

$$2An + 2MnO_2(s) + 2\,HCO_3^- + 2H^+ = AzB + 2MnCO_3(s) + 4H_2O \quad \text{(K-22)}$$

is given by

$$\Delta G_{reaction} = -nF\Delta E \qquad (12\text{-}87)$$

where n = 4 and ΔE is the difference between the E_H-values of the half reactions Eq. K-20 and Eq. K-17 under the given conditions. These E_H-values have been calculated above. Hence,

$$\Delta G_{reaction} = -(4)(96.49)[(+0.48)-(-0.04)] = -200.7 \text{ kJ} \cdot \text{mol}^{-1}$$

Under these conditions, the reaction Eq. K-22 occurs spontaneously from the left to the right.

Assessing the Equilibrium Speciation of Redox Couples under Given Redox Conditions

Problem

Consider a 1 μM solution of 2-hydroxy-1,4-naphthoquinone (lawsone, "LAW") in 1 mM aqueous hydrogen sulfide solutions at pH 6.0 and 8.0 at 25°C. Assume that hydrogen sulfide is oxidized to elemental sulfur. The E_H^o-value of the half-reaction K-23 is + 0.35 V (*Clark, 1960*).

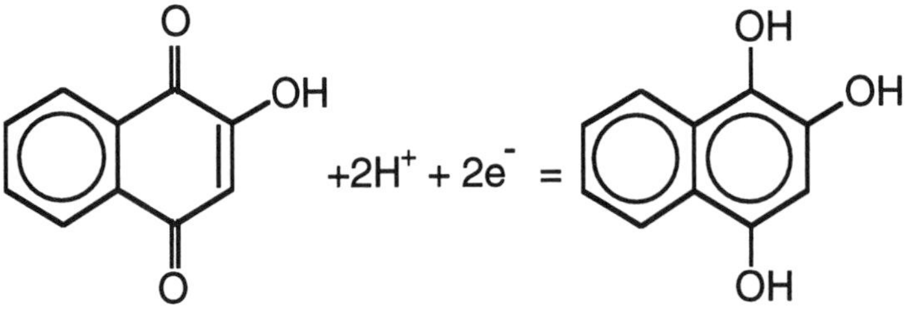

$$\text{LAW} + 2H^+ + 2e^- = H_2\text{LAW} \qquad (\text{K-23})$$

LAW

$pK_a^{(ox)} = 3.98$

H_2LAW

$pK_{a1}^{(red)} = 8.68$

$pK_{a2}^{(red)} = 10.71$

The Nernst equation for the lawsone couple at 25°C is (see also *Eq. 12-120):*

$$E_H = E_H^o + 0.03 \log \frac{[LAW]}{[H_2LAW]}$$

$$+ 0.03 \log \frac{[H^+]^3 + K_{a1}^{(red)}[H^+]^2 + K_{a1}^{(red)} K_{a2}^{(red)}[H^+]}{[H^+] + K_a^{(ox)}} \qquad (K\text{-}24)$$

where [LAW] and $[H_2LAW]$ are the total concentrations of the oxidized and reduced species, respectively. Calculate the equilibrium concentrations of the various oxidized and reduced lawsone species present in solution at the two pH-values. (Note that both LAW and H_2LAW are weak acids.)

Answer

Assume that the E_H of the hydrogen-sulfide buffer solution is given by *Eq. 12-99* with $K_a = 10^{-7}$ and $[H_2S]_{tot} = 10^{-3}$ M:

$$E_H = +0.14 + 0.03 \log \frac{[H^+]\{[H^+] + (10^{-7})\}}{(10^{-3})}$$

Hence, at pH 6.0 $\{[H^+] = 10^{-6}\}$, E_H is – 0.13 V, and at pH 8.0 $\{[H^+] = 10^{-8}\}$, E_H is – 0.22 V.

Insert these values into Eq. K-24 and solve for ($[LAW]/[H_2LAW]$) using the K_a-values given above. For pH = 6.0, the result is

$$\log \frac{[LAW]}{[H_2LAW]}$$

$$= \frac{(-0.13) - (0.35)}{0.03}$$

$$- \log \frac{(10^{-6})^3 + (10^{-8.68})(10^{-6})^2 + (10^{-8.68})(10^{-10.71})(10^{-6})}{(10^{-6}) + (10^{-3.98})}$$

$$= -16.0 + 14.0 = -2.0$$

or

$$\frac{[\text{LAW}]}{[\text{H}_2\text{LAW}]} = 0.01$$

Thus, at equilibrium at pH 6.0, 99% of the lawsone is present in the reduced, nondissociated form (i.e., as H_2LAW; note that $pK_{a1}^{(red)}$ is 8.68). The 1% oxidized lawsone is present predominantly in the dissociated (anionic) form ($K_a^{(ox)}$ = 3.98).

At pH 8.0, however,

$$\log \frac{[\text{LAW}]}{[\text{H}_2\text{LAW}}) = -19.0 + 20.0 = 1.0$$

or

$$\frac{[\text{LAW}]}{[\text{H}_2\text{LAW}]} = 10$$

That is, more than 90% of the lawsone is present in the oxidized (dissociated) form, while only about 10% is present as reduced lawsone. Of this 10%, the fraction of nondissociated H_2LAW is 0.83 (*Eq. 8-16*.)

This example illustrates that pH may have a considerable effect on the speciation of compounds that may undergo reversible redox reactions in natural waters (e.g., constituents of natural organic matter, see *Dunnivant et al., 1992*).

PROBLEMS

● *K-1 Hydrolysis of an Insecticide in a River*

After a fire in a chemical storehouse at Schweizerhalle, Switzerland, in November 1986, several tons of various pesticides, solvents, dyes, and other raw and intermediate chemicals were flushed into the Rhine River (*Capel et al., 1988; Wanner et al., 1989*). Among these chemicals was the insecticide disulfoton, of which 3500 kg were introduced into the river water (11°C, pH 7.5). During the 8 days "travel time" from Schweizerhalle to the Dutch border, 2500 kg of this compound were "eliminated" from the river water. Somebody wants to know how much of this elimination was due to abiotic hydrolysis. Since in the literature you do not find any good kinetic data for the hydrolysis of disulfoton, you make your own measurements in the laboratory. Under all selected experimental conditions, you observe (pseudo) first-order kinetics, and you get the following results:

temperature	k_{obs} (s^{-1}) for $(C_2H_5O)_2P(=S)-SCH_2CH_2SCH_2CH_3$		
°C	pH 6.0	pH 11.98	pH 11.72
20		1.3×10^{-5}	
30	4.0×10^{-7} [a]		3.6×10^{-5}
40	9.6×10^{-7}		
45	1.5×10^{-6}		
50	2.9×10^{-6}		

[a] A similar k_{obs}-value was obtained at pH 4.0 and 30°C.

Determine the k_{obs}-value for the conditions in the river (11°C, pH 7.5), and calculate how much disulfoton was transformed by hydrolysis over the 8 days. What is (are) the most likely hydrolysis product(s)?

● *K-2 Chemical Transformation of Polychlorinated Ethanes in a Lake*

Assume that the three polychlorinated ethanes, 1,1,2,2-tetrachloroethane, 1,1,1,2-tetrachloroethane, and pentachloroethane are introduced into a lake by an accident. Calculate the half-life for chemical transformation of each of the three compounds in (a) the epilimnion of the lake (T = 25°C, pH 8.5) and (b) the hypolimnion of the lake (T = 5°C, pH 7.5). Furthermore, indicate for each compound the pH (for the epilimnion and for the hypolimnion) at which the neutral and the base-catalyzed reaction would be equally important. What is (are) the transformation product(s) of these compounds? Try to explain the different reactivities of the three compounds. The kinetic data summarized below are from *Haag and Mill (1988)* and *Jeffers et al. (1989)*.

	neutral reaction		base-catalyzed reaction	
compound	$k_N(s^{-1})$ [a]	$E_a(kJ \cdot mol^{-1})$	$k_B(M^{-1} \cdot s^{-1})$ [a]	$E_a\ (kJ \cdot mol^{-1})$
$CHCl_2 - CHCl_2$	1.6×10^{-11}	93	2.0×10^{8}	78
$CCl_3 - CH_2Cl$	4.3×10^{-10}	95	3.6×10^{-4}	100
$CCl_3 - CHCl_2$	8.2×10^{-10}	95	$2.2 \times 10^{+1}$	81

[a] At 25°C

● *K-3 What Happens to Trimethylphosphate in Seawater and in a Leachate?*

The hydrolysis half-life of trimethylphosphate (TMP)

$$CH_3O-\overset{\overset{\displaystyle O}{\|}}{\underset{\underset{\displaystyle OCH_3}{|}}{P}}-OCH_3$$

(TMP)

in pure water is 438 days at 25°C and pH 7.0 (*Table 12.14*). Your colleague in oceanography, with whom you already had a small bet earlier (see Problem B-5), claims that in sterile seawater, he observed a half-life for TMP of only 44 days at 25°C and pH 7. Is this result reasonable? What are the major products of the abiotic transformation of TMP in seawater?

Because you are more interested in groundwater contamination, you wonder how fast TMP would be transformed by chemical reactions at 10°C and pH 8.0 in a leachate from a waste disposal site containing 0.25 M Cl^-, 0.05 M Br^-, and 10^{-4} M CN^-. Calculate the approximate half-life of TMP under these conditions by trusting your colleague's measurements and by assuming that all relevant reactions exhibit about the same activation energy of 90 $kJ \cdot mol^{-1}$.

● *K-4 More Questions About the Chemical Transformation of Methylbromide in a Leachate*

A leachate of a dump site (15°C; pH 7.0; $[Cl^-] = 5 \times 10^{-2}$ M; $[Br^-] = 10^3$ M; $[H_sS]_{tot} = 10^{-6}$ M, $pK_a(H_2S) = 7.0$) contains 10^{-6} M methyl bromide (CH_3Br). After how much time has 95% of the CH_3Br reacted when considering only abiotic nucleophilic substitution reactions? What are the major products, and what are their concentrations after this time period? Recall from the illustrative examples above that in pure water at pH 7.0 and 25°C, the half-life of CH_3Br is 20 days. Furthermore, for all possible reactions, assume an activation energy, E_a, approximately 100 $kJ \cdot mol^{-1}$. Finally, recall from your basic mathematics courses that the solution of a linear inhomogeneous first-order differential equation of the form

$$\frac{dy}{dt} = x + \lambda y$$

with a time-variable parameter x(t), but with a constant λ, is given by

$$y(t) = y^o e^{\lambda t} + \int_0^t e^{\lambda(t-t')} x(t') dt'$$

● *K-5 Synthesizing the "Right" Carbamates*

You work in the chemical industry and you are asked to synthesize two different carbamates of either Type I or Type II (see below). One carbamate should have a hydrolysis life of approximately 1 month at 25°C and pH 8.0, while the hydrolysis half-life of the other one should be about 10 months at 25°C and pH 9.0. You assume that only the base-catalyzed reaction is important at the pH's of interest, and you search the literature for k_B-values for these type of compounds. For some Type I compounds k_B-values are summarized in Table K.1, and for some Type II compounds you find the data given below. What are the structures of the molecules that you are going to synthesize in order to get the desired half-lives?

Type I

Type II

Second-Order Rate Constants k_B at 25°C for the Hydrolysis of Some Subsituted Phenyl N-Methyl-N-Phenyl Carbamates (Type II)[a]

R	$k_B (M^{-1} \cdot s^{-1})$	R	$k_B (M^{-1} \cdot s^{-1})$
	7.5×10^{-5}	NO_2	3.9×10^{-4}
NH_2	2.8×10^{-5}	NO_2, CH_3	3.3×10^{-4}
$\oplus N(CH_3)_3$	2.5×10^{-4}[b]		

[a] Data from references given in *Table 12.13.*

[b]The Hammett σ_{meta}-value for - $\overset{\oplus}{N}(CH_3)_3$ is 0.88; for an extensive compilation of Hammett Substituent Constants see *Harris and Hayes (1990).*

● *K-6 Are All of These Reactions Really Occurring Spontaneously?*

Somebody claims that all of the redox reactions (a-c) involving organic compounds occur spontaneously at 25°C in aqueous solutions under the indicated conditions. Is this person correct? Use the information given below and summarized in *Tables 12.16* and *12.17.*

(a) The oxidation of cysteine to cystine by FeOOH(s) (assuming that $FeCO_3$(s) is formed) under the following conditions:

$$pH = 7.0, \; [HCO_3^-] = 5 \times 10^{-3} M$$

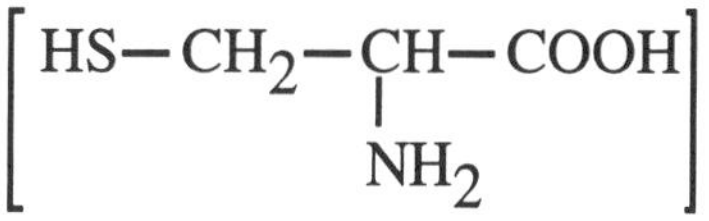

$$\left[HS{-}CH_2{-}\underset{\underset{NH_2}{|}}{CH}{-}COOH \right] = 10^{-6} M$$

cysteine

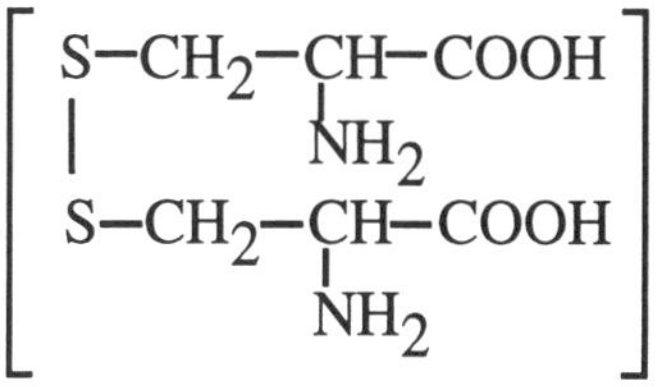

$$\left[\begin{array}{l} S{-}CH_2{-}\underset{\underset{NH_2}{|}}{CH}{-}COOH \\ | \\ S{-}CH_2{-}\underset{\underset{NH_2}{|}}{CH}{-}COOH \end{array} \right] = 10^{-4} M$$

cystine

(b) The reduction of azobenzene to hydrazobenzene ($E_H^o(w)$ = – 0.11V) by hydrogen sulfide (assuming that S(s) is formed) under the following conditions:

$pH = 6.0$

$[H_2S]_{tot} = 10^{-3}M$, $pK_a(H_2S) = 7.0$

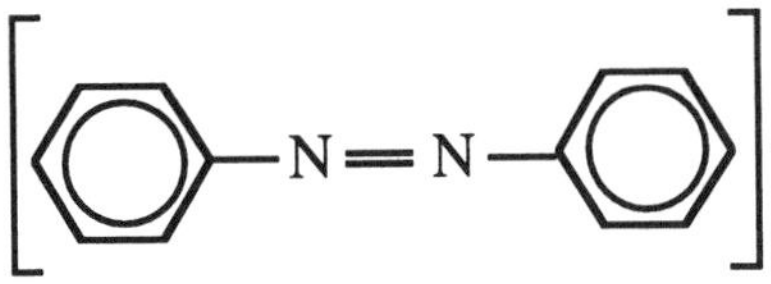

$[C_6H_5{-}N{=}N{-}C_6H_5] = 10^{-6}M$

azobenzene

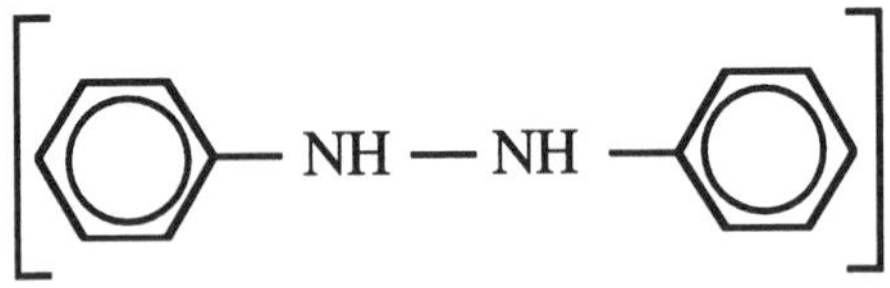

$[C_6H_5{-}NH{-}NH{-}C_6H_5] = 10^{-5}M$

hydrazobenzene

(c) The reduction of nitrobenzene (NB) to aniline (An) by reduced juglone (H_2JUG, for pH-dependence of the E_H-value of the JUG/H_2JUG couple see *Eq. 12-120* and *Table 12.20*) at pH 8.5 and the following concentrations of the species involved:

$[NB] = 10^{-9}M$

$[An] = 10^{-5}M$

$[JUG] = 10^{-3}M$

$[H_2JUG] = 10^{-4}M$

● *K-7 Estimating Reduction Rates of Nitroaromatic Compounds in Aqueous Solutions Containing Hydrogen Sulfide and Natural Organic Matter*

Dunnivant et al. (1992) have demonstrated that natural organic matter (NOM) from a variety of sources mediates the reduction of nitroaromatic compounds (NACs) in aqueous solution containing hydrogen sulfide (H_2S) as bulk electron donor. They found that in the absence of NOM, the reduction of the NACs investigated occurred extremely slowly. For a given NOM and for given system conditions of pH, T, and $[H_2S]_{tot}$ (i.e.,

E_H), they defined a carbon-normalized second-order rate constant, k_{NOM}, which describes the reduction rate of a series of substituted benzenes ($ArNO_2$) by the second-order rate law:

$$-\frac{d(ArNO_2)}{dt} = k_{NOM} \cdot [ArNO_2]\,[TOC]$$

where [TOC] is the total organic carbon concentration. For "Hyde County NOM", a set of k_{NOM}-values of 10 monosubstituted nitrobenzenes is summarized below together with the one-electron reduction potentials of the compounds. Try to solve the following questions and problems:

(a) Is the actual electron transfer between reduced NOM moieties and the NACs rate-determining?

(b) Estimate the $E_H^{1'}(w)$-value of 2,4-dinitrotoluene.
<u>Hint</u>: Use Hammett substituent constants (*Table 8.4*). But be careful! Watch out for ortho-effects. Base your estimate on $E_H^{1'}(w)$ of 2-Me.

(c) Estimate the half-life of 2,4-dinitrotoluene in an aqueous solution (25°C, 5mM H_2S, pH 7.2) containing 5 $mg \cdot L^{-1}$ "Hyde County NOM".

(d) Why is the rate of reduction of NACs by H_2S so slow?

Compound	abr.	$E_H^{1'}(ArNO_2)$ [a] (V)	k_{NOM} [b] $(h^{-1}(mgC \cdot L^{-1})^{-1})$
nitrobenzene	H	- 0.485	4.5×10^{-5}
2-methylnitrobenzene	2-Me	-0.590	3.0×10^{-6}
3-methylnitrobenzene	3-Me	- 0.475	6.6×10^{-5}
4-methylnitrobenzene	4-Me	-0.500	2.9×10^{-5}
2-chloronitrobenzene	2-Cl	- 0.485	1.6×10^{-4}
3-chloronitrobenzene	3-Cl	- 0.405	8.1×10^{-4}
4-chloronitrobenzene	4-Cl	- 0.450	3.6×10^{-4}
2-acetylnitrobenzene	2-$COCH_3$	- 0.475	3.8×10^{-4}
3-acetylnitrobenzene	3-$COCH_3$	- 0.405	1.0×10^{-3}
4-acetylnitrobenzene	4-COCH3	- 0.360	3.3×10^{-2}

[a] From *Dunnivant et al. (1992)*; $E_H^{1'}(w)$ is the reduction potential of the half-reaction $ArNO_2 + e^- = ArNO_2^{\bullet -}$ at pH 7 and 25°C.

[b] Hyde County NOM, pH 7.2.

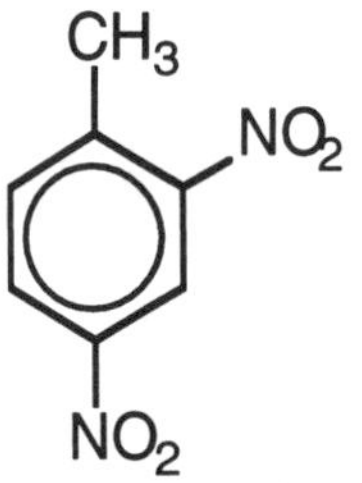

CH_3 / NO_2 / NO_2

2,4-dinitrotoluene

$E_H^1(w) = ?$

$k_{NOM} = ?$

L. PHOTOCHEMICAL TRANSFORMATION REACTIONS

ILLUSTRATIVE EXAMPLES

Determining Decadic Molar Extinction Coefficients of Organic Pollutants

Somebody measures the electronic absorption spectrum of an 0.1 mM solution of nitrobenzene in pure water. For wavelengths below 310 nm, the spectrum is recorded using a 1 cm cuvette; for higher wavelengths, a 5 cm cuvette is used. The following absorbances are recorded:

Wavelength λ(nm)	Absorbance (1cm) A
250	0.54
265 (λ_{max})	0.76
280	0.60
290	0.43
300	0.25

Wavelength λ(nm)	Absorbance (5cm) A
310	0.70
320	0.40
330	0.29
340	0.28
350	0.14
360	0.07
370	0.01

$$\text{C}_6\text{H}_5\text{—NO}_2$$

nitrobenzene

Problem
Calculate the decadic molar extinction coefficients of nitrobenzene for the wavelengths indicated above.

Answer

Rearrange *Eq. 13-5* to calculate $\varepsilon(\lambda)$:

$$\varepsilon(\lambda) = \frac{A(\lambda)}{C \cdot l}$$

where C = 10^{-4} M and l = 1 cm or 5 cm, respectively. The results are summarized in Table L.1.

Table L.1 Decadic Molar Extinction Coefficients of Nitrobenzene at Various Wavelengths

λ (nm)	$\varepsilon(M^{-1} \cdot cm^{-1})$	λ (nm)	$\varepsilon(M^{-1} \cdot cm^{-1})$
250	5400	320	800
265 (λ_{max})	7600	330	580
280	6000	340	560
290	4300	350	280
300	2500	360	140
310	1400	370	20

Estimating Light Penetration Into a Natural Water Body

Consider a well-mixed, nonturbid water body with a dissolved organic carbon concentration (DOC) of 4 $mgC \cdot L^{-1}$. The decadic beam attenuation coefficients, $\alpha(\lambda)$, determined for a water sample at five wavelengths are the following (see also other examples given in *Fig. 13.11*):

λ (nm)	$\alpha(\lambda)^a$ (m^{-1})
300	4.2
350	1.5
400	0.6
450	0.3
500	0.2

[a]Values taken from *Table 13.6* for epilimnic water of Greifensee, Switzerland.

Problem

Calculate for each wavelength indicated above the thickness of the water layer required to attenuate sunlight by a factor of 2. At what depth is 99% of the incoming light of a given wavelength absorbed by the water body? (Assume an average value for the distribution function, $D(\lambda)$, in *Eq. 13-11* of 1.2.)

Answer

Rearrange *Eq. 13-10* to calculate $z_{mix,0.5}$ and $z_{mix,0.01}$, respectively:

$$z_{mix} = \frac{1}{\alpha_D(\lambda)} \log \frac{W(\lambda)}{W(z_{mix}, \lambda)}$$

where $\alpha_D(\lambda) \simeq 1.2 \cdot \alpha(\lambda)$. Setting $W(\lambda) / W(z_{mix},\lambda) = 2$ and 100, respectively, yields

λ (nm)	$z_{mix, 0.5}$ (m)	$z_{mix, 0.01}$ (m)
300	0.06	0.4
350	0.17	1.1
400	0.42	2.8
450	0.83	5.6
500	1.25	8.3

Note that light attenuation and thus light penetration in a natural water body are strongly wavelength-dependent.

Estimating Specific Light Absorption Rates and Direct Photolysis Half-Lives of Organic Pollutants in Natural Waters

Problem
Estimate the near-surface total specific light absorption rate of nitrobenzene present at low concentration in a natural water body on a clear midsummer day (at noon and averaged over 24 h) at 40°N latitude and sea level.

Answer
Use *Eq. 13-22* to calculate the near-surface total specific light absorption rate of nitrobenzene:

$$k_a^o \simeq 2.3 \, \Sigma \, Z(\lambda) \cdot \varepsilon(\lambda) \qquad (13\text{-}22)$$

Take the Z(noon, λ) and Z(24h, λ) values, respectively, given in *Table 13.3* for the appropriate wavelength range (i.e., 297.5 - 370 nm). Use the $\varepsilon(\lambda)$ values calculated in Table L.1. Calculate ε values by linear interpolation for those wavelengths for which no value is given. The resulting k_a^o values are (see calculation in Table L.2):

$$k_a^o \text{(noon)} = 8.7 \times 10^{-3} \text{ einstein} \cdot (\text{mol NB})^{-1} \cdot s^{-1}$$

and

$$k_a^o \text{(24h)} = 3.2 \times 10^{2} \text{ einstein} \cdot (\text{mol NB})^{-1} \cdot d^{-1}$$

Note that calculated on a daily basis, k_a^o(noon) (= 7.5×10^2 einstein$\cdot$(mol NB)$^{-1}\cdot$d^{-1}) is about 2.5 times larger than k_a^o (24h).

Table L.2 Calculation of the Near-Surface Total Specific Light Absorption Rate of Nitrobenzene (NB) on a Clear Mid-summer Day at Noon and Averaged over 24h at 40°N Latitude and Sea Level

					$k_a^o(\lambda) = 2.303\ Z(\lambda)\ \varepsilon(\lambda)$	
λ(Center) (nm)	λ-Range ($\Delta\lambda$) (nm)	Z(noon,λ)[a] (millieinstein $\cdot cm^{-2} \cdot s^{-1}$)	Z(24h,λ)[a] (millieinstein $\cdot cm^{-2} \cdot d^{-1}$)	$\varepsilon(\lambda)$ ($M^{-1} \cdot cm^{-1}$)	noon (einstein·mol NB)$^{-1} \cdot s^{-1}$) $10^3 k_a^o$(noon,λ)	24h-average (einstein·(mol NB)$^{-1} \cdot d^{-1}$) k_a^o(24h,λ)
297.5	2.5	1.19(−9)	2.68(−5)	2950	0.008	0.18
300.0	2.5	3.99(−9)	1.17(−4)	2500	0.023	0.67
302.5	2.5	1.21(−8)	3.60(−4)	2225	0.062	1.84
305.0	2.5	3.01(−8)	8.47(−4)	1950	0.135	3.80
307.5	2.5	5.06(−8)	1.62(−3)	1675	0.195	6.25
310.0	2.5	8.23(−8)	2.68(−3)	1400	0.265	8.64
312.5	2.5	1.19(−7)	3.94(−3)	1250	0.342	11.34
315.0	2.5	1.60(−7)	5.30(−3)	1100	0.405	13.42
317.5	2.5	1.91(−7)	6.73(−3)	950	0.418	14.72
320.0	2.5	2.24(−7)	8.12(−3)	800	0.413	14.96
323.1	3.75	4.18(−7)	1.45(−2)	730	0.702	24.43
330.0	10	1.41(−6)	5.03(−2)	580	1.883	67.18
340.0	10	1.60(−6)	6.34(−2)	560	2.063	81.75
350.0	10	1.71(−6)	7.03(−2)	280	1.102	45.32
360.0	10	1.83(−6)	7.77(−2)	140	0.590	25.05
370.0	10	2.03(−6)	8.29(−2)	20	0.093	3.82
					$k_a^o = \Sigma\ k_a^o(\lambda) = 8.7 \times 10^{-3}$	321.4

[a] Value in parentheses indicate powers of 10.

Problem

Estimate the total specific light absorption rate, k_a, of nitrobenzene present at low concentration in the epilimnion of a lake (z_{mix} = 5 m) on a clear midsummer day at noon at 40°N latitude and sea level. The average $\alpha(\lambda)$ values of the water from the epilimnion of the lake are given in Table L.3.

Answer

The $\alpha(\lambda)$ value of the highest wavelength at which nitrobenzene absorbs sunlight (i.e., 370 nm) is 1.0 m^{-1}. Hence, at z_{mix} = 5 m (see *Eq. 13-10*):

$$\log \frac{W(370 \text{ nm})}{W(z_{mix}, 370 \text{ nm})} = z_{mix} \cdot \alpha(370 \text{ nm}) = 5$$

and, therefore, it can be assumed that virtually all light is absorbed in the epilimnion. Thus, use *Eq. 13-27* to calculate the total specific light absorption rate of nitrobenzene in the epilimnion:

$$k_a^t = \frac{1}{z_{mix}} \sum \frac{W(\lambda) \cdot \varepsilon(\lambda)}{\alpha(\lambda)} \qquad (13\text{-}27)$$

Using the W(noon,λ) and $\varepsilon(\lambda)$ values given in *Table 13.3* and Table L.2, respectively, one obtains a k_a^t(noon)-value of (see Table L.3):

$$k_a^t(\text{noon}) = 3.3 \times 10^{-4} \text{ einstein} \cdot (\text{mol NB})^{-1} \cdot \text{s}^{-1}$$

Inspection of Tables L-2 and L-3 shows that the rate of light absorption of nitrobenzene is greatest in the wavelength range between 320 and 360 nm with a maximum at about 340 nm. Thus, k_a^t(noon) could also have been calculated from the k_a^o(noon) value derived in Table L.2 using an average light-screening factor $S(\lambda_m)$ (see *Eq. 13-32)*:

$$k_a^t(\text{noon}) = S(\lambda_m) \cdot k_a^o(\text{noon}) \qquad (13\text{-}32)$$

where $S(\lambda_m)$ is equal to (see *Eq. 13-31*)

Table L.3 Calculation of the Total Specific Light Absorption Rate of Nitrobenzene (NB) in the Epilimnion of a Lake (z_{mix} = 5 m, $\alpha(\lambda)$ values given below) on a Clear Midsummer Day at Noon at 40°N Latitude and Sea Level

λ (Center) (nm)	λ-Range ($\Delta\lambda$) (nm)	$W(noon,\lambda)$[a] (millieinstein $\cdot cm^{-2} \cdot s^{-1}$)	$\alpha(\lambda)$ (m^{-1})	$\varepsilon(\lambda)$ ($M^{-1} \cdot cm^{-1}$)	$k_a^t(\lambda) = W(\lambda) \cdot \varepsilon(\lambda) / (z_{mix} \cdot \alpha(\lambda))$ (einstein$\cdot$(mol NB)$^{-1} \cdot s^{-1}$) $10^6\, k_a^t(noon,\lambda)$
297.5	2.5	1.08 (−9)	4.30	2950	0.15
300.0	2.5	3.64 (−9)	4.15	2500	0.44
302.5	2.5	1.10 (−8)	3.95	2225	1.24
305.0	2.5	2.71 (−8)	3.75	1950	2.82
307.5	2.5	4.55 (−8)	3.55	1675	4.29
310.0	2.5	7.38 (−8)	3.35	1400	6.17
312.5	2.5	1.07 (−7)	3.20	1250	8.36
315.0	2.5	1.34 (−7)	3.05	1100	9.67
317.5	2.5	1.71 (−7)	2.90	950	11.20
320.0	2.5	2.01 (−7)	2.75	800	11.69
323.1	3.75	3.75 (−7)	2.60	730	21.06
330.0	10	1.27 (−6)	2.20	580	66.96
340.0	10	1.45 (−6)	1.85	560	87.78
350.0	10	1.56 (−6)	1.50	280	58.24
360.0	10	1.66 (−6)	1.25	140	37.18
370.0	10	1.86 (−6)	1.00	20	7.44
				$k_a^t(noon) = \Sigma\, k_a^t(noon,\lambda) =$	3.3×10^{-4}

[a] Values in parentheses indicate powers of 10.

$$S(\lambda_m) = \frac{[1 - 10^{-(1.2) \cdot \alpha(\lambda_m) \cdot z_{mix}}]}{(2.303) \cdot (1.2) \cdot \alpha(\lambda_m) \cdot z_{mix}} \tag{L-1}$$

The $\alpha(\lambda_m)$ value for λ_m = 340 nm is 1.85 m^{-1} (see Table L.3). Hence,

$$S(340\text{ nm}) = \frac{1}{(2.303)\,(1.2)\,(1.85\text{ m}^{-1})\,(5\text{m})} = 0.039$$

This result indicates that, on the average, only about 4% of the light is available for direct photolysis of NB. Therefore,

$$k_a^t\,(\text{noon}) = (0.039)\,(8.7 \times 10^{-3})\text{ einstein}\cdot(\text{mol NB})^{-1}\cdot\text{s}^{-1}$$

$$= 3.4 \times 10^{-4}\text{ einstein}\cdot(\text{mol NB})^{-1}\cdot\text{s}^{-1},$$

a value that is very similar to the one calculated in Table L.3.

Problem

Estimate the 24h-averaged near-surface direct photolysis half-life of nitrobenzene present at low concentration in a natural water body on a clear midsummer day at 40° N latitude and sea level.

Answer

Express the rate of the near-surface direct photolysis of 4-NB by

$$-\frac{d[4\text{-NB}]}{dt} = k_p^o\,[4\text{-NB}]$$

with a half-life of

$$t_{1/2}^o = \frac{\ln 2}{k_p^o}$$

and

$$k_p^o = \Phi_r \cdot k_a^o \tag{13-36}$$

The reaction quantum yield Φ_r of nitrobenzene is 2.9 x 10^{-5} (see *Table 13-7*). Using the k_a^o(24h) value calculated in Table L.2, a k_p^o(24h) value of

$$k_p^o(24h) = 2.9 \times 10^{-5} \frac{(\text{mol NB})}{\text{einstein}} \left(3.2 \times 10^{2} \frac{\text{einstein}}{(\text{mol NB})\cdot \text{d}}\right) = 9.3 \times 10^{-3}\ \text{d}^{-1}$$

is obtained and, therefore,

$$t_{1/2}^o = \frac{0.693}{9.3 \times 10^{-3}\ \text{d}^{-1}} = 75\ \text{d}$$

Problem

Estimate the 24h-averaged direct photolysis half-life of 4-nitrophenol (4-NP) (a) near the surface, and (b) in the well-mixed epilimnion of a lake (pH=7.5, z_{mix}=5 m, $\alpha(\lambda)$ values given in *Table 13.6*) on a clear midsummer day at 40° N latitude and sea level. The following data are available:

4-NP is a weak acid with a pK_a of 7.11 (see Chapter E):

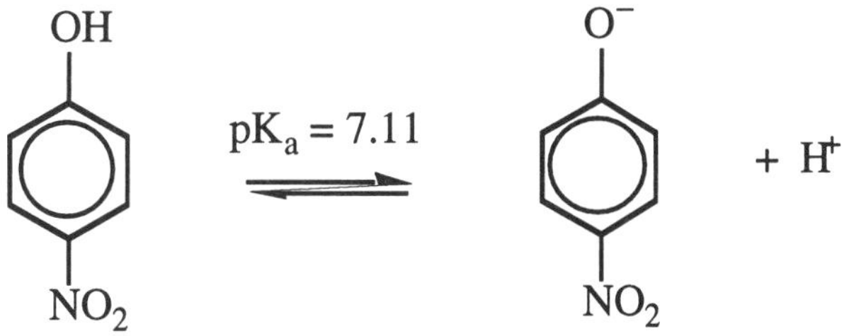

OH — NO_2 (HA) $\underset{}{\overset{pK_a = 7.11}{\rightleftharpoons}}$ O^- — NO_2 (A^-) + H^+

The estimated 24h-averaged near-surface total specific light absorption rates of the nondissociated (HA) and the dissociated (A^-) species are

$$k_a^o(24h, \text{HA}) = 4.5 \times 10^{3}\ \text{einstein}\cdot(\text{mol HA})^{-1}\cdot\text{d}^{-1}\ (\lambda_m \simeq 330\ \text{nm})$$

$$k_a^o(24h, \text{A}^-) = 3.2 \times 10^{4}\ \text{einstein}\cdot(\text{mol A}^-)^{-1}\cdot\text{d}^{-1}\ (\lambda_m \simeq 400\ \text{nm})$$

The corresponding quantum yields can be found in *Table 13.7:*

$$\Phi_r\,(\text{HA}) = 1.1 \text{ x } 10^{-4}\,(\text{mol HA})\cdot\text{einstein}^{-1}$$

$$\Phi_r\,(\text{A}^-) = 8.1 \text{ x } 10^{-6}\,(\text{mol A}^-)\cdot\text{einstein}^{-1}$$

Answer (a)

The near-surface rate of direct photolysis of 4-NP (HA + A$^-$) is given by

$$-\frac{d[\text{4-NP}]_{tot}}{dt} = k_p^o\,(24\text{h, 4-NP}) \cdot [\text{4-NP}]_{tot}$$

k_p^o (24h, 4-NP) is a "lumped" rate constant that can be expressed as the sum of the k_p^o values for the nondissociated and dissociated species, respectively:

$$k_p^o\,(24\text{h, 4-NP}) = \alpha_a \cdot k_p^o\,(24\text{h, HA}) + (1-\alpha_a)\cdot k_p^o\,(24\text{h, A}^-) \qquad \text{(L-2)}$$

$$= \alpha_a \cdot \Phi_r\,(\text{HA}) \cdot k_a^o\,(24\text{h, HA}) + (1-\alpha_a)\cdot \Phi_r(\text{A}^-)\cdot k_a^o\,(24\text{h, A}^-)$$

with

$$\alpha_a = \frac{1}{1+10^{(\text{pH}-\text{p}K_a)}}$$

At pH 7.5, $\alpha_a = 0.29$. Substitution of the corresponding values into Eq. L-2 yields

$$k_p^o(24\text{h,4-NP}) = (0.29)(1.1\text{x}10^{-4})(4.5\text{x}10^{3})\text{d}^{-1}$$

$$+(0.71)(8.1\text{x}10^{-6})(3.2\text{x}10^{4})\text{d}^{-1}$$

$$= (0.14\text{ d}^{-1} + 0.18\text{ d}^{-1}) = 0.32\text{ d}^{-1}$$

and

$$t_{1/2}^o\,(24\text{h, 4-NP}) = \frac{0.693}{0.32\text{ d}^{-1}} = 2.2\text{ d}$$

Note that at pH 7.5, the direct photolytic transformations of HA and A$^-$, respectively, contribute about equally to the overall near-surface direct photolysis rate of 4-NP.

Answer (b)

For the well-mixed epilimnion, the overall rate of direct photolysis of 4-NP can be expressed as

$$-\frac{d[\text{4-NP}]_{tot}}{dt} = k_p\,(24h,\,\text{4-NP}) \cdot [\text{4-NP}]_{tot}$$

with

$$k_p\,(24h,\,\text{4-NP}) = \alpha_a \cdot \Phi_r(HA) \cdot k_a(24h,\,HA) + (1-\alpha_a) \cdot \Phi_r\,(A^-) \cdot k_a(24h,\,A^-) \qquad \text{(L-3)}$$

Using appropriate screening factors $S(\lambda_m)$, $k_p(24h,\,\text{4-NP})$ can be approximated by (see *Eq. 13-32* and Eq. L-1)

$$k_p(24h,\,\text{4-NP}) = \alpha_a \cdot \Phi_r(HA) \cdot S(\lambda_m^{HA}) \cdot k_a^o\,(24h,\,HA) + (1-\alpha_a) \cdot \Phi_r(A^-) \cdot S(\lambda_m^{A^-}) \cdot k_a^o\,(24h, A^-) \qquad \text{(L-4)}$$

The α values for λ_m^{HA} = 330 nm and $(\lambda_m^{A^-})$ = 400 nm are 2.20 m^{-1} and 0.55 m^{-1}, respectively (see *Table 13.6*). Insertion of these α values into Eq. L-1 with z_{mix} = 5 m yields $S(\lambda_m)$ values of 0.033 and 0.13, respectively. Thus,

$$k_p(24h,\,\text{4-NP}) = (0.29)\,(1.1 \times 10^{-4})\,(0.033)\,(4.5 \times 10^3)\,d^{-1} + (0.71)\,(8.1 \times 10^{-6})\,(0.13)\,(3.2 \times 10^4)\,d^{-1}$$

$$= (0.0047\,d^{-1} + 0.0239\,d^{-1}) = 0.029\,d^{-1}$$

and

$$t_{1/2}(24h,\text{4-NP}) = \frac{0.693}{0.029\,d^{-1}} \simeq 24\,d$$

Note that in contrast to the near-surface situation, because of the very different screening factors, the reaction of the dissociated species is much more important in determining the overall direct photolysis rate of 4-NP in the well-mixed epilimnion.

Estimating Indirect Photolysis Half-Lives of Organic Pollutants in Natural Waters

Problem

Estimate the half-life of atrazine in a nitrate-rich shallow pond under clear skies during summertime at 47.5° N latitude, assuming that the reaction with hydroxyl radical (HO•) is the only elimination mechanism. For estimating the HO• steady-state concentration in the pond, use the midday near-surface rate of HO•production given by *Eq. 13-61*. Note that in a field study (*Kolpin and Kolkhoff, 1993*), the observed elimination of atrazine during daytime in a small river in Iowa was attributed primarily to reaction with HO•.

$$r^{o}_{f,HO\cdot}\,(\text{noon}) = (3 \times 10^{-7})\,[NO_3^-] \quad (M \cdot s^{-1}) \qquad (13\text{-}61)$$

In addition, the following information is available:

Pond:			
Mean depth	1 m	α(320 nm)	1 m^{-1}
Water temperature	20°C	$[NO_3^-]$	0.5 mM
pH	7.8	$[HCO_3^-]$	1 mM
[DOC]	2.0 $mg \cdot L^{-1}$	$[CO_3^{2-}]$	0.003 mM

CH_3—CH_2—HN–(chloro-triazine ring: Cl, N, N, N)–NH—CH(CH_3)$_2$

Atrazine

$k'_{HO\cdot} \simeq 5 \times 10^9\ M^{-1}s^{-1}$

Answer

Insert the concentrations of DOC, NO_3^-, HCO_3^-, CO_3^{2-} into *Eq. 13-62* to estimate the midday near-surface HO• steady-state concentration:

$$[HO\bullet]^o_{ss}\,(\text{noon}) = \frac{(3 \times 10^{-7})\,(5 \times 10^{-4})}{(1.5 \times 10^{7})\,(10^{-3}) + (4.2 \times 10^{8})\,(3 \times 10^{-6}) + (2.5 \times 10^{4})\,(2)}$$

$$\simeq 2 \times 10^{-15}\ \text{M}$$

The near-surface 24h-averaged HO• steady-state concentration can then be roughly estimated using *Eq. 13-55.*:

$$[HO\bullet]^o_{ss}(24\text{h}) \simeq [HO\bullet]^o_{ss}(\text{noon})\ \frac{\Sigma\, Z(24\text{h},\lambda)}{86400\ \Sigma\, Z(\text{noon},\lambda)} \qquad \textit{(13-55)}$$

Nitrate absorbs sunlight in the wavelength range between 290-340 nm. Since in the summer, light intensities do not change dramatically with latitude (see *Fig. 13.10*), insert the $Z(\lambda)$ values given in *Table 13.3* into Eq. 13-55. The result is

$$[HO\bullet]^o_{ss}\,(24\text{h}) \simeq (2 \times 10^{-15})\,\frac{(1.6 \times 10^{-1})}{(86400)\,(4.3 \times 10^{-6})} \simeq 9 \times 10^{-16}\ \text{M}$$

The 24h-averaged HO• steady-state concentration in the pond can then be approximated by (see *Eq. 13-56)*

$$[HO\bullet]_{ss}(24\text{h}) \simeq [HO\bullet]^o_{ss}(24\text{h}) \cdot S(\lambda_{max})$$

where λ_{max} = 320 nm (wavelength of the maximum light absorption rate of nitrate). S(320 nm) is given by (see Eq. L-1)

$$S(320\ \text{nm}) = \frac{[1 - 10^{-(1.2)\,(1)\,(1)}]}{(2.303)\,(1.2)\,(1)\,(1)} = 0.34$$

and thus

$$[\mathrm{HO}\bullet]_{ss}(24\mathrm{h}) \simeq (9 \times 10^{-16})\,(0.34) \simeq 3 \times 10^{-16}\ \mathrm{M}$$

Using the $k'_{\mathrm{HO}\cdot}$ value given above, the indirect photolysis half-life of atrazine in the pond is calculated as

$$t_{1/2} = \frac{\ln 2}{k'_{\mathrm{HO}\cdot}[\mathrm{HO}\bullet]_{ss}(24\mathrm{h})} = \frac{0.693}{(5 \times 10^{9})\,(3 \times 10^{-16})} = 4.6 \times 10^{5}\ \mathrm{s}$$

$$= 5.3\ \mathrm{d}$$

Note that atrazine itself does not absorb light significantly above 290 nm and that, therefore, direct photolysis can be neglected.

PROBLEMS

● L-1 *Is Everything O.K. with These Light Penetration Measurements?*

During the summer, somebody measures how much light coming into a small mesotrophic lake is absorbed in the water column as a function of depth. He gets the following results:

		% of incoming light intensity of wavelength λ		
Depth (m)	T (°C)	300 nm	400 nm	500 nm
0	20.3	100	100	100
1	20.2	10	60	90
2.5	20.3	0.3	30	75
5	16.5	< 0.1	3	40
7.5	10.2		0.3	24
10	8.4		< 0.1	18

Calculate the diffuse light attenuation coefficients, $\alpha_D(\lambda)$, as a function of depth for the three indicated wavelengths. Do the results of these light penetration measurements make sense?

● L-2 *Photolysis or Hydrolysis? What Process(es) is (are) Primarily Responsible for the Elimination of an Insecticide from a Shallow Water Body?*

Consider a well-mixed oligotrophic shallow water body (z_{mix} = 1 m; α(300 nm) = 0.2 m^{-1}; α(320 nm) = 0.15 m^{-1}, α(350 nm) = 0.1 m^{-1}; pH = 7.0; T = 15°C) exposed to sunlight on a clear summer day at 40°N latitude. Due to spraying of the insecticide, carbaryl, in the surroundings, there is a significant input of this compound into the water body. As an employee of the company that manufactures this insecticide, you are asked how persistent this compound is in this water body. Calculate the half-life of carbaryl in the water under the given conditions by assuming that photolysis and abiotic hydrolysis are the dominant elimination mechanisms. Is this assumption correct? What other processes could be important for the elimination of carbaryl from this water body. Can you say anything about the half-lives of the hydrolysis products of carbaryl? In the literature you find the following data for carbaryl (*Roof, 1982; Table 12.13*):

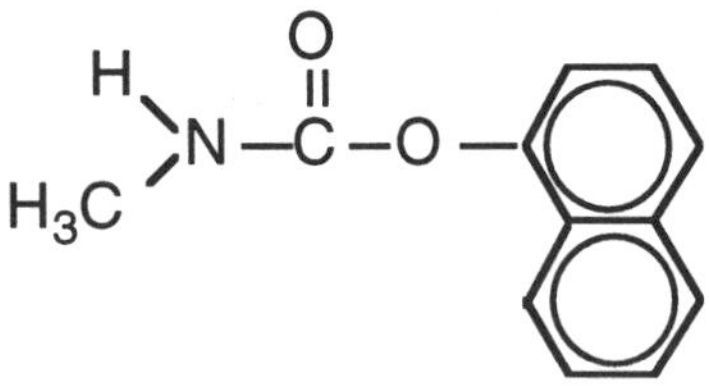

carbaryl

$k_N(25°C) = 9.0 \times 10^{-7} s^{-1}$

$k_B(25°C) = 5.0 \times 10^{1} M^{-1} s^{-1}$

Φ_r (313 nm) = 0.006

and the decadic molar extinction coefficients

λ(nm)	ε($M^{-1} \cdot cm^{-1}$)	λ(nm)	ε($M^{-1} \cdot cm^{-1}$)
297.5	1480	315	261
300	918	317.5	235
302.5	741	320	101
305	532	323	45
307.5	427	330	11
310	356	340	< 1
312.5	288		

● *L-3 How Important are Reactions of Substituted Phenols with $^1O_2$ in the Epilimnion of a Lake?*

In an extensive study on the kinetics of the reaction of substituted phenols with singlet oxygen ($^1O_2$), *Tratnyek and Hoigné (1991)* have derived quantitative structure-reactivity relationships for both the non-dissociated and the dissociated phenol species. For compounds with substituents that do not exhibit a strong negative resonance effect (-R effect, see *Table 8.3*), they found the following correlations:

$$\log k_{^1O_2}^{ArOH} \text{ (in } M^{-1} \cdot s^{-1}) = 8.9 - 3.5\, E_{1/2} \text{ (in V)}$$

$$\log k_{^1O_2}^{ArO^-} \text{ (in } M^{-1} \cdot s^{-1}) = 9.6 - 2.0\, E_{1/2} \text{ (in V)}$$

where $k_{^1O_2}^{ArOH}$ and $k_{^1O_2}^{ArO^-}$ are the second-order rate constants for the reactions of the phenol and the phenolate, respectively with $^1O_2$, and $E_{1/2}$ is the half-wave potential (relative to the standard calomel electrode!) determined by anodic voltametry. $E_{1/2}$ reflects the potential for the first one-electron oxidation step.

Using the information given below, calculate the half-lives of 2,4-dichlorophenol and 2,6-dimethoxyphenol for the reaction with $^1O_2$ in the epilimnion of Greifensee, (T = 25°C, pH = 8.5, z_{mix} = 5 m, $\alpha(\lambda)$ see *Table 13.6*), Switzerland under light conditions typical of a clear midsummer day (see *Table 13.6*). Assume a midday, near-surface steady-state concentration of $^1O_2$ ($[^1O_2]_{ss}^o$) of 8 x 10^{-14} M, with a maximum $^1O_2$ production at a wavelength of 400 nm. For the two compounds, the following information is available:

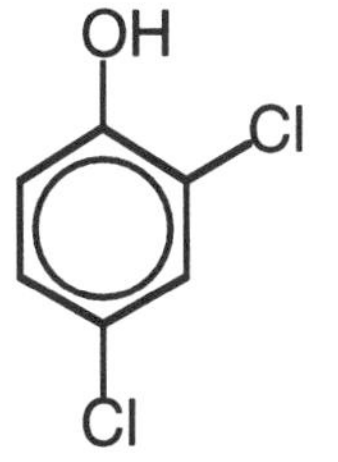

OH, Cl, Cl (structure)

2,4-dichlorophenol
$E_{1/2}$ = 0.645 V

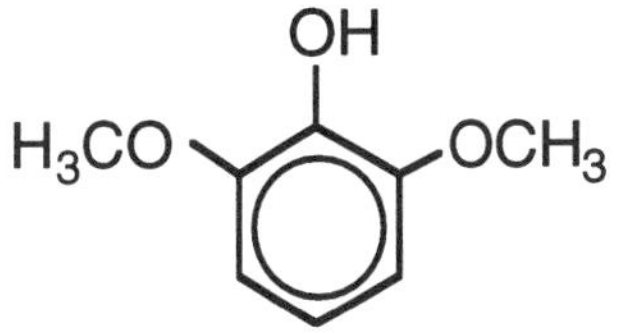

OH, H_3CO, OCH_3 (structure)

2,6-dimethoxyphenol
$E_{1/2}$ = 0.317 V

Are there other indirect photolysis processes that are important for these two compounds?

● *L-4 Does Direct Photolysis Affect the Phenanthrene to Anthracene Ratio in Aerosol Droplets?*

Gschwend and Hites (1981) observed that the two closely related polycyclic aromatic hydrocarbons, phenanthrene and anthracene, occur in a ratio of about 3-to-1 in urban air. In contrast, sedimentary deposits obtained from remote locations (e.g., the Adirondack mountain ponds) exhibited phenanthrene to anthracene ratios of 15-to-1. You hypothesize that these chemicals are co-carried in aerosol droplets from midwestern USA urban environments via easterly winds to remote locations (like the Adirondacks) where the aerosol particles fall out of the atmosphere and rapidly accumulate in the ponds' sediment beds without any further compositional change (i.e., the phenanthrene-to-anthracene ratio stops changing after the aerosols leave the air). If summertime direct photolysis was responsible for the change in phenanthrene-to-anthracene ratio, estimate how long the aerosols would have to have been in the air. Comment on the assumptions that you make. What are your conclusions?

Hints and Help
All necessary data can be found in the textbook (*Appendix, Figure 13.3, Table 13.7*).

M. BIOLOGICAL TRANSFORMATION REACTIONS

ILLUSTRATIVE EXAMPLES

Establishing Mass Balances For Oxygen and Nitrate in a Given System

As discussed in *Sections 14.2* and *14.3*, oxygen plays an important role in the microbial metabolism of organic pollutants in the environment. On the one hand, oxygen is the energetically most favorable among the most abundant natural electron acceptors (see *Table 12.16*); on the other hand, it is a pivotal cosubstrate in many biochemical reactions. Therefore, it is in many cases relevant to know whether there is sufficient oxygen available in a given system, or whether anaerobic conditions have to be expected. Also, when bioremediating a contaminated site under aerobic conditions, it is necessary to estimate how much oxygen is required to mineralize a given amount of organic waste.

Another important electron acceptor is nitrate. Although in the absence of molecular oxygen, microorganisms have to use a different strategy to degrade organic pollutants (particularly for the initial transformation step(s)), there are quite a few bacteria that may grow or at least survive under both aerobic and nitrate reducing conditions. Furthermore, a frequently applied alternative to the use of oxygen as electron acceptor in bioremediation is to use nitrate, because the latter is much more soluble in water and may, therefore, be supplied to the microorganisms at much higher concentrations (*Hunkeler et al., 1995*). It is, therefore, reasonable to consider these two electron acceptors together, in as much as after removal of oxygen and nitrate (and possibly manganese oxide), the reduction potential drops dramatically (see *Table 12.16*).

Problem
Consider a situation in which bank filtrate of a polluted river is used for drinking water supply. Among other water constituents, dissolved organic material (measured as dissolved organic carbon, DOC), ammonia, oxygen, and nitrate are monitored continuously in the river and in a well located at a distance of 10 meters from the river bank. The average values obtained for the four parameters are

Species (measured parameter)	Concentration in the river	Concentration in the well
Organic material (CH_2O) ([DOC])	4.2 mg $C{\cdot}L^{-1}$	1.2 mg $C{\cdot}L^{-1}$
Ammonia (NH_4^+) ($[NH_4^+ - N]$)	2.1 mg $N{\cdot}L^{-1}$	< 0.1 mg $N{\cdot}L^{-1}$
Dissolved Oxygen (O_2) ($[O_2]$)	9.6 mg $O_2{\cdot}L^{-1}$	< 0.1 mg $O_2{\cdot}L^{-1}$
Nitrate (NO_3^-) ($[NO_3^- - N]$	2.1 mg $N{\cdot}L^{-1}$	1.6 mg $N{\cdot}L^{-1}$

Inspection of the data field shows that a significant portion of the organic material and virtually all NH_4^+ and O_2 are eliminated by microbial processes during infiltration, but that the infiltrated water still exhibits 75% of the nitrate concentration observed in the river. Are these findings reasonable when assuming that no additional (water) input occurs during infiltration, and that the organic material is oxidized to CO_2, NH_4^+ is oxidized to NO_3^-, and that NO_3^- is reduced to N_2?

Answer
Use the half reactions given in *Table 12.16* to establish electron balances for the various processes:

(i) Oxidation of organic material:

$$CH_2O + H_2O = CO_2 + 4H^+ + 4e^- \qquad \text{(M-1)}$$

(ii) Oxidation of ammonia (nitrification):

$$NH_4^+ + 3H_2O = NO_3^- + 10H^+ + 8e^- \qquad \text{(M-2)}$$

(iii) Reduction of oxygen:

$$O_2 + 4H^+ + 4e^- = 2H_2O \qquad \text{(M-3)}$$

(iv) Reduction of nitrate (denitrification):

$$NO_3^- + 6H^+ + 5e^- = \frac{1}{2}N_2 + 3H_2O \qquad \text{(M-4)}$$

Calculate how many electrons are produced and consumed, respectively, by the various processes during infiltration:

(i) $\Delta[CH_2O]$	$= -3.0$ mg C·L^{-1}	$= -0.25$ mM CH_2O·L^{-1}	$= +1.0$ mM e^-
(ii) $\Delta[NH_4^+]$	$= -2.1$ mg N·L^{-1}	$= -0.15$ mM NH_4^+·L^{-1}	$= +1.2$ mM e^-
(iii) $\Delta[O_2]$	$= -9.6$ mg O_2·L^{-1}	$= -0.3$ mM O_2·L^{-1}	$= -1.2$ mM e^-
Total electrons without considering denitrification			$= +1.0$ mM e^-

Thus, in order to balance the electrons, 1 mM e^- have to be consumed by denitrification. Hence, the calculated consumption of nitrate is 1 mM e^- = 0.2 mM NO_3^- = 2.8 mg N·L^{-1}, which is more than is present in the river water. Note, however, that reaction (ii) produces 0.15 mM = 2.1 mg N·L^{-1} nitrate. Thus, one would expect a net decrease in nitrate of only 0.7 mg N·L^{-1}, which compares well with the observed 0.5 mg N·L^{-1} decrease. The measured concentration changes of the four water constituents are, therefore, reasonable.

Problem

For remediation of an aquifer that has been contaminated with toluene, groundwater is pumped through the contaminated zone, then pumped back to the surface, saturated with air or supplied with nitrate, and finally introduced into the ground again. The idea of this quite widely applied procedure is to stimulate those indigenous microorganisms that are capable of mineralizing a given substrate, in this case toluene, by supplying the necessary oxidants. Calculate how much water is at least required to supply sufficient O_2 or NO_3^-, respectively, for degradation of 1 kg of toluene, when assuming (i) that toluene is not mobilized by this procedure, (ii) that it is completely mineralized to CO_2 and H_2O, and (iii) that the water contains either 10 mg $O_2 \cdot L^{-1}$ or 100 mg $NO_3^- \cdot L^{-1}$, respectively. Note that much more NO_3^- could, in principle, be dissolved in the water, but that the maximum allowed concentration is commonly limited by the water authorities.

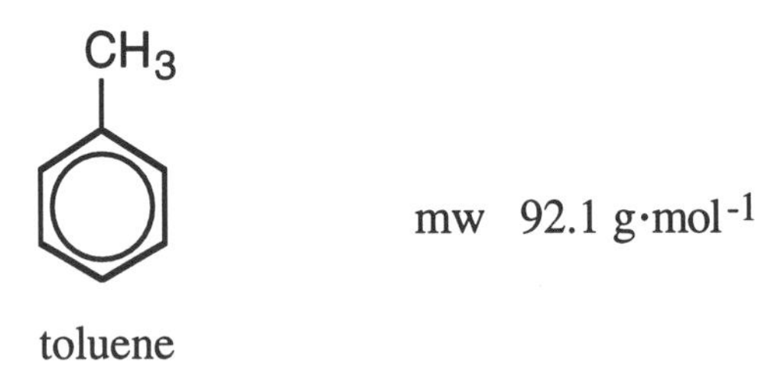

Answer

To calculate how many electrons have to be transferred to O_2 or respectively, when oxidizing 1 mole of toluene (C_7H_8) to CO_2, determine first the average oxidation state of the carbon atoms present in toluene (see also examples given in *Table 2.4.*). Since this compound is made up only of carbon and hydrogen atoms (the oxidation state of H is +I), you can just consider the hydrogen/carbon ratio, which yields an average carbon oxidation state of −8/7). Considering that the oxidation state of carbon in CO_2 is +IV, it is easy to see that a total of (4−(−8/7)) x 7 = 36 moles of electrons have to be transferred. The overall reactions are, therefore,

$$C_7H_8 + 9\,O_2 = 7\,CO_2 + 4\,H_2O \qquad \text{(M-5)}$$

if O_2 is the oxidant (4 electrons per O_2, see *Table 12.16*) , and, with NO_3^- as the electron acceptor (5 electrons per NO_3^-, *Table 12.16*):

$$C_7H_8 + \frac{36}{5}NO_3^- + \frac{216}{5}H^+ = 7\,CO_2 + \frac{18}{5}N_2 + \frac{108}{5}H_2O \qquad \text{(M-6)}$$

Consequently, 9 moles of O_2 or 7.2 moles of NO_3^- are required to mineralize 1 mole of toluene. Since 1 kg of toluene corresponds to 10.86 moles, this means that at least 97.72 moles O_2 or 78.12 moles NO_3^- have to be provided by the water that is pumped through the contaminated zone. Thus, in the case of O_2 (10 g $O_2 \cdot m^{-3}$), the total water volume required is

$$V = \frac{97.72 \text{ mol}}{(10 \text{ g} \cdot \text{m}^{-3})/(32 \text{ g} \cdot \text{mol}^{-1})} = 312.7 \text{ m}^3$$

If NO_3^- (100 g $NO_3^- \cdot m^{-3}$) is used, the calculated water volume is only

$$V = \frac{78.1 \text{ mol}}{(100 \text{ g} \cdot \text{m}^{-3})/(62 \text{ g} \cdot \text{mol}^{-1})} = 48.5 \text{ m}^3$$

Evaluating the Biodegradation of Organic Compounds in a Well-Mixed Reactor

Suppose you need to purify wastewater stream containing a readily degradable compound in fairly high concentrations. If one can deliver that waste stream into a well-mixed reactor, which is simultaneously receiving any other "supplies" (e.g., oxygen, nutrients) needed by the microorganisms for growth, then one may be able to build up a microbial population capable of transforming the organic waste to innocuous substances like CO_2 and H_2O before the water is discharged. For setting up such a reaction, one needs information about the rate of growth of the initial microbial innoculum (i.e., the maximum growth rate μ_{max}, and the Monod constant K_M; *Eq. 14-118*) and about the "yield" of the microbial process (Y, *Eq. 14-119*), that is, the relationship between biomass production and substrate consumption.

Problem

You have a well-mixed tank of 10 m^3 volume, which you want to set up as a bioreactor to remove glycerol from an industrial waste stream flowing at 2 $m^3 \cdot h^{-1}$. You also have a microbial culture of *Aerobacter sp*. that has been characterized as being able to grow on glycerol with a maximum doubling rate, μ_{max}, of 1.2 per hour and a Monod half-saturation constant, K_M, for glycerol consumption of 120 μM (see *Table 14.10*). This bacterium has also been found to convert glycerol to biomass at a yield, Y, of 10^{14} · cells (mol glycerol)$^{-1}$. Calculate the steady-state output glycerol concentration after the microorganisms have increased their biomass to some steady-state level, by assuming that the input concentration of glycerol is 100 μM (9 ppm). What is the bacterial cell density at steady state?

```
CH2— CH — CH2        glycerol
|     |     |
OH    OH    OH
```

Answer

The tank will reach steady-state conditions when the number of bacteria present has increased to some constant level, dictated by the balance of their ability to grow on glycerol against their rate of loss due to being flushed from the tank. Said in mathematical terms:

$$\frac{d[\text{bacteria}]}{dt} = -k_{flush}\,[\text{bacteria}] + \mu[\text{bacteria}] \qquad \text{(M-7)}$$

$$= 0 \text{ at steady state}$$

Therefore, the growth rate for this bacterial culture, after an initial phase of very rapid exponential growth, will ultimately settle down to a value fixed by the rate of washout! (Note: such a culture apparatus is widely used in the laboratory and is referred to as a "chemostat".) Since the rate of tank flushing can be readily deduced from its volume and the wastewater input rate, one obtains

$$\mu_{\text{steady state}} = k_{flush} = \text{input flow rate / tank volume} \qquad \text{(M-8)}$$

$$= (2\ m^3/h)\ /\ (10\ m^3)$$

$$= 0.2\ h^{-1}$$

Now use *Eq. 14-118* and the information on the *Aerobacter sp.* growth characteristics to calculate the corresponding steady-state glycerol concentration:

$$\mu = \frac{\mu_{max}\,[\text{glycerol}]}{K_M + [\text{glycerol}]} \qquad (14\text{-}118)$$

Hence, at steady-state growth

$$0.2\ \text{h}^{-1} = \frac{1.2\ \text{h}^{-1}\,[\text{glycerol}]_{\text{steady state}}}{120\ \mu\text{M} + [\text{glycerol}]_{\text{steady state}}}$$

so that

$$[\text{glycerol}]_{\text{steady state}} = 24\ \mu\text{M}$$

Note that this result is independent of the concentration of glycerol in the input stream as long as the other necessary nutrients for growth are sufficient! This result is also independent of the original size of the inocculum. The bacteria simply increase in number until their use of the glycerol (100 μM – 24 μM = 76 μM) corresponds to the bacterial biomass they form. You can also use this result together with the information on metabolic yield (10^{14} cells·mol^{-1}) to estimate the steady-state bacterial cell density:

$$(76 \times 10^{-6}\ \text{mol}\cdot\text{L}^{-1}) \times (10^{14}\ \text{cells}\cdot\text{mol}^{-1}) = 7.6 \times 10^{9}\ \text{cells}\cdot\text{L}^{-1}$$

Note that the oxidation of glycerol to CO_2 requires 3.5 moles of O_2 for every mole of glycerol consumed. Thus, the degradation of 76 μmol glycerol per liter can just be accomplished with the O_2 present initially at saturation (ca. 280 μM).

Problem

You have a wastewater stream flowing at 2 $m^3 \cdot h^{-1}$ and containing 10 ppm trichloroethylene (TCE). Realizing that methane-utilizing bacteria (or methanotrophs) have been found capable of oxidizing TCE (*Fogel et al., 1986*) you decide to set up a closed bioreactor (10 m^3) in which such methanotrophs are encouraged to grow by supplying the water with methane (0.1 mM) and oxygen (0.3 mM):

$$CH_4 + 2O_2 = CO_2 + 2H_2O \qquad \text{(M-9)}$$

The methanotrophs are found to grow with a μ_{max} of 2 h^{-1}, a K_M (methane) of 10 μM, and a yield of 10^{13} cells$\cdot$(mol methane)$^{-1}$.

The methane mono-oxygenase of this species exhibits the following enzymatic properties: K_{MM}(methane) = 1 μM and K_{MM}(TCE) = 5 μM; V_{max}(methane) = 10 V_{max}(TCE).

In order to get an idea about the performance of such a reactor, you have to answer the following questions:

(a) What will be the steady-state methane concentration exiting the bioreactor?

(b) What will be the concentration of methanotrophs in the reactor at steady state?

(c) What will be the biodegradation rate constant for TCE?

(d) What will be the steady-state TCE concentration exiting the bioreactor?

Answer

(a) To determine the concentration of methane exiting the bioreactor, first recognize that the growth rate of methanotrophs will be determined by the flushing rate of the reactor (see previous problem). Thus,

$$\mu_{\text{steady state}} = \frac{\mu_{max}\,[\text{methane}]}{K_M + [\text{methane}]}$$

or

$$(0.2\ h^{-1}) = \frac{(2\ h^{-1})\,[\text{methane}]}{10\ \mu M + [\text{methane}]}$$

which gives

$$[\text{methane}]_{\text{steady state}} = 1.1\ \mu\text{M}$$

(b) Realizing that the influent methane concentration of 100 μM is reduced to 1.1 μM in the reactor, the methanotrophs consume about 99 μmol methane per liter. Given the reported yield of 10^{13} cells per mol methane consumed, calculate the steady-state cell density:

$$[\text{B}] = (\text{d}[\text{methane}] / \text{dt})\ (\text{Y}) / \mu \qquad (14\text{-}121)$$

$$= (99 \times 10^{-6}\ \text{mol}\cdot\text{L}^{-1})\ (10^{13}\ \text{cells}\cdot\text{mol}^{-1})\ (0.2\ \text{h})^{-1}$$

$$= 10^{9}\ \text{cells}\cdot\text{L}^{-1}$$

(c) The methanotrophs degrade TCE incidentally to their primary substrate, methane. Thus the rate of TCE biotransformation is dictated by the relative effectiveness of the enzyme, methane mono-oxygenase, for interacting with TCE rather than methane. The enzymatic processing of both substrate may be described by a Michaelis-Menten-type expression:

$$\frac{\text{d}[\text{substrate}]}{\text{dt}} = -\frac{V_{\text{max}}\ [\text{substrate}]}{K_{\text{MM}} + [\text{substrate}]} \qquad (14\text{-}91)$$

From the preceding discussions concerning methane use by the methanotrophs, and using data on the mono-oxygenase's half saturation constant for methane, K_{MM}(methane) = 1 μM, *Eq. 14-91* can be used to calculate the steady state system's V_{max}:

$$(100\ \mu\text{M} - 1.1\ \mu\text{M}) / 5\text{h} = -\frac{V_{\text{max}}\ [1.1\ \mu\text{M}]}{(1\ \mu\text{M} + 1.1\ \mu\text{M})}$$

or

$$V_{\text{max}}(\text{methane}) = -38\ \mu\text{M}\cdot\text{h}^{-1}$$

This maximal methane consumption rate is set by both the enzymatic properties and the cell density achieved in this particular bioreactor. Normalizing to the cell density, one finds

$$V_{max}(\text{methane}) \simeq 4 \times 10^{-14}\ \text{mol}\cdot\text{L}^{-1}\cdot\text{h}^{-1}\ \text{cell}^{-1}$$

From experience, the maximal processing of TCE is reported to be one tenth the rate of oxidation of methane (as given above, $V_{max}(\text{methane}) = 10 \times V_{max}(\text{TCE})$), so

$$-\frac{d[\text{TCE}]}{dt} = \frac{0.1\ V_{max}(\text{methane})\ [\text{TCE}]}{(K_{MM}(\text{TCE}) + [\text{TCE}])} = \frac{0.1 \times 38\ \mu\text{M}\cdot\text{h}^{-1}\ [\text{TCE}]}{(5\ \mu\text{M} + [\text{TCE}])}$$

If the incoming [TCE] (10 ppm = $10^{-2}\text{g}\cdot\text{L}^{-1}$ = 76 μM) is biodegraded to a steady-state value, which is small relative to K_{MM} (TCE) of 5 μM, then the overall rate expression becomes:

$$-\frac{d[\text{TCE}]}{dt} = 0.1 \times 38\ \mu\text{M}\cdot\text{h}^{-1}\ [\text{TCE}]\ /\ (5\ \mu\text{M})$$

$$= 0.76\ \text{h}^{-1}\ [\text{TCE}]$$

and the "first-order biodegradation rate constant" is $0.76\ \text{h}^{-1}$. However, if the steady-state concentration of TCE proves to be large relative to K_{MM} (TCE), the biodegradation rate constant will be zero-order in [TCE]:

$$-\frac{d[\text{TCE}]}{dt} = 0.1 \times 38\ \mu\text{M}\cdot\text{h}^{-1}$$

$$= 3.8\ \mu\text{M}\cdot\text{h}^{-1}$$

(d) Set up a mass balance equation for TCE:

$$\frac{d[\text{TCE}]}{dt} = +\frac{\text{input}}{\text{reactor volume}} - \text{biodegradation} - \frac{\text{flushing}}{\text{reactor volume}} \qquad \text{(M-10)}$$

$$\begin{aligned} = &+ (2000\ \text{L}\cdot\text{h}^{-1})\ (76\ \mu\text{mol}\cdot\text{L}^{-1}\ )\ /\ (10000\ \text{L}) \\ &- (3.8\ \mu\text{M}\cdot\text{h}^{-1})\ [\text{TCE}]\ /\ (5\ \mu\text{M} + [\text{TCE}]) \\ &- ((2000\ \text{L}\cdot\text{h}^{-1})\ [\text{TCE}]\ /\ (10000\ \text{L}) \end{aligned}$$

$$= + 15.2 \ \mu mol \cdot L^{-1} \cdot h^{-1} - (3.8 \ \mu M \cdot h^{-1}) [TCE] / (5 \ \mu M + [TCE]) - 0.2 \ h^{-1} [TCE]$$

At steady state, $\frac{d[TCE]}{dt} = 0$; and by solving the mass balance equation, one finds

$$[TCE]_{steady\ state} = \sim 59 \ \mu M$$

This result indicates that the methanotrophs are "co-metabolizing" the TCE at a concentration well above their monooxygenase's half saturation constant (5 µM), and thus, the effective biodegradation rate is zero-order and has a value of only

$$k_{bio} = V_{max}(TCE) = 3.8 \ \mu M \cdot h^{-1}$$

Since the water detention in the tank is only 5 hours, this result implies the methanotrophs could only process (5 h) x ($3.8 \ \mu M \cdot h^{-1}$) = 19 µM of TCE (2.5 ppm).

Estimating Biotransformation Rates of Organic Pollutants in Natural System

As already pointed out in *Chapter 14*, it is extremely difficult to make sound predictions of biotransformation rates of organic pollutants in a natural system. One major reason for this difficulty is the lack of information on the abundance and on the activities of the microorganisms present that are capable of metabolizing the compound of interest. Furthermore, other factors such as the availability of nutrients, trace elements, oxidants, or the presence of competing substrates may strongly influence the rate of biotransformation. Also, particularly when dealing with strongly sorbing compounds, the availability of the chemical to the microorganisms, e.g., the rate of desorption, may determine the overall biotransformation rate.

Nevertheless some simple calculations using information from laboratory systems may provide a "first feeling" for whether this process may or may not be important in a given system.

Problem

You are concerned about the fate of 2-nitrophenol (2-NP) found at a concentration of 10 ppb (0.07 µM) in some groundwater. According to the literature (*Zeyer and Kocher, 1988*), this compound can be transformed aerobically by a soil pseudomonad. The biodegradation pathway appears to involve an oxygenase that first converts the nitrophenol to catechol, and then the catechol is further metabolized (see *Table 14.1*).

OH NO_2 → oxygenase OH OH

2-nitrophenol (2-NP) catechol

In order to anticipate the *in situ* rate of removal of 2-NP, you grow a culture of the pseudomonad, isolate and enrich (40x) a cell protein fraction containing the oxygenase, and study the ability of this enzyme to execute the first transformation step shown above. In a first experiment, you vary the concentration of 2-NP in the culture medium to assess the oxygenase's Michaelis-Menten parameters (see *Fig. 14.13*). The data are

2-nitrophenol concentration (μM)	rate of degradation by oxygenase ($\mu mol \cdot g_{protein}^{-1} \cdot min^{-1}$)
2	1040
3	1420
4	1920
5	2150
10	2480
20	2530

(At still higher 2-nitrophenol concentrations, the specific rate of transformation to catechol actually slows, indicating that some sort of inhibition effect becomes important. Neglect, however, this phenomenon for the following calculations.)

Answer now the following questions:

(a) What are the V_{max} and K_{MM} for this enzymatic reaction?

(b) Assuming a soil pseudomonad used this enzyme to transform 2-NP without growing on this compound, what biodegradation rate constant, k_{bio} would you estimate for 2-NP removal?

(c) Assuming the 2-NP exhibits a K_d for the aquifer solids of 5 $L \cdot kg^{-1}$, what is the overall half-life of this phenol derivative in the aquifer assuming biodegradation is the chief removal mechanism?

Answer

(a) To deduce the approximate enzyme parameters, fit the data from the partially purified enzyme to the Michaelis-Menten equation (*Eq. 14-91* and *Fig. 14.13*). This can be done by inverting this hyperbolic expression to yield a linear form:

$$1/v = \frac{K_{MM}}{V_{max}\,[\text{substrate}]^{-1}} + \frac{1}{V_{max}} \qquad \text{(M-11)}$$

Performing a least squares fit of the transformed data

(2-nitrophenol concentration)$^{-1}$ (μM^{-1})	(rate of degradation by oxygenase)$^{-1}$ ($g_{protein} \cdot min \cdot \mu mol^{-1}$)
0.50	0.000962
0.33	0.000704
0.25	0.000521
0.20	0.000465
0.10	0.000403
0.05	0.000395

yields

$$1/v \text{ (in } g_{protein} \cdot min \cdot \mu mol^{-1}) = 0.00131 \text{ [2-NP (in } \mu M)]^{-1} + 0.000263$$

$$R = 0.97$$

From the coefficients, the parameters can be derived:

$$1/V_{max} = 0.000263 \; g_{protein} \cdot min \cdot \mu mol^{-1}$$

Thus,

$$V_{max} = 3800 \; \mu mol \cdot g_{protein}^{-1} \cdot min^{-1}$$

or considering the 40 x concentration factor,

$$V_{max} \simeq 100 \; \mu mol \cdot g_{protein}^{-1} \cdot min^{-1}$$

If these bacterial cells are taken to be 1 μm in diameter (volume $\simeq 5 \times 10^{-13} \cdot cm^3$), to consist of 90% water (i.e., 5×10^{-14} g dry weight each), and to contain 20% of their organic content as protein, one estimates each bacterial cell to contain 10^{-14} g of protein. Using this estimate, the V_{max} calculated above can be normalized per cell:

$$V_{max} \approx 10^{-12} \; \mu mol \cdot cell^{-1} \cdot min^{-1}$$

Such "per cell or per protein" maximum rates may be tunable to situations of interest by multiplying by the abundance of relevant cells present or the concentration of protein, which serves as a surrogate measure of the microbial abundance. Also,

$$(K_{MM} / V_{max}) = 0.00131$$

and, therefore,

$$K_{MM} = 0.00131 \times 3800 = 5\ \mu M \text{ (or about 700 ppb)}$$

(b) To estimate an <u>in situ</u> biodegradation rate, one needs to "tune" for the abundance of microorganisms participating in the 2-NP removal; unfortunately, as already stated above, such data are typically unavailable. Thus, here we estimate the rate for two "plausible" soil pseudomonad abundances assuming an aquifer r_{sw} of 10 kg/L (see *Eq. 11-8*):

(i) about 10^6 cells $\cdot kg^{-1}$ or 10^7 cells$\cdot L^{-1}$ and
(ii) about 10^8 cells$\cdot kg^{-1}$ or 10^8 cells$\cdot L^{-1}$

In case (i), one estimates V_{max} to be

$$V_{max} \text{ (for } 10^7 \text{ cells}\cdot L^{-1}) \simeq (10^{-12}\ \mu mol\cdot cell^{-1}\cdot min^{-1})\ (10^7\ cells\cdot L^{-1})$$
$$\simeq 10^{-5}\ \mu mol\cdot L^{-1}\cdot min^{-1}$$

Using *Eq. 14-91* and the data provided, write for the <u>dissolved</u> 2-NP:

$$\frac{d[2\text{-NP}]_d}{dt} = -\frac{V_{max}\ [2\text{-NP}]_d}{K_{MM} + [2\text{-NP}]_d}$$

where the subscript d indicates the assumption that only dissolved 2-NP molecules are "available" for biodegradation at any particular moment (see *Section 15.4*).

Note that in the usual pseudo-first order modeling formulation:

$$\frac{d(2\text{-NP})_d}{dt} = -k_{bio}\ [2\text{-NP}]_d \qquad \text{(M-12)}$$

the parameter reflecting the rate of biodegradation includes three factors:

$$k_{bio} = \frac{V_{max}}{K_{MM} + [2\text{-NP}]_d} \qquad \text{(M-13)}$$

For case (i) one obtains:

$$k_{bio} = \frac{(10^{-5}\ \mu\text{mol}\cdot\text{L}^{-1}\cdot\text{min}^{-1})}{(5\ \mu\text{mol}\cdot\text{L}^{-1} + 0.07\ \mu\text{mol}\cdot\text{L}^{-1})} = 2 \times 10^{-6}\ \text{min}^{-1}$$

and for case (ii):

$$k_{bio} = \frac{(10^{-3}\ \mu\text{mol}\cdot\text{L}^{-1}\cdot\text{min}^{-1})}{(5\ \mu\text{mol}\cdot\text{L}^{-1} + 0.07\ \mu\text{mol}\cdot\text{L}^{-1})} = 2 \times 10^{-4}\ \text{min}^{-1}$$

Obviously, a key factor influencing the magnitude of these rates is the abundance of microorganisms capable of transforming 2-NP.

(c) In light of the degradation rates estimated in (b), one may now estimate the overall biodegradation half-life of 2-NP in this subsurface system. First, assume that the transformation only occurs in proportion to the fraction of 2-NP accessible to the microorganissm, here taken as the fraction dissolved. Next, assume that desorption is fast relative to biodegradation so that the overall removal is limited by the biotransformation step. Finally, neglect "flushing" of the aquifer as a competing sink. In such a case, the governing mass balance equation is (see *15-53*):

$$\frac{d[2\text{-NP}]_{total}}{dt} = -f_w \cdot (k_{bio,d}) \cdot [2\text{-NP}]_{total} \qquad \text{(M-14)}$$

For the case of $K_d = 5\ \text{L}\cdot\text{kg}^{-1}$ and $r_{sw} = 10\ \text{kg}\cdot\text{L}^{-1}$

$$f_w = (1 + r_{gw}\cdot K_d)^{-1} = (1 + 10 \times 10 \times 5)^{-1} = 0.02$$

Thus, the "half-life" of 2-NP in this groundwater for the "low" pseudomonad abundance and other specified conditions is

$$t_{1/2} = \frac{\ln 2}{(f_w \cdot k_{bio,d})} = \frac{0.693}{(0.02 \cdot 2 \cdot 10^{-6}\ \text{min}^{-1})} = 1.7 \times 10^{7}\ \text{min}\ (= 40\text{y})$$

while the result for the "high" pseudomonad abundance is

$$t_{1/2} = \frac{\ln 2}{f_w \cdot k_{bio.d}} = \frac{0.69}{(0.02 \times 2 \times 10^{-4}\ min^{-1})} = 2 \times 10^5\ min\ (= 120\ d)$$

Especially for the low abundance case, one may worry that the rate of biological removal is slow relative to subsurface transport, and thus one may want to reconsider whether neglecting groundwater-mediated transport was a good idea!

Problem

Nitrilotriacetic acid (NTA) is used in some detergents as a complexing agent. As a result of this usage, NTA is introduced into sewage and may be released into the environment along with treated sewage effluent or sludge if it is not biodegraded during sewage treatment. Estimate the biodegradation rate constant, k_{bio}, for NTA introduced into a lake such that 1 to 10 nM (ca. 0.2-2 ppb) levels result.

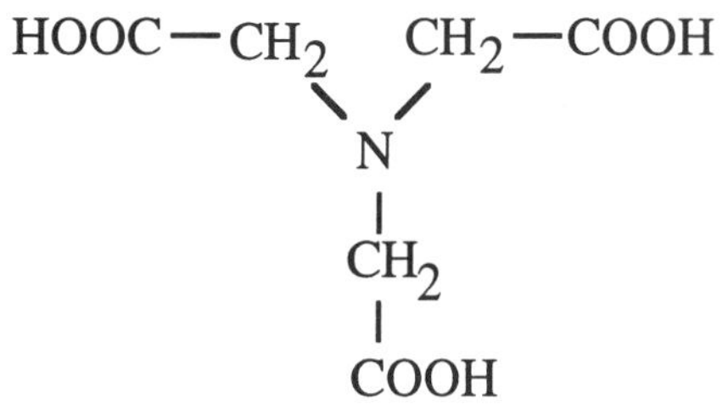

$HOOC{-}CH_2 \quad CH_2{-}COOH$ (both bonded to N); N bonded to $CH_2{-}COOH$

mw 191.1 $g \cdot mol^{-1}$

nitrilotriacetic acid
(NTA)

Answer

Given the low concentrations involved, assume that the NTA is not present at a sufficient level to allow growth by a specific microorganism; rather assume that it is metabolized incidentally by constitutive (i.e., always present) enzymes. In this case, a Michaelis-Menten-type kinetic description may be used to estimate k_{bio}. Hence, search the literature for studies that are representative for the system considered. In a field study (*Bartholomew et al., 1983)* of NTA degradation in an open water body (an estuarine site), analysis of the NTA biodegradation rates using the Michaelis-Menten equation *Eq. 14*-91 yielded

$$V_{max} = 0.3 \text{ to } 2.6 \quad \text{nmol}\cdot\text{L}^{-1}\cdot\text{h}^{-1}$$

and

$$K_{MM} = 290 \text{ to } 580 \text{ nmol}\cdot\text{L}^{-1}$$

Since the lake considered here only contains NTA in the range of 1 to 10 nM, the biodegradation rate constant may be estimated by:

$$k_{bio} = \frac{V_{max}}{K_{MM} + [\text{NTA}]} \simeq \frac{V_{max}}{K_{MM}} \simeq \frac{0.3 \text{ to } 2.6 \text{ nmol.L}^{-1}\cdot\text{h}^{-1}}{290 \text{ to } 580 \text{ nmol}\cdot\text{L}^{-1}}$$

$$\simeq 0.002 \text{ h}^{-1} \text{ or } 0.05 \text{ d}^{-1} \quad \text{(using the geometric means of the ranges)}$$

Of course, this estimate should be considered with caution because it assumes (a) that the "biological reagent" concentration (controlling V_{max}) in the lake of interest is the same as that in the estuarine sites studied by *Bartholomew et al., (1983)*, and (b) the effectiveness of the organisms in processing NTA, especially as reflected by the half-saturation constant, K_{MM}, is the same in both environmental systems.

It is interesting to note, however, that *Ulrich et al. (1994)* reported that the k_{bio} values necessary to explain the NTA cycling in a Swiss lake ranged from 0.02 to 0.05 d^{-1}, consistent with the value estimated above. These authors note that a much larger k_{bio} has been found for NTA in a shallow river system where higher operative biomass per volume of water is probably present (*Giger et al., 1991*) due to biofilms on the bed surface, thus enabling a higher "system V_{max}".

Assessing the Rate of Biodegradation in a Shallow Stream

In some environmental systems, the organisms responsible for most of the transformations of organic chemicals are not evenly distributed throughout the volume of the system. A good example of this involves shallow stream systems in which most of the microbial activity is located on/in the stream bed. In some cases, where rocky surfaces are present, the microorganisms may even be present in a thin layer referred to as a biofilm. To describe the biodegradation rate in this type of system, one must consider that the rate may be limited (a) by the rate of delivery of the organic compound (or necessary "reagents" like oxygen) from the flowing stream to the bed or (b) by the rate of microbial processing of the chemical of interest.

Problem

Groundwater leachate discharging from a waste site into a drainage ditch is seen to cause the water flowing overland along the ditch to have 4 μM (ca. 400 ppb) toluene decreasing to about 10 nM 400 meters downstream (*Kim et al., 1995*). To evaluate the *in situ* toluene biodegradation rate, k_{bio}(toluene) you perform the following experiments:

(a) Using shake flasks containing gravel from the stream bottom, you determine the biodegradation of toluene added to the flasks at concentrations ranging from 1 to 10 μM. Fitting the data gives a half-saturation constant, K_{MM}, of 2 μM (*Cohen et al., 1995*) .

(b) Collecting rocks and leaves from the bottom of the stream, which is 10 cm deep, you gently scrape the surfaces and find a release of 18 $mg_{biomass}\ cm^{-2}_{rock/leaf}$ (*Kim, 1995*). You also find that the stream bed is covered by enough leaf litter that at any one point there is an average of three leaves lying on top of one another (i.e., the biologically effective bottom area of the stream is six times larger than the apparent stream bottom area.)

(c) Suspending the rock/leaf scrapings so as to have a biomass density of 2300 $mg_{biomass} \cdot L^{-1}$, you find that toluene added at 0.2 μM disappears at a rate of 1.2 h^{-1}.

Using this information, estimate the k_{bio}(toluene) in the ditch.

Answer

First, assume that the microorganisms degrading toluene (a) are not growing on this substrate and (b) that mass transfer of toluene into the films of organisms coating the rocks and leaves on the stream bottom is not limiting toluene biodegradation. In such cases, enzyme kinetics are taken to describe k_{bio}(toluene):

$$k_{bio}(\text{toluene}) = \frac{V_{max}}{K_{MM} + [\text{toluene}]} \qquad \text{(M-13)}$$

In light of experiment (c) above, find the biomass-normalized V_{max} by substituting into this expression for k_{bio}:

$$1.2\ \text{h}^{-1} = \frac{(V_{max}\ \text{per unit biomass})\ (2300\ \text{mg}_{\text{biomass}}\cdot\text{L}^{-1})}{(2\ \mu\text{M} + 0.2\ \mu\text{M})}$$

Hence,

$$V_{max}\ \text{per unit biomass} = 0.0011\ \mu\text{mol}\cdot\ (\text{mg}_{\text{biomass}})^{-1}\cdot \text{h}^{-1}$$

To determine the effective biomass in the stream, use the data from experiment (b) and the stream water depth (10 cm):

$$\text{biomass} = (18\ \text{mg}_{\text{biomass}}\cdot 6\ \text{cm}^{2}_{\text{rock/leaf}}\cdot \text{cm}^{-2}_{\text{stream area}}\ /10\ \text{cm depth}$$

$$= 11\ \text{mg}_{\text{biomass}}\cdot \text{cm}^{-3}\ \text{ or }\ 11000\ \text{mg}_{\text{biomass}}\cdot\text{L}^{-1}$$

Finally, to estimate the in-stream k_{bio}, use the above results and the K_{MM} from experiment (a):

$$k_{bio}(\text{toluene}) = \frac{(0.0011\ \mu\text{mol}\cdot\text{mg}^{-1}_{\text{biomass}}\cdot \text{h}^{-1})\ (11000\ \text{mg}_{\text{biomass}}\cdot\text{L}^{-1})}{(2\ \mu\text{mol}\cdot\text{L}^{-1} + [\text{toluene}])}$$

$$= \frac{12\ \mu\text{mol}\cdot\text{L}^{-1}\cdot\text{h}^{-1}}{(2\ \mu\text{mol}\cdot\text{L}^{-1} + [\text{toluene}])}$$

For portions of the ditch near the groundwater discharge (with 4 µM toluene), the estimated biodegradation rate is, therefore,

$$k_{bio}(\text{toluene}) = \frac{12\ \mu\text{mol}\cdot\text{L}^{-1}\cdot\text{h}^{-1}}{(2\ \mu\text{mol}\cdot\text{L}^{-1} + 4\ \mu\text{mol}\cdot\text{L}^{-1})} = 2\ \text{h}^{-1}$$

Further downstream, where the toluene concentrations have diminished (<< 2 µM, due to dilution, volatilization, and biodegradation), the rate becomes

$$k_{bio}(\text{toluene}) = \frac{12\ \mu\text{mol}\cdot\text{L}^{-1}\cdot\text{h}^{-1}}{2\ \mu\text{mol}\cdot\text{L}^{-1}} = 6\ \text{h}^{-1}$$

Several comments should be made. First, these biodegradation rate estimates have been made assuming that delivery of the toluene (or other necessary constituents like oxygen), is not limiting the biotransformation process. In the in-stream studies of *Kim et al. (1995),* the value of k_{bio} corresponding to the conditions described above was 1.7 h^{-1}. Part of the reason the estimated values exceed this field result may be that mass transfer of the toluene into the biofilms on the leaves and rocks, not all of which were fully exposed to the overlying water, could be limiting the degradation.

Furthermore, the characteristic biodegradation half-life ($t_{1/2,bio} \simeq \ln 2$ / 1.7 $h^{-1} \simeq$ 20 minutes) in this drainage ditch is very fast as contrasted with what is reported for other deeper water bodies. For example, *Reichardt et al. (1981)* observed toluene biodegradation in seawater at a very slow rate (characteristic time of about 1 year!) A large part of this difference is due to the enormous difference in the number of microorganisms (per volume) involved in toluene degradation. This is reflected by the huge difference in V_{max}; the V_{max} (toluene, ditch) = 10^7 x V_{max}(toluene, seawater)! Thus as noted in the NTA problem above, application of open water data to shallow systems, and vice versa, is likely to fail unless biomass differences can be considered.

PROBLEMS

● *M-1 What Redox Zones Can Be Expected in This Laboratory Aquifer Column?*

You work in a research laboratory and your job is to investigate the microbial degradation of organic pollutants in laboratory aquifer column systems. You supply a column continuously with a synthetic groundwater containing 0.3 mM O_2, 0.5 mM NO_3^-, 0.5 mM SO_4^{2-} and 1 mM HCO_3^-, as well as 0.1 mM benzoic acid butyl ester, which is easily mineralized to CO_2 and H_2O. The temperature is 20°C and the pH is 7.5 (well buffered). Would you expect sulfate reduction or even methanogenesis to occur in this column? Establish an electron balance to answer this question.

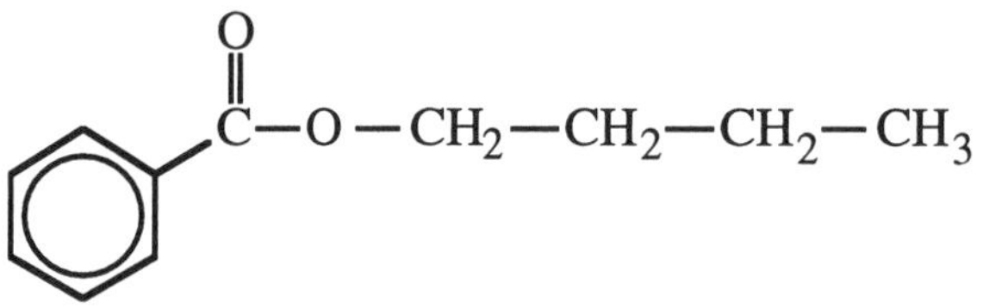

$C_6H_5-C(=O)-O-CH_2-CH_2-CH_2-CH_3$ mw 178.2 $g \cdot mol^{-1}$

benzoic acid butyl ester

● *M-2 Some Additional Questions Concerning the Bioremediation of Contaminated Aquifers*

You are involved in the remediation of an aquifer that has been contaminated with 2-methylnaphthalene. Similar to the toluene case discussed above (see Illustrative Examples), the aquifer is flushed with air-saturated water that is pumped into the ground at one place and withdrawn nearby. Calculate how much water is at least required to supply sufficient oxygen for the microbial mineralization of 1 kg of 2-methylnaphthalene assuming that the water contains 10 mg $O_2 \cdot L^{-1}$.

In order to check whether naphthalene is indeed microbially degraded at the site, after several weeks of operation, you get some samples of the contaminated soil and you try to enrich for 2-methylnaphthalene-degrading microorganisms. After some initial difficulties, you finally manage to obtain a liquid culture in which, at 25°C and pH 7, 2-methylnaphthalene is rapidly mineralized in the laboratory. However, at the field site, there are several indications that 2-methylnaphthalene is degraded only very slowly. How could this be explained? Give various possible reasons.

CH_3

mw 142.2 $g \cdot mol^{-1}$

2-methylnaphthalene

● M-3 *Optimizing a Bioreactor*

Suppose you are interested in improving the degradation of glycerol by *Aerobacter sp.* in the 10 m^3 bioreactor described in the Illustrative Example given above (see page M5). One option is to vary the volume of the reactor, thereby influencing the washout rate (k_{flush}). If you double the reactor volume, what improvement in glycerol removal do you achieve? (Note that in the reactor described in the Illustrative Example, 76% of the glycerol was degraded.) How does this improvement compare to that achieved by feeding the effluent from the first 10 m^3 bioreactor into a second similar 10 m^3 tank (with additional O_2, nutrients, etc., added as necessary)?

● M-4 *Estimating the Biodegradability of Herbicides by a Hydrolase*

You are concerned about the longevity of two herbicides, linuron and diuron, leaching into a river. Given the structure of these urea derivatives (see below), you suspect they should be biodegraded via a hydrolysis mechanism. You recall a report of a hydrolase enzyme, isolated from a common bacterium, that exhibited a half-saturation constant, K_{MM}(linuron), of 2 μM and a maximal degradation rate, V_{max}(linuron), of 2.5 $nmol \cdot mg_{protein}^{-1} \cdot s^{-1}$.

(a) If you found that linuron at 100 nM disappeared from the river water with a half-life of 60 days, what would the hydrolase enzyme concentration ($mg_{protein} \cdot L^{-1}$) in the river water have to be, if biodegradation via this enzyme accounted for all of linuron's removal?

(b) What would be the initial products of such a hydrolysis?

(c) If diuron exhibits a K_{MM} which is half that of linuron and a V_{max} which is twice that of linuron, what half-life (in days) would you expect for diuron initially present at 100 nM in the river water if biodegradation by the same hydrolase was the only removal mechanism? What would the half-life (in days) be if diuron was initially present at 30 μM?

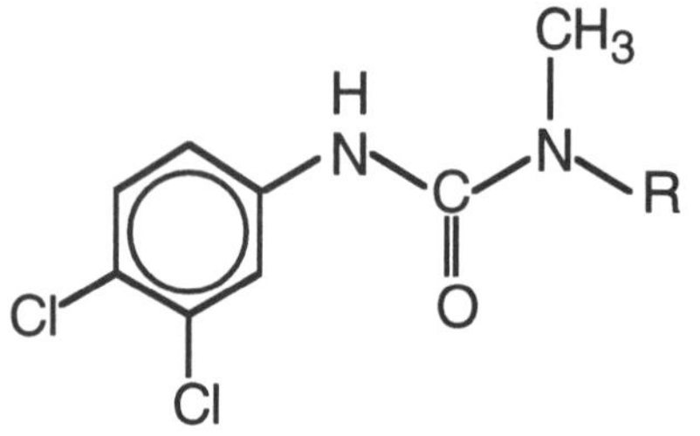

R = OCH_3 linuron
R = CH_3 diuron

● *M-5 Evaluating the Relative Transformation Rates of Substituted Anilines*

Paris and Wolfe (1987) reported the following rates of microbial oxidation of a series of substituted anilines at very low concentrations:

(a) Are the relative rates of transformation of these compounds consistent with the hypothesis that the rate was limited by an enzymatic reaction involving (i) an enzyme having a hydrophobic reaction site, and (ii) using an electrophilic form of oxygen?

(b) What rate ($hour^{-1}$)do you expect for 3-aceto-aniline (3-$COOCH_3$)?

compound	pseudo first-order rate constant (hour^{-1})
aniline	2.1
3-methyl-aniline	0.17
3-chloro-aniline	0.23
3-bromo-aniline	0.37
3-methoxy-aniline	0.050
3-nitro-aniline	0.012
3-cyano-aniline	0.0021

NH_2
1
2
3
aniline

● *M-6 A Case of Oxygen-Limited Biodegradation*

Isopropanol (rubbing alcohol) is being continuously discharged into a shallow (2 meters) pond. As a result, a bacterial species (with $\mu_{max} = 0.3$ h^{-1}, K_M(isopropanol) = 100 μM, and Y(isopropanol) = 2×10^{14} cells·mol^{-1}) has increased in numbers throughout the pond, depleting the isopropanol. After a time, the isopropanol concentration becomes constant at 300 μM (i.e., inputs are exactly balanced by biodegradation.)

(a) If the rate of isopropanol bio-oxidation is ultimately rate-limited by the rate of input of oxygen from the atmosphere (the weather is constantly overcast, winds out of the northwest at 2 m sec^{-1}, and cold at 10°C), what will be the maximum rate of biodegradation (in mol·L^{-1}·d^{-1})? (Hint: The solubility of O_2 in water at 10°C is 350 μM.)

(b) For this biodegradation rate, what bacterial cell density is necessary?

$$CH_3\diagdown CH — OH / CH_3$$

isopropanol

● *M-7 Evaluating Whether an Intermediate Product Accumulates in a Given System*

There are cases in which a xenobiotic substrate, S, is transformed by a first enzyme, E_1, to an intermediate product, IP, which may accumulate before it further reacts with a second enzyme, E_2, to yield a product P. An example is the reductive transformation of 3,5-dichlorobenzoate to benzoate by a methanogenic consortium (*Eq. 14-4;* for details see *Suflita et al., 1983*):

$$\mathrm{S} \xrightarrow[K_{MM1},\,V_{max1}]{E_1} \mathrm{IP} \xrightarrow[K_{MM2},\,V_{max2}]{E_2} \mathrm{P}$$

COO⁻ Cl Cl S; COO⁻ Cl IP; COO⁻ P

Consider a batch reactor with an initial concentration of S=200 $\mu mol \cdot L^{-1}$. Assume that K_{MM1} and K_{MM2} are 43 and 55 $\mu mol \cdot L^{-1}$, and V_{max1} and V_{max2} are 8 and 15 $\mu mol \cdot L^{-1} \cdot h^{-1}$, respectively. You want to know to what extent the intermediate product, IP, accumulates during the course of the reaction when assuming that (a) the educt S does not inhibit the activity of the enzyme E_2, and (b) S inhibits E_2 with a K_i of 0.1 $\mu mol \cdot L^{-1}$. Derive first the rate expressions for the consumption and production, respecively, of S, IP, and P. Plot the time courses of [S], [IP], and [P] as a function of time using a computer calculation program (e.g., EXCEL). What maximum concentration of IP (in μM) is reached in the reactor for the two cases (a) and (b)?

Hints and Help

Note that if the activity of E_2 is inhibited by S, the turnover rate of IP is given by

$$v(\mathrm{IP}) = \frac{V_{max2}\,[\mathrm{IP}]}{[\mathrm{IP}] + K_{MM2}\,(1 + [\mathrm{S}]\,/\,K_i)}$$

Try to derive this expression yourself.

N. ORGANIC COMPOUNDS IN PONDS AND LAKES - CASE STUDIES

ONE-BOX MODELS WITHOUT SEDIMENT-WATER EXCHANGE

In this first of three chapters devoted to case studies, you are invited to exercise the integrative modeling of organic pollutant behavior in systems that, with respect to transport and mixing processes, are relatively easy to handle. As pointed out in *Chapter 15*, from a didactic point of view, ponds and lakes are very well suited to demonstrate how the interplay of transport and reaction processes determines the spatial and temporal distribution of a given organic compound in the natural environment. Furthermore, for such systems, simple back-of-the-envelope calculations using one- or two-box models may already help to answer important questions, including a quick assessment of which processes are the most important ones. To this end, it is useful to express each process by a first-order or a pseudo-first-order rate law exhibiting a characteristic rate constant (for details see *Imboden and Schwarzenbach, 1985,* as well as *Chapter 15*.) By doing so, the relative importance of each of the processes may be assessed immediately by a direct comparison of the characteristic (pseudo-) first-order rate constants. Figure N.1 illustrates such an approach for a well-mixed shallow water body, such as a shallow pond or the epilimnion of a small lake. Note that in the latter case, it is assumed that the total water input occurs into the epilimnion and that exchange of water with the hypolimnion is neglected. Furthermore, in both cases, exchange with the sediments (diffusion, resuspension) is neglected, and the concentration in the air is assumed to be zero.

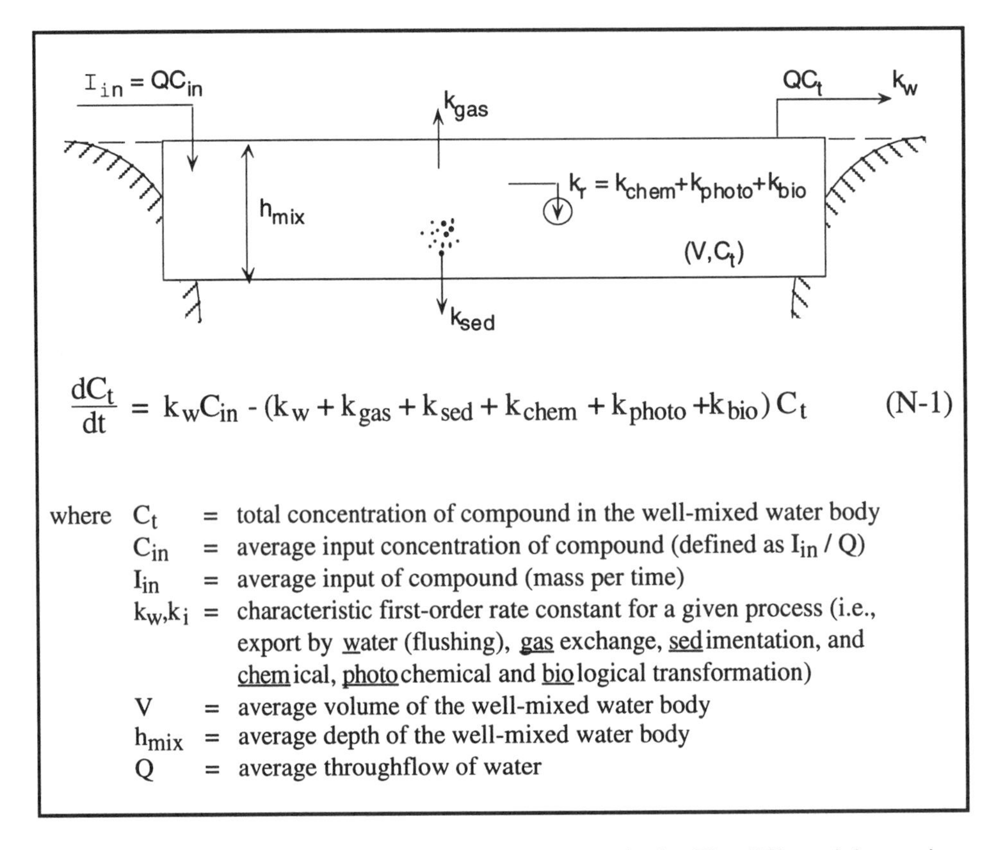

$I_{in} = QC_{in}$
QC_t
k_w
k_{gas}
h_{mix}
$k_r = k_{chem} + k_{photo} + k_{bio}$
(V, C_t)
k_{sed}
$$\frac{dC_t}{dt} = k_w C_{in} - (k_w + k_{gas} + k_{sed} + k_{chem} + k_{photo} + k_{bio}) C_t \quad \text{(N-1)}$$
where C_t = total concentration of compound in the well-mixed water body
C_{in} = average input concentration of compound (defined as I_{in} / Q)
I_{in} = average input of compound (mass per time)
k_w, k_i = characteristic first-order rate constant for a given process (i.e., export by water (flushing), gas exchange, sedimentation, and chemical, photochemical and biological transformation)
V = average volume of the well-mixed water body
h_{mix} = average depth of the well-mixed water body
Q = average throughflow of water

Figure N.1 One-box model for a well-mixed water body. The differential equation (Eq. N-1) holds for a constant I_{in}, Q, and V and time-invariable characteristic rate constants (adapted from *Ulrich et al., 1994)*

As discussed in *Section 15.3*, the solution of the first-order linear inhomogeneous differential equation (FOLIDE) with constant coefficients (see Eq. N-1 in Figure N.1) is given by (*Table 15.3*)

$$C_t(t) = C_t^{\infty} + (C_t^{o} - C_t^{\infty}) \cdot e^{-(k_w + \Sigma k_i)\, t} \qquad \text{(N-2)}$$

with the initial concentration C^o and the steady-state concentration

$$C_t^{\infty} = \frac{k_w}{k_w + \Sigma k_i} C_{in} \qquad \text{(N-3)}$$

The "time to steady state" is defined as (see also *Eq. 15-16*):

$$t_{ss} = \frac{3}{k_w + \Sigma k_i} \tag{N-4}$$

The characteristic rate constant for elimination by flushing, k_w, is calculated by

$$k_w = \frac{Q}{V} \tag{N-5}$$

Because elimination by flushing is equally important for all compounds (dissolved and particulate species), k_w can be considered as a reference point for assessing the importance of all other processes. Note that the characteristic rate constants given in Figure N.1 refer to the total concentration. Consequently, for processes that affect only the dissolved (e.g., gas exchange) or particulate (e.g., sedimentation) species, respectively, the corresponding rate constants have to be multiplied by the fraction of the compound present in that form. Thus, for example, k_{gas} is expressed as

$$k_{gas} = k_g \cdot f_w \tag{N-6}$$

with (see also *Eq. 15.11*)

$$k_g = \frac{v_{tot}}{h_{mix}} \tag{N-7}$$

and, f_w, the fraction in dissolved form (see *Eq. 11-5*):

$$f_w = \frac{1}{1 + [S] \cdot K_d} \tag{N-8}$$

Note that in this expression the solid-to-water phase ratio r_{sw} is approximated by the particle concentration [S]. In analogy, k_{sed} is given by

$$k_{sed} = k_s\,(1\text{-}f_w) \tag{N-9}$$

with (*Eq. 15-51*)

$$k_s = \frac{v_s}{h_{mix}} \tag{N-10}$$

where v_s is the average particle settling velocity, which is defined as the ratio of the flux of particles (F_s) per unit area and time out of the epilimnion and the average particle concentration [S] in the epilimnion (see *Section 15.4*):

$$v_S = \frac{F_S}{[S]} \qquad \text{(N-11)}$$

For the transformation processes, appropriate assumptions for the reaction rates in the dissolved and particulate form have to be made.

● *Case Study N-1 A Vinyl Acetate Spill into a Pond*

Due to an accident during the summer holidays, an unknown amount of vinyl acetate is introduced into a small, well-mixed pond located in the center of a city. Working in a consulting firm, you are asked (1) to determine the concentration of vinyl acetate in the pond, (2) to estimate how much vinyl acetate has entered the pond, and, most importantly, (3) to say something about the half-life of this compound in the pond.

Because your laboratory technician is out of town, it takes you 10 days until you are ready to make the first measurement. At this time (i.e., 10 days after the input), the measured vinyl acetate concentration in the pond water is 50 $\mu g \cdot L^{-1}$. Immediately, one wants to know from you, how long it will take until the concentration will have dropped below 1 $\mu g \cdot L^{-1}$. Try to answer this question by making the (worst-case) assumption that vinyl acetate is not biodegraded in the pond. With this assumption, calculate also how much vinyl acetate has at least been introduced into the pond. Base your answers on the data given below.

Since you are interested in checking whether your predictions were right, you measure the vinyl acetate concentration in the pond again 20 days after the accident. The value that you obtain is now 4 $\mu g \cdot L^{-1}$. Compare this result with your predictions. Try to explain any discrepancies, and, if necessary, revise your answers given 10 days after the accident.

Hints and Help

Note that in this particular case, there is only one input at the begin of the time period considered leading to an initial concentration, C_t^o. Hence, I_{in}, and thus C_{in}, are equal to zero, which simplifies Eq. N-2 to

$$C_t(t) = C_t^o \cdot e^{-(k_w + \Sigma k_i)\, t}$$

Compound, System, and Process Data for a Vinyl Acetate Spill in a Pond

$CH_3-C(=O)-O-CH=CH_2$	mw	86.1
	T_m	−93.2°C
	T_b	72.5°C
vinyl acetate	P^o(25°C)	1.4×10^{-1} atm
	C_w^{sat}	2.3×10^{-1} mol·L^{-1}
	K_{ow}(25°C)	5
Hydrolysis:	k_A(25°C)	1.4×10^{-4} M^{-1}·s^{-1}
	k_N(25°C)	1.1×10^{-7} s^{-1}
	k_B(25°C)	10 M^{-1}·s^{-1}

Pond:

Volume	V	10^4 m^3
Surface area	A	2×10^3 m^2
Water throughflow	Q	10^2 m^3·d^{-1}
Water temperature	T	25°C
	pH	7.0
Dissolved org. carbon	[DOC]	4 mg C·L^{-1}
Nitrate	$[NO_3^-]$	10^{-4} mol·L^{-1}
Bicarbonate	$[HCO_3^-]$	10^{-3} mol·L^{-1}
Particle concentration	[S]	5×10^{-6} kg_s·L^{-1}
Fraction of org. material	f_{om}	0.4 $kg_{om} \cdot kg_s^{-1}$
Average settling velocity of particles	v_s	2 m·d^{-1}
Average wind speed 10 m above the pond	u_{10}	1.5 m·s^{-1}
Geographic latitude		47.5°N[a]

[a] Note that for this latitude typical $^1O_2$ and OH• concentrations are given in *Section 13.4*. Assume clear skies during the time period considered.

● *Case Study N-2 What Steady-State Concentration of Anthracene Is Established in the Epilimnion of This Lake?*

With a leachate from a coal tar site, each day, 2 kg of anthracene are introduced continuously into the epilimnion of a eutrophic lake. The lake is stratified between April and November. As an employee of the state water authorities you are asked to monitor the anthracene concentration in the epilimnion of this lake. In order to get an idea about how sensitive your analytical technique has to be, you wonder what anthracene concentration (order of magnitude) you have to expect. To this end, answer the following questions using the average epilimnion characteristics and conditions given below. Neglect any water exchange between the epilimnion and the hypolimnion.

(a) What would be the anthracene concentration at steady-state if anthracene would show conservative behavior in the epilimnion of the lake?

(b) What anthracene steady-state concentration can actually be expected when assuming that biodegradation can be neglected?

(c) What is the time required to reach this steady state?

Average Epilimnion Characteristics and Conditions

Volume	V	5×10^7 m^3
Surface area	A	1×10^7 m^2
Mean water residence time	τ_w	100 days
Average particle concentration	$[S]$	5×10^{-3} kg$_s \cdot$m^{-3}
Fraction of org. material	f_{om}	0.4 kg$_{oc} \cdot$kg$_s^{-1}$
Average particle flux	F_s	5×10^{-3} kg$_s \cdot$m$^{-2} \cdot$d^{-1}
Average wind speed 10 m above the lake surface	u_{10}	1.5 m$\cdot$s^{-1}
Average water temperature	T	20°C
	pH	8.2
Dissolved Organic Carbon	[DOC]	4 mgC$\cdot$L^{-1}

Average Epilimnion Characteristics and Conditions (cont.)		
Decadic beam attenuation coefficients $\alpha(\lambda)$ of the water at various wavelengths (λ)	λ (nm)	$\alpha(\lambda)$ (m^{-1})
	300	4.0
	325	2.5
	350	1.5
	375	1.0
	400	0.6
	450	0.3

Hints and Help

The structure of anthracene is

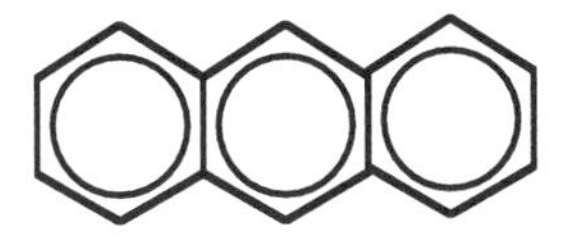

You can find all relevant physical-chemical properties of anthracene in the *Appendix* of the textbook. Note that ΔH_{Henry} is about 40 kJ·mol^{-1} (*Table 6.1*). Furthermore, when inspecting *Figure 13.3* and *Table 13.7* you realize that direct photolysis could be an important process. Before going through tedious calculations, you remember that a friend of yours who works at the EPA laboratory in Athens (Georgia) told you that the near-surface photolysis half-life (averaged over 24h) of anthracene under clear skies in the summer at 40°N latitude (by accident exactly the same latitude of your lake!) is about 5 days. You also remember from a course in environmental photochemistry, that the average light intensity between April and October is about 50% of that observed on a clear midsummer day. This estimate includes both the seasonal variation in light intensity as well as the effect of clouds.

● *Case Study N-3 How Fast Is Benzene Biodegraded in This Pond?*

In the same well-mixed pond in which you already had to deal with a vinyl acetate spill (see case study N-1), somebody monitors the benzene concentration during the summer. Interestingly, the concentration of benzene in the pond water does not vary much over time, and is always in the order of 0.05 $\mu g \cdot L^{-1}$. A colleague of yours who is responsible for air pollution measurements in the area tells you that the average benzene concentration in the air is 0.04 $\mu g \cdot L^{-1}$. He claims that input of benzene by gas exchange from the air is the predominant source of this compound in the pond. You remember vaguely that you already dealt with such a problem in Section H. Assuming he is right, and assuming that biodegradation is the only transformation process for benzene in the pond, calculate the characteristic biodegradation rate constant for benzene by using the pond characteristics and conditions given in Case Study N-1.

TWO-BOX MODELS WITHOUT SEDIMENT-WATER EXCHANGE

The next step towards a more sophisticated lake model is the two-box model that takes into account that water bodies such as lakes may be seasonally or even permanently stratified. The goal of this section is to illustrate that, when neglecting sediment-water exchange and by making some simplifying assumptions, it is still possible to carry out back-of-the-envelope calculations, even when using a two-box model for describing a lake. Furthermore, more complex models such as the one-dimensional vertical transport reaction model are based, in principle, on the concepts used for the two-box model (see *Figure 15.15*). Thus for a better understanding and proper application of these models, for which computer programs are necessary and available (*Ulrich , 1991; Ulrich et al., 1994*), it is very useful to spend some time with some simple basic two-box model calculations.

As discussed in *Section 15.3* and illustrated by the mass balance equations shown in Figure N.2, the major difference between the one-box- and two-box model is that one has to deal with two systems that are coupled by exchange processes. Even when expressing each of the processes by a pseudo-first order rate law exhibiting a characteristic rate constant (see also Figure N.1), one has to deal with a system of two coupled first-order linear inhomogeneous differential equations for which the solutions are already significantly more complicated as compared to the one-box model (see *Table 15.5*). Nonetheless, when considering only steady-state situations, as will be the case in the following problems, this task becomes manageable.

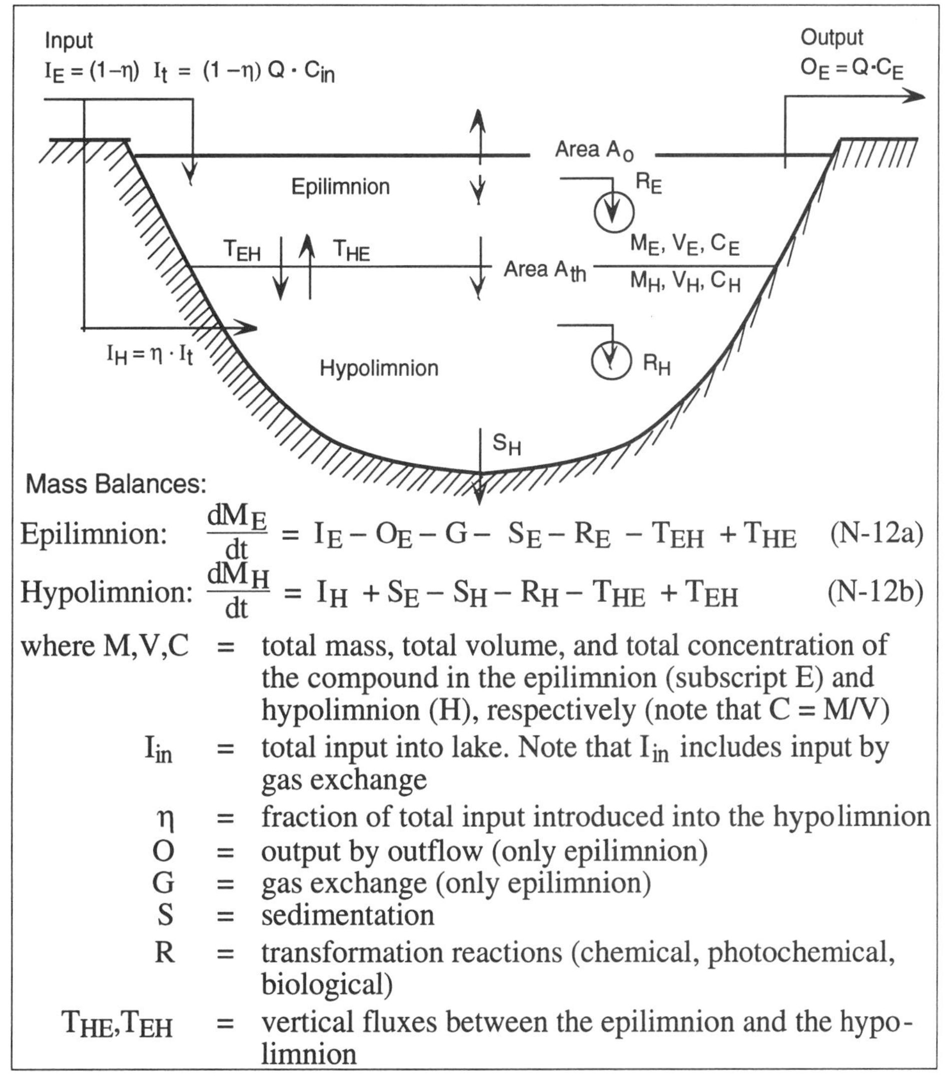
Input
$I_E = (1-\eta)\ I_t = (1-\eta)\ Q \cdot C_{in}$
Output
$O_E = Q \cdot C_E$
Area A_o
Epilimnion
R_E
T_{EH}
T_{HE}
Area A_{th}
M_E, V_E, C_E
M_H, V_H, C_H
$I_H = \eta \cdot I_t$
Hypolimnion
R_H
S_H
Mass Balances:
Epilimnion: $\frac{dM_E}{dt} = I_E - O_E - G - S_E - R_E - T_{EH} + T_{HE}$ (N-12a)
Hypolimnion: $\frac{dM_H}{dt} = I_H + S_E - S_H - R_H - T_{HE} + T_{EH}$ (N-12b)
where M,V,C = total mass, total volume, and total concentration of the compound in the epilimnion (subscript E) and hypolimnion (H), respectively (note that C = M/V)
I_{in} = total input into lake. Note that I_{in} includes input by gas exchange
η = fraction of total input introduced into the hypolimnion
O = output by outflow (only epilimnion)
G = gas exchange (only epilimnion)
S = sedimentation
R = transformation reactions (chemical, photochemical, biological)
T_{HE}, T_{EH} = vertical fluxes between the epilimnion and the hypolimnion

Figure N.2 Two-box model for (seasonally or permanently) stratified lakes. The mass balance equations N-12a and N-12-b hold for constant volumes. Water inflow and outflow is assumed to occur only in and out of the epilimnion.

Starting with the mass balance Equations given in Figure N.2 (Eqs. N-12a and b), one can derive two equations for the mass total concentrations in the epilimnion and hypolimnion, respectively:

$$\frac{dC_E}{dt} = (1-\eta)k_{w,E} \cdot C_{in} + k_{g,E}(C_a/K'_H) - (k_{w,E} + k_{gas,E} + k_{sed,E} + k_{r,E} + k_{ex,E})\, C_E + k_{ex,E} \cdot C_H \qquad \text{(N-13a)}$$

$$\frac{dC_H}{dt} = \eta \cdot k_{w,H} \cdot C_{in} - (k_{sed,H} + k_{r,H} + k_{ex,H})C_H + (k'_{sed.H} + k_{ex,H})C_E \qquad \text{(N-13b)}$$

with the following definitions of the various characteristic rate constants (all pseudo-first-order!):

$k_{w,E} = \frac{Q}{V_E}$	Flushing rate constant of epilimnion
$k_{w,H} = \frac{Q}{V_H} = k_{w,E}\,(V_E/V_H)$	Used for quantification of input into hypolimnion
$k_{g,E} = \frac{A_o}{V_E} v_{tot} = \frac{v_{tot}}{h_E}$	Rate constant for air-water exchange ($h_E = V_E/A_o$ = mean depth of epilimnion); note that $k_{gas} = k_g \cdot f_w$ (Eq. N-6), and that gas exchange has been split into an input term ($k_{g,E}(C_a/K'_H)$) and an output term described by k_{gas}, (see also *Eq. 15-4*)
$k_{sed,E} = \frac{A_o}{V_E} v_{s,E}\,(1-f_{w,E})$	Rate constant for sedimentation out of the epilimnion
$k'_{sed.H} = \frac{A_{th}}{V_H} v_{s,E}\,(1-f_{w,E})$	Rate constant for sedimentation into the hypolimnion
$k_{sed,H} = \frac{A_{th}}{V_H} v_{s,H}\,(1-f_{w,H})$	Rate constant for sedimentation out of the hypolimnion
$k_{r,E}$, $k_{r,H}$	Sum of the rate constants for chemical, photochemical, and biological transformation in the epilimnion and hypolimnion, respectively; $k_r = k_{chem} + k_{photo} + k_{bio}$

$k_{ex,E} = \frac{v_{th} \cdot A_{th}}{V_E}$	Rate constant for water exchange of epilimnion by internal mixing with hypolimnion; v_{th} is the "piston velocity" for turbulent diffusion across the thermocline (*Eq. 15-27*) and has the dimension of a velocity
$k_{ex,H} = \frac{v_{th} \cdot A_{th}}{V_H} = k_{ex,E} \cdot \frac{V_E}{V_H}$	Rate constant for water exchange of hypolimnion by internal mixing with epilimnion

For calculating the steady-state concentrations in the epilimnion (C_E^{∞}, box 1) and in the hypolimnion (C_H^{∞}, box 2), respectively, it is useful to calculate first the following terms:

$J_1 = I_E/V_E =$ $(1-\eta)\, k_{w,E} \cdot C_{in} + k_{g,E}\,(C_a/K_H')$	Total input into the epilimnion other than by mixing (mass per volume and time!)
$J_2 = I_H/V_H = \eta \cdot k_{w,H} \cdot C_{in}$	Total input into the hypolimnion other than by sedimentation or mixing (mass per volume and time!)
$k_{11} =$ $k_{w,E} + k_{gas,E} + k_{sed,E} + k_{r,E} + k_{ex,E}$	Sum of rate constants of processes leading to an elimination from the epilimnion
$k_{22} = k_{sed,H} + k_{r,H} + k_{ex,H}$	Sum of rate constants of processes leading to an elimination from the hypolimnion
$k_{12} = k_{ex,E}$	Rate constant for input of compound from the hypolimnion into the epilimnion (turbulent mixing)
$k_{21} = k'_{sed.H} + k_{ex,H}$	Sum of rate constants for input of compound from the epilimnion into the hypolimnion (sedimentation, turbulent mixing).

Using these terms, the steady-state concentrations are then given by (see *Table 15.5*):

$$C_E^{\infty} = \frac{k_{22} \cdot J_1 + k_{12} \cdot J_2}{k_{11} \cdot k_{22} - k_{12} \cdot k_{21}} \quad \text{(N-14a)}$$

$$C_H^{\infty} = \frac{k_{21} \cdot J_1 + k_{11} \cdot J_2}{k_{11} \cdot k_{22} - k_{12} \cdot k_{21}} \quad \text{(N-14b)}$$

The time-dependent solution can be found in *Table 15.5*. For calculating epilimnic and hypolimnic concentrations as a function of time, the use of a computer program (e.g., the one developed by *Ulrich, 1991*) is recommended. However, in order to estimate the time needed for the system to approach steady-state, t_{ss}, only the two "eigenvalues" k_1 and k_2 of the differential equation have to be calculated:

$$k_1 = \frac{1}{2}[(k_{11} + k_{22} + q)]$$

$$k_2 = \frac{1}{2}[(k_{11} + k_{22} - q)] \quad \text{(N-15)}$$

with $q = \left((k_{11} - k_{22})^2 + 4\, k_{12} \cdot k_{21}\right)^{1/2}$

$$t_{ss} = \frac{3}{\min\{k_i\}} \quad \textit{(15-31)}$$

where $\min\{k_i\}$ is the smaller of the eigenvalues k_1 and k_2.

It is illustrative to consider a special case, i.e., that of a conservative chemical (e.g., Cl^-), for which the various rate terms in Eqs. 14a and b simplify to

$$k_{11} = k_{w,E} + k_{ex,E}$$

$$k_{22} = k_{ex,H}$$

$$k_{12} = k_{ex,E}$$

$$k_{21} = k_{ex,H}$$

Inserting into Eqs. N-14 yields

$$C_E^\infty = \frac{J_1}{k_{w.E}} + \frac{k_{ex.E} \cdot J_2}{k_{ex.H} \cdot k_{w.E}}$$

$$= \frac{(1-\eta)\, Q \cdot C_{in}}{V_E \cdot Q/V_E} + \frac{V_H \cdot \eta \cdot Q \cdot C_{in}}{V_E \cdot V_H \cdot Q/V_E} = C_{in}$$

and

$$C_H^\infty = \frac{k_{ex.H} \cdot J_1 + (k_{ex.E} + k_{w.E})\, J_2}{k_{ex.H} \cdot k_{w.E}}$$

$$= (1 - \eta)\, C_{in} + \eta \frac{k_{w.E} + k_{ex.E}}{k_{ex.E}} C_{in}$$

Thus, independent of where the input of a conservative chemical occurs, the steady-state concentration in the epilimnion will always be equal to the average input concentration C_{in}. Recall that C_{in} has been defined as total compound input per unit time divided by total water input (which has been assumed to occur only into the epilimnion, see Fig. N.2):

$$C_{in} = \frac{I_E + I_H}{Q} = \frac{I_{in}}{Q}$$

For the hypolimnion, the situation is quite different. Only if the total input occurs into the epilimnion, i.e., $\eta = 0$, and/or if $k_{ex,E} >> k_{w,E}$, then $C_H^\infty = C_E^\infty = C_{in}$. For a substance such as Cl^-, this is a common situation in lakes, and is reflected in the uniform vertical concentration profiles often observed for such species. If, however, a substantial input occurs into the hypolimnion, and if $k_{w,E}$ is of similar size as $k_{ex,E}$ or larger, then much higher steady-state concentrations can be expected in the hypolimnion as compared to the epilimnion. An interesting example is the continuous introduction of salt into the hypolimnion of a lake which, due to the establishment of large density differences between epilimnion and hypolimnion, may lead to a permanent stratification (*Imboden and Wüest, 1995*). Finally, without any calculations, one can intuitively foresee that the input of chemicals into the hypolimnion that exhibit much higher reactivities in the epilimnion as compared to the hypolimnion will lead to large differences in their given steady-state concentrations in the two lake compartments (see example given in *Fig. 15.8*).

● *Case Study N-4 Extending the Anthracene Case*

In Case Study N-2 you were asked to calculate the steady-state concentration of anthracene in the epilimnion of a lake under the assumption that 2 kg of this compound are introduced into the epilimnion every day. For your calculation you neglected any water exchange between the epilimnion and the hypolimnion. Extend your calculation now to the whole lake, and consider various input scenarios. In particular, answer the following questions:

(a) What are the anthracene steady-state concentrations in the epilimnion and in the hypolimnion, respectively, if (i) all input (i.e., $2\ \text{kg} \cdot \text{d}^{-1}$) occurs into the epilimnion, (ii) all input occurs into the hypolimnion, and (iii) half of the input occurs in each the epilimnion and the hypolimnion? Assume that during stratification all water input occurs into the epilimnion.

(b) Considering that the lake is stratified between the middle of April and the end of November, is there enough time to reach steady-state? Discuss each of the three cases.

(c) Take case (i) and assume that steady-state is reached at the end of November. Furthermore, assume that at this time, the whole lake mixes instantaneously. What anthracene concentration do you expect at the end of March? How close is this concentration to the steady-state concentration that would establish under typical winter conditions?

Table N.1 Average Lake Characteristics and Conditions

Total volume	V	$15 \times 10^7\ m^3$
Average volume of epilimnion	V_E	$5 \times 10^7\ m^3$
Surface area	A_o	$1 \times 10^7\ m^2$
Area at thermocline	A_{th}	$8 \times 10^6\ m^2$
Throughflow of water	Q	$5 \times 10^5\ m^3 \cdot d^{-1}$
Summer		
Turbulent exchange velocity between epilimnion and hypolimnion	v_{th}	$0.05\ m \cdot d^{-1}$
Average particle concentration		
in the epilimnion	$[S]_E$	$5 \times 10^{-3}\ kg_s \cdot m^{-3}$
in the hypolimnion	$[S]_H$	$2 \times 10^{-3}\ kg_s \cdot m^{-3}$
Fraction of organic material	$f_{om,E}$; $f_{om,H}$	$0.4\ kg_{oc} \cdot kg_s^{-1}$
Average setting velocity of particles	$v_{s,E}$; $v_{s,H}$	$1\ m \cdot d^{-1}$
Average wind speed 10 m above the lake surface	u_{10}	$1.5\ m \cdot s^{-1}$
Average water temperature	T_E	20°C
	T_H	5°C
Average pH	pH_E	8.2
	pH_H	7.5
Dissolved Organic Carbon	$[DOC]_E$	$4\ mgC \cdot L^{-1}$
	$[DOC]_H$	$2\ mgC \cdot L^{-1}$
Decadic beam attenuation coefficients	$\alpha(\lambda)_E$	see Case Study N-2
Winter (whole lake)		
Average particle concentration	$[S]$	$2 \times 10^{-3} kg_s \cdot m^{-3}$
Fraction of organic material	f_{om}	$0.2\ kg_{oc} \cdot kg_s^{-1}$
Average settling velocity of particles	v_s	$1\ m \cdot d^{-1}$
Average wind speed 10 m above the lake surface	u_{10}	$2\ m \cdot s^{-1}$
Average water temperature	T	5 °C
Average pH	pH	7.5
Dissolved Organic Carbon	$[DOC]$	$2\ mgC \cdot L^{-1}$

Hints and Help

Note that some of the characteristic rate constants calculated in Case Study N-2 are also valid for this case study. For estimating the rate of direct photolysis in the winter, assume that $\alpha(\lambda)$ is proportional to [DOC] and that the average light intensity is about 3-4 times lower as compared to the summer (compare 24-hour averaged light intensity values in *Table 13.3* with those in *Table 13.4*). If, for some reason, you have not worked on Case Study N-2, it would be useful for you to read at least the Hints and Help section of that problem.

● *Case Study N-5 Determining In-Situ Biodegradation Rates of NTA in a Lake*

Nitrilotriacetate (NTA) is widely used in detergents as complexing agent. It enters the aquatic environment primarily through the sewage system. Since there is some concern that NTA (like other complexing agents such as EDTA) may mobilize heavy metals in the environment, its rate of biodegradation in natural waters is of great interest. In a field study, *Ulrich et al. (1994)* have measured inputs and vertical concentration profiles of NTA in a small lake in Switzerland. They used a one-dimensional vertical model to derive in-situ elimination rates for this compound in the water column. However, before applying this more sophisticated model, a quick back-of-the-envelope calculation was made to estimate the order of magnitude of the elimination processes. You are also invited to perform this calculation for the summertime by assuming steady-state and by making the following additional assumptions:

(i) Besides flushing, biodegradation is the only elimination process for NTA in the lake.

(ii) The average input of NTA is 5 $kg \cdot d^{-1}$ of which 80% is introduced into the epilimnion, and 20% into the hypolimnion.

(iii) The average NTA concentrations are 2$\mu g \cdot L^{-1}$ and 0.5 $\mu g \cdot L^{-1}$ in the epilimnion and hypolimnion, respectively.

Calculate now the average $k_{bio,E}$ and $k_{bio,H}$ values for NTA in the lake, using a two-box model and the average lake characteristics and conditions given in Case Study N-4.

HOOC—CH_2 CH_2—COOH
\ /
N
|
CH_2
|
COOH

NTA
(acid form)

SEDIMENT-WATER INTERACTION

Elimination to the sediments by settling particles may be an important process in removing a chemical from a lake or pond, especially if the fraction of the compound sorbed to particles, $(1\text{-}f_w)$, is clearly greater than zero (see Eq. N-9). In contrast to the elimination by flushing, sedimentation is not necessarily an irreversible process. Under certain conditions the chemicals buried in the sediments may be transported back into the water body. Such a situation may arise if, following a period of significant loading, the input of a pollutant is stopped or at least significantly reduced. As a consequence, the exchange between water and sediment can reverse its direction, i.e., the chemical can be transported from the polluted sediments back into the water column.

In order to describe the dynamical behavior of the joint water/sediment system, a mass balance equation for the compound concentration in the sediments has to be developed that can be combined with the dynamic equation of the concentration in the free water column. As discussed in *Section 15.4*, the simplest approach is to introduce the concept of a "surface mixed sediment layer" (SMSL). The relevant processes are shown in *Fig. 15.10*. They can be integrated into the linear differential equation for C_{ss}, the compound concentration per dry sediment mass:

$$\frac{dC_{ss}}{dt} = \frac{v_s}{m} f_s C_t - \beta \frac{r_{sw} \cdot v_s}{m} C_{ss} + k_{ex} (f_w \cdot K_{ds} \cdot C_t - C_{ss}) - k_{r,ss} \cdot C_{ss} \qquad (15\text{-}64)$$

The terms on the right hand side of *Eq. 15-64* are

(1) Input to SMSL by settling particles (v_s: average particle settling velocity, $f_s = 1\text{-}f_w$: fraction of compound on solids), where

$$m = z_{mix}\ (1\text{-}\phi)\ \rho_s \qquad (15\text{-}58)$$

is the solid mass per unit area in the SMSL of thickness z_{mix}, porosity ϕ, and solid mass density ρ_s.

(2) Removal from SMSL into the "permanent sediment layer" (PSL) (β: preservation factor, i.e., relative fraction of settled particulate matter reaching the PSL). Note that r_{sw} is the solid-to-solution phase ratio of the water column; in the following calculation r_{sw} is approximated by [S], the suspended particle concentration.

(3) Exchange of compound between the surface of the sediment particles and the overlying water either by resuspension or/and direct sorption/dessorption at the sediment surface (k_{ex}: sediment/water exchange rate, see *Eq. 15-63)*. Note that the distribution coefficient of the sediment particles, K_{ds}, may be different from the one in the water column since the chemical composition of the particles (e.g., f_{om}) is not necessarily the same.

(4) Chemical or biochemical degradation in the SMSL with first-order rate constant $k_{r,ss}$.

Since term (3) describes a two-way process, the mass balance equation developed for the water body (Eq. N-1 in Fig. N.1) has to be completed by a corresponding term (see last term of *Eq. 15-65)*:

$$\frac{dC_t}{dt} = k_w C_{in} - (k_w + \Sigma k_i) \cdot C_t - k_{ex}^{*}\,(f_w \cdot C_t - \frac{C_{ss}}{K_{ds}}) \qquad (N\text{-}16)$$

where (h: mean depth of water body)

$$k_{ex}^{*} = k_{ex} \cdot \frac{m \cdot K_{ds}}{h} \qquad (15\text{-}66)$$

The solution of the combined system of Eqs. *15-64* and N-16 is obtained by the scheme given in *Table 15.5.* If the water column is given the role of box 1, the SMSL the role of box 2, then the coefficients of *Eq. (a)* in *Table 15.5* are defined by

$J_1 = k_w \cdot C_{in} + k_{gas} (C_a/K_H^{'})$	Input into the water column other than from the SMSL
$J_2 = 0$	No input into SMLS other than from the water column
$k_{11} = k_w + \Sigma k_i + k_{ex}^* \cdot f_w$	Sum of rate constants of processes leading to an elimination from the water column
$k_{12} = k_{ex}^* / K_{ds}$	Rate constant for input of compound from the SMSL into the water column
$k_{21} = \frac{v_s}{m} \cdot f_s + k_{ex} \cdot f_w \cdot K_{ds}$	Sum of rate constants for input of compound from the water column into the SMSL
$k_{22} = \beta \frac{[S] \cdot v_s}{m} + k_{ex} + k_{r,ss}$	Sum of rate constants of processes leading to an elimination from the SMSL (Note: r_{sw} replaced by [S])

The steady-state concentrations, C_t^∞ and C_{ss}^∞, are then given by expressions as Eqs. N-14. The eigenvalues (Eq. N-15) serve to estimate the time to reach steady-state.

Note that the mixed coefficients, k_{12} and k_{21}, have units of $kg_s \cdot m^{-3} \cdot d^{-1}$ and $m^3 \cdot kg_s^{-1} \cdot d^{-1}$, respectively, i.e., not the usual dimension $(\text{time})^{-1}$ as expected for a rate. This is because the two dynamic variables, C_t and C_{ss}, do not have the same units. Convince yourself that the solution given in *Table 15.5* remains dimensionally correct.

● *Case Study N-6 Determining the Fate of Hexachlorobenzene in a Pond*

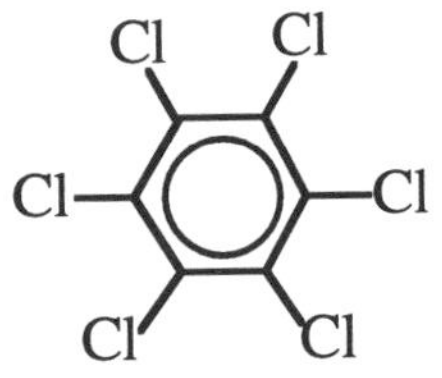

hexachlorobenzene (HCB)

mw 284.8 $g \cdot mol^{-1}$

It remained unnoticed for several years that the herbicide hexachlorobenzene (HCB) had continuously entered the small well-mixed pond that you already had to deal with in Case Study N-1. One day, your colleague determines the HCB concentration in the surface sediments of the pond and finds a C_{ss} value of 1.3 $\mu mol \cdot kg_s^{-1}$. You are alarmed by this value, because you fear that the HCB concentration in the water column could be dangerously high. Since you cannot get a water sample, at once, you first try to calculate the concentration in the water, C_t, by assuming a steady-state situation. From earlier investigations the following data on the sediments are available.

Sediment Characteristics

Depth of SMSL	z_{mix}	2 cm
Particle density	ρ_s	2.5 $g_s \cdot cm^{-3}$
Porosity in SMSL	ϕ	0.92
Preservation factor	β	0.8
Organic matter content of particles in SMSL	f_{oms}	0.1 $kg_{om} \cdot kg_s^{-1}$
Sediment/water exchange rate	k_{ex}	0.011 d^{-1}

(a) Calculate C_t^{∞}.

(b) Estimate the ratio between the total HCB mass in the water column and in the SMSL, respectively.

(c) What fraction of the HCB entering the pond at any given time is eventually leaving the pond by gas exchange and through the outlet? What happens to the rest?

(d) Estimate the input of HCB per unit time into the pond. After a thorough investigation you locate the HCB source and stop it.

(e) Estimate the time needed to eliminate 95% of the HCB from the combined water column/SMSL system.

● *Case Study N-7 The Long-Term Memory of a Lake to Former Pollution by Heptachlorobiphenyl*

Over a long time period, a lake has been exposed to pollution by different polychlorinated biphenyl (PCB) congeners. As discussed in *Section 15.4*, a fraction of the PCBs introduced into a lake is stored in the sediments.

You are responsible for the PCB monitoring program in the lake. Despite the fact that all external PCB input has been stopped, you still find significant PCB concentrations in the water column. A detailed survey of the sediments shows that in the top 10 cm of the sediment layer the mean concentration C_{ss} of 2,2',3,4,5,5',6-heptachlorobiphenyl (HCBP) is 4.0 $nmol \cdot kg_s^{-1}$.

(a) Calculate the mean concentration C_t in the water column that is in equilibrium with C_{ss}.

(b) How long does it take until C_{ss} and C_t have dropped to 95% of their present values?

The characteristic data describing the lake are summarized in the following table. Note that the physico-chemical properties of HCBP can be found in *Table 15.9.*

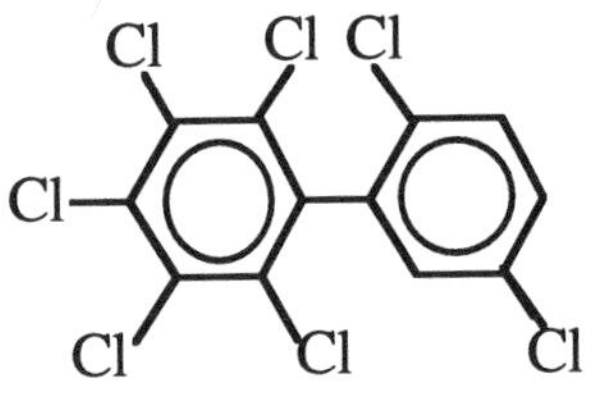

Cl Cl Cl Cl– Cl Cl Cl

mw 395.3 $g \cdot mol^{-1}$

2,2',3,4,5,5',6-heptachlorobiphenyl
(HCBP)

Lake		
Total volume	V	$7.5 \times 10^{8}\ m^3$
Surface area	A	$30 \times 10^{6}\ m^2$
Throughflow of water	Q	$7.5 \times 10^{6}\ m^3 \cdot d^{-1}$
Average settling velocity of particles	v_s	$0.7\ m \cdot d^{-1}$
Average particle concentration	$[S]$	$1 \times 10^{-3}\ kg \cdot m^{-3}$
Fraction of organic material	f_{om}	0.4
Average total air/water exchange velocity of HCBP	v_{tot}	$0.4\ m \cdot d^{-1}$
Sediment		
Depth of surface mixed sediment layer (SMSL)	z_{mix}	0.1 m
Particle density	ρ_s	$2'500\ kg \cdot m^{-3}$
Porosity of SMSL	ϕ	0.9
Preservation factor	β	0.8
Relative organic matter content of particles in SMSL	f_{oms}	0.1
Sediment/water exchange rate for HCBP	k_{ex}	$2.6 \times 10^{-5}\ d^{-1}$

Hints and Help

When calculating the characteristic coefficients, k_{ij}, you will find that the response velocity of the sediment reservoir is much smaller than the one of the open water column. In order to predict the decrease of both concentration, C_{ss} and C_t, you can assume a quasi-steady state between the two concentrations in which the system is controlled by the decrease of the slowly reacting sediment reservoir.

After you have done your model calculation you realize that the measured C_t-value is nearly twice as big as the value which you have calculated. You remember that colloidal organic matter can significantly change the behavior of the PCBs in the water column (see *Table 15.10)*. Your measurements give an average concentration of colloidal organic matter of [COM] = 3.2×10^{-3} kg·m^{-3}. Modify the above calculations and check whether the colloids can explain the discrepancy between observation and model calculation.

Personal Notes

O. ORGANIC COMPOUNDS IN RIVERS

TRANSPORT, MIXING, AND REACTIONS

Introduction

Unlike lakes, which, as a first approximation, can be described as completely mixed boxes *(Section 15.2)*, transport processes in rivers and streams are dominated by the unidirectional flow of the water, which forces all chemicals to spread downstream. The goal of this section is to describe the dynamic behavior of organic pollutants, either dissolved or sorbed on suspended particles, as they are transported along the river undergoing all kinds of transformations and exchange processes between the water and the adjacent environmental compartments, the atmosphere and the sediments.

Unless certain simplifications are made, the description of river flow and mixing can be extremely complicated. The following discussion will be confined to the case of *stationary flow*, that is, to a situation in which the water discharge Q at every cross section of the river, A(x), is constant with time (Fig. O.1). In reality, Q(x) along the longitudinal axis of the river, x, is not necessarily constant because of merging rivers and water exchange with the groundwater. Nevertheless, a river can always be looked at as consisting of sections in between river junctions for which the above assumption holds.

Stationary flow does not necessarily imply that the concentration of a chemical along the river is stationary as well. To the contrary, often one has to assess the fate of chemicals that are accidentally spilled into a river and are transported downstream as a concentration cloud. One then may need to predict the temporal change of the concentration of the chemicals at a given location downstream of the spill, especially the time when the concentration starts to increase and the maximum concentration that will be reached when the spill passes by the location.

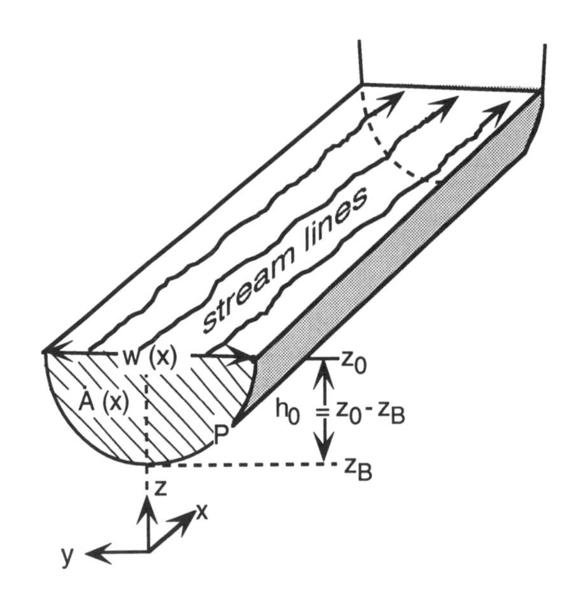

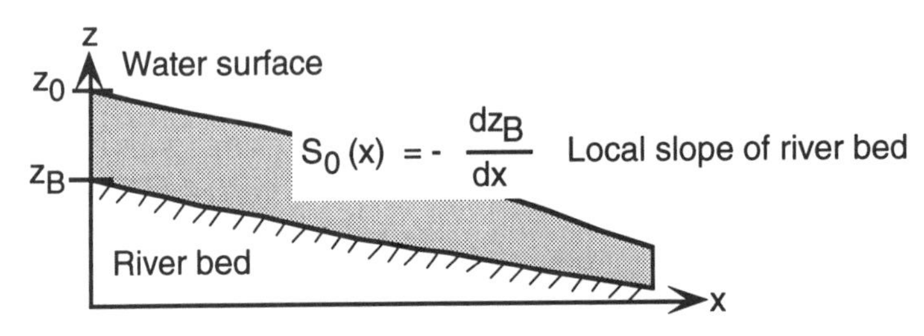

Figure O.1 Coordinate system and geometric parameters used to characterize mixing processes in a river. See Table O.1 for definitions of symbols.

The following explanations, of course, cannot replace the large literature on hydraulics and mixing processes in rivers (see, e.g., *Fischer et al., 1979; Rutherford, 1994*). As a first step towards "full-scaled" river modeling we shall explain and describe the following processes:

(a) Advective transport along the river caused by the mean flow of the water
(b) First-order chemical reaction in the water ("volumetric transformation")
(c) Linear exchange process occurring at the boundary of the water body, e.g., air/water-exchange ("surface reaction")
(d) Vertical mixing by turbulence

(e) Lateral mixing by turbulence
(f) Longitudinal mixing by dispersion
(g) Sedimentation and sediment-water interaction

Note that this list does not mention longitudinal mixing *by turbulence*; its influence is masked by longitudinal *dispersion*. As described below, dispersion, an inevitable attendant of unidirectional advective transport, is commonly much larger than turbulent diffusion along the direction of flow.

Figure O.1 and Table O.1 give all the necessary definitions. The coordinate system at any location is chosen such that x always points in the direction of the main flow, y across the river, and z vertically upwards.

Table O.1 Definition of Characteristic River Parameters (Fig. O.1)

Choice of local coordinates	x-axis	main direction of flow (horizontal)
	y-axis	lateral direction (horizontal)
	z-axis	vertical, positive upwards
Geometry and Discharge		
$A(x)$	m^2	Cross-section of river filled by water
$w(x)$	m	Width at water surface
$z_B(x)$	m	Height of deepest point of river bed at cross-section A(x), relative to arbitrary base
$z_0(x)$	m	Height of water surface at cross-section A(x)
$h_0 = z_0 - z_B$	m	Maximum depth
$h = A/w$	m	Mean depth
$S_0 = -dz_B/dx$	-	Local slope of river bed
P	m	Wetted perimeter (wet area per unit length)
$R = A/P$	m	Hydraulic radius
Q	$m^3 \cdot s^{-1}$	Discharge of water
$\bar{u} = Q/A$	$m^3 \cdot s^{-1}$	Mean flow velocity along the x-axis
f	-	friction factor
$g = 9.81$	$m\ s^{-2}$	Acceleration due to gravity
Mixing Parameters		
E_y, E_z	$m^2 \cdot s^{-1}$	Turbulent diffusion coefficients
E_{dis}	$m^2 \cdot s^{-1}$	Coefficient of longitudinal dispersion

Note: All variables are subjected to changes if Q changes with time.

Transport by the Mean Flow

The mathematical description of the time-dependent, three-dimensional concentration distribution of a chemical in a river, C(x,y,z,t), is not simple. However, depending on the actual situation and the questions to be answered with the model, significant simplifications can be made. In the following discussion, complexity is added step by step in order to allow the reader to build up to what is otherwise a confusing picture. As a starting point, the concentration change with time due to the mean advective transport in a short river section, bounded by two cross-sections at x and x + Δx, is considered as sketched in the following figure:

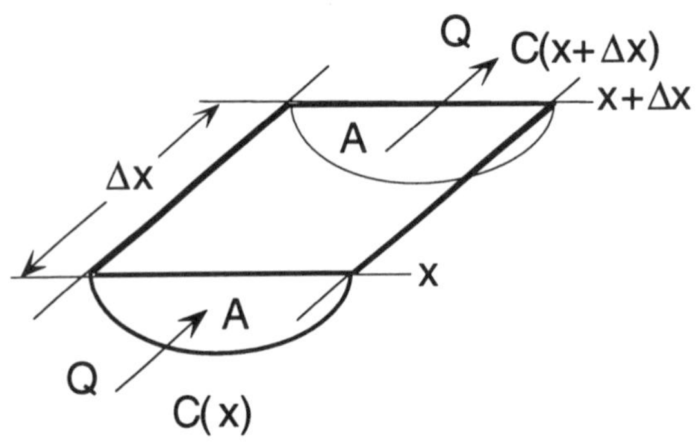

The concentration C(x,t) is the cross-sectional average; lateral and vertical deviations from the average are considered to be small. At this point, all other processes are excluded. The mass balance for a small section (length Δx) is given by

$$A \cdot \Delta x \left(\frac{\partial C}{\partial t}\right)_{ad} = Q \cdot C(x) - Q \cdot C(x+\Delta x) \qquad \text{(O-1)}$$

The discharge Q is assumed to be constant from x to x+Δx. Division by A·Δx and taking the usual limit Δx → 0 yields (see also *Eq. 9-33*)

$$\left(\frac{\partial C}{\partial t}\right)_{ad} = -\frac{Q}{A} \cdot \frac{\partial C}{\partial x} = -\bar{u} \cdot \frac{\partial C}{\partial x} \qquad \text{(O-2)}$$

where

$$\bar{u} = Q/A \text{ is the mean flow velocity.} \qquad \text{(O-3)}$$

Prediction or measurement of $\bar{u}$ is of paramount importance to assessing transport of chemicals in rivers. The following discussion will be restricted to the special case of *stationary uniform flow*. This means that at a fixed location the discharge Q is constant. Furthermore, the cross-section A does not change in size or shape in the x-direction, and the bed and surface slope are equal and constant. An equilibrium between the gravitational force pushing the water downhill and the frictional force, which increases in proportion to $\bar{u}^2$, can be formulated for this case. The first force is represented by the product of the slope of the river bed, S_o, the water volume per unit length (that is equal to the cross-sectional area A), and the acceleration due to gravity, g. The second force is expressed by the product of the nondimensional *friction factor,* f, the square of the mean velocity, and the contact area per unit length of the water with the river bed, P (the wetted perimeter). This leads to the *Darcy-Weisbach* equation for stationary uniform flow (see, e.g., *Chow, 1959*):

$$\bar{u} = \left(\frac{8g \cdot A}{f \cdot P} S_o\right)^{1/2} = \left(\frac{8g}{f} R \cdot S_o\right)^{1/2} \qquad \text{(O-4)}$$

(see Fig. O.1 and Table O.1 for definitions). If the river is wide, i.e., if its width, w, is much larger than its mean depth, h = A/w, P is about equal to w, and Eq. O-4 becomes

$$\bar{u} = \left(\frac{8g}{f} h \cdot S_o\right)^{1/2} \quad \text{wide river} \qquad \text{(O-4a)}$$

The friction factor typically lies between f = 0.02 (smooth river bed, like a man-made channel) and f = 0.1 (rough river bed, like a mountain stream, or a small river with large dunes or sand bars). For a given river bed, f decreases with increasing depth h, i.e., with increasing discharge Q.

Problem

Consider the River G, which has a slope of 0.4 m per kilometer, a width of w = 10 m, and is bounded by rather steep slopes (inclination 3:1). The usual discharge is Q = 0.8 $m^3 \cdot s^{-1}$ (Situation I) and reaches maximum values of 10 $m^3 \cdot s^{-1}$ during heavy rain (Situation II). The friction factor is assumed to be f = 0.04 for Situation I and f = 0.026 for Situation II. Calculate the mean flow velocity $\bar{u}$ and the mean depth h for both situations.

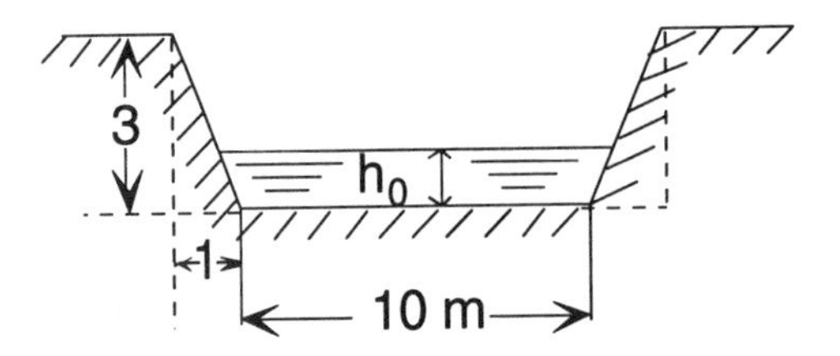

Answer

As a first approximation the river bed is described as a "wide" rectangular channel. *Note:* Using this result, it can then be checked in a second step whether the original assumption was justified. If not, the calculation can iteratively be continued with more realistic river cross-sections. This means that $A = h \cdot w$ and $Q = \bar{u} \cdot A = \bar{u} \cdot h \cdot w$. The geometric relationship is used in the shallow river Darcy-Weisbach Equation O-4a to replace h by $Q/(\bar{u} \cdot w)$:

$$\bar{u} = \left(\frac{8g}{f} h \cdot S_o\right)^{1/2} = \left(\frac{8g}{f} \frac{Q}{\bar{u} \cdot w} S_o\right)^{1/2}$$

Solving for $\bar{u}$ yields

$$\bar{u} = \left(\frac{8g \cdot Q \cdot S_o}{f \cdot w}\right)^{1/3} \qquad \text{(O-5)}$$

With $S_o = 0.4$ m/1 km $= 4 \times 10^{-4}$ and $h = \frac{Q}{\bar{u} \cdot w}$ one obtains

Situation I	$Q = 0.8\ m^3 \cdot s^{-1}$	$\bar{u} = 0.40\ m \cdot s^{-1}$	$h = 0.20$ m
Situation II	$Q = 10\ m^3 \cdot s^{-1}$	$\bar{u} = 1.06\ m \cdot s^{-1}$	$h = 0.94$ m

Situation I does not pose any problems with respect to the rectangular channel assumption. Note that for Situation II due to the slope of 3:1 on both sides of the river, the surface width w increases to about $[10+0.94 \cdot (2/3)]$ m = 10.6 m. Thus the cross-section is slightly larger than 10 [m] $\cdot$ h. Yet, this effect is more than compensated for by the fact that the contact length P is greater than w = 10 m (P = 12.0 m). From Eq. O-4 it can be concluded that the true $\bar{u}$ is slightly smaller and thus h larger than the values deduced from the rectangular wide river approximation. In fact, the corresponding correction of $\bar{u}$ is – 6% ($\bar{u} = 1.00\ m \cdot s^{-1}$), not much in comparison with the uncertainty caused by assuming an appropriate friction factor f. The following simple values will be used in all the calculations below:

Situation I	$Q = 0.8\ m^3 \cdot s^{-1}$	$\bar{u} = 0.40\ m \cdot s^{-1}$	$h = 0.20\ m$	$w = 10\ m$
Situation II	$Q = 10\ m^3 \cdot s^{-1}$	$\bar{u} = 1.0\ m \cdot s^{-1}$	$h = 1.0\ m$	$w = 10\ m$

Mean Flow and Chemical Transformation

As a next step Eq. O-2 describing transport by the mean flow is combined with a first-order volumetric reaction of the form

$$\left(\frac{\partial C}{\partial t}\right)_{react} = - k_r \cdot C \qquad \text{(O-6)}$$

where k_r is a first-order reaction rate constant. Combining Eqs. O-2 and O-6 yields

$$\frac{\partial C}{\partial t} = \left(\frac{\partial C}{\partial t}\right)_{ad} + \left(\frac{\partial C}{\partial t}\right)_{react} = - \bar{u}\frac{\partial C}{\partial x} - k_r \cdot C \qquad \text{(O-7)}$$

For stationary conditions with respect to flow ($\bar{u}(t)$ = const.) *and* concentration ($\partial C/\partial t = 0$) one gets

$$\frac{dC}{dx} = - \frac{k_r}{\bar{u}} C \qquad \text{(O-8)}$$

Since the derivative with respect to time, $\partial C/\partial t$, no longer appears in the above expression, the partial derivative was replaced by the normal derivative, dC/dx. In order to solve this ordinary differential equation and to calculate the concentration variation along the river, C(x), it should be remembered that $\bar{u}(x)$ is not necessarily constant along x – even if Q is – since the cross-section A(x) may change (see Eq. O-3). By separating the variables, C and x, one gets

$$\frac{dC}{C} = - k_r \frac{dx}{\bar{u}} \qquad \text{(O-9)}$$

which after integration between x = 0 and x, or C(0) and C(x), respectively, turns into

$$\int_{C(0)}^{C(x)} \frac{dC}{C} = \ln C(x) - \ln C(0) \tag{O-10}$$

$$= \ln\left(\frac{C(x)}{C(0)}\right) = -\,k_r \int_0^x \frac{dx}{\bar{u}} = -\,k_r \cdot T(x)$$

where $T(x) = \int_0^x \frac{dx}{\bar{u}}$ is the mean flow time from x = 0 to x.

Thus, by taking the exponential on both sides of Eq. O-10 one obtains

$$C(x) = C(0)\, e^{-k_r \cdot T(x)} \tag{O-11}$$

Not so surprisingly this result shows that in the flowing water parcel the concentration is exponentially decaying with flow time T(x), precisely as it should be for a linear decay reaction.

Note that for the case of constant $\bar{u}(x)$ along the river, i.e. for constant cross-section A(x), T(x) increases linearly with x:

$$T(x) = \frac{x}{\bar{u}} \tag{O-12}$$

Thus, the concentration drops exponentially along x with the spatial length scale, $\bar{u}/k_r$, in much the same way as C(t) in an individual water parcel drops exponentially with time:

$$C(x) = C(0)\, e^{-(k_r/\bar{u})x} \quad \text{for } \bar{u} = \text{const.} \tag{O-13}$$

Note: Incomplete lateral mixing does not affect a volumetric *first-order* reaction as described by Eq. O-6. The reaction rate simply depends on the mean concentration and is independent of the exact distribution of the chemical within the water body. However, this would not be true for a nonlinear reaction, e.g., for a Michaelis-Menten-type reaction (see *Section 14.4*).

Mean Flow and Air-Water Exchange

Air-water exchange serves to exemplify the influence of a boundary process on the concentration, C(x). Again it is assumed that in a given river cross-section, deviations of the real concentrations from the mean are small in the lateral and vertical direction. Then the expressions derived for lakes in *Section 15.3* are also applicable to rivers :

$$\left(\frac{\partial C}{\partial t}\right)_{air/water} = k_g\,(C_{eq} - C), \quad k_g = \frac{v_{tot}}{h} \tag{O-14}$$

where $C_{eq} = C_a/K_H'$ is the concentration in the water in equilibrium with the atmospheric concentration C_a, K_H' is the nondimensional Henry's Law coefficient, and h is the mean depth of the river. Thus, in analogy to Eq. O-7 we can write

$$\frac{\partial C}{\partial t} = \left(\frac{\partial C}{\partial t}\right)_{ad} + \left(\frac{\partial C}{\partial t}\right)_{air/water} = -\bar{u}\,\frac{\partial C}{\partial x} + k_g\,(C_{eq} - C) \tag{O-15}$$

As before, for stationary conditions ($\partial C/\partial t = 0$) this becomes an ordinary differential equation of the form

$$\frac{dC}{dx} = \frac{k_g}{\bar{u}}\,(C_{eq} - C) = \frac{v_{tot}}{h \cdot \bar{u}}\,(C_{eq} - C) \tag{O-16}$$

Note that $h = A/w$, where w is the width of the river, $\bar{u} = Q/A$, and thus $h \cdot \bar{u} = Q/w$. Equation O-16 can be rewritten as

$$\frac{dC}{dx} = \frac{v_{tot}}{Q}\,w\,(C_{eq} - C) \tag{O-17}$$

In fact, the appropriate coordinate to describe the change of C along the river due to an interfacial process – like air-water exchange – is the contact area S(x) of the river measured from some (arbitrary) point, where S(0) = 0, to the coordinate x (Fig. O.2). The infinitesimal increment of S, dS, is related to dx by

$$dS = w \cdot dx, \qquad \text{thus} \quad \frac{dx}{dS} = \frac{1}{w} \,, \tag{O-18}$$

and therefore solving for the concentration change per area using Eqs. O-17 and O-18:

$$\frac{dC}{dS} = \frac{dC}{dx} \cdot \frac{dx}{dS} = \frac{v_{tot}}{Q}(C_{eq} - C) \tag{O-19}$$

Thus, even if depth, width and cross-section of the river are changing, Eq. O-19 can still be integrated according to the FOLIDE recipe (see *Section 15.3*) as long as v_{tot} and Q are constant:

$$C[S(x)] = C_{eq} + [C(0) - C_{eq}] \exp[- \frac{v_{tot}}{Q} S(x)] \tag{O-20}$$

In this case, the surface area S(x) takes the role of a coordinate along the river (Fig. O.2).

For constant width, this yields

$$C(x) = C_{eq} + [C(0) - C_{eq}] \exp[- \frac{v_{tot} \cdot w}{Q} x] \tag{O-21}$$

Note that incomplete lateral mixing (in combination with *complete vertical mixing*) does not alter the mean surface concentration and thus does not affect the total air-water exchange.

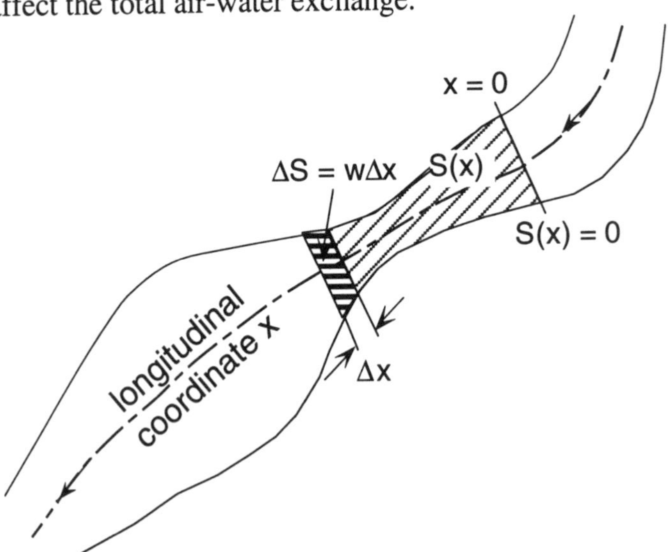

Figure O.2 The surface area of the river can serve as a coordinate along the the river

Problem

Wastewater is introduced into the River G (see preceding problem) at location x = 0. It causes a continuous input of 0.25 $mol \cdot h^{-1}$ of benzyl chloride (BC) and 0.08 $mol \cdot h^{-1}$ of tetrachloroethene (PER).

CH_2Cl (structure of benzyl chloride) — $Cl_2C{=}CCl_2$ (structure of tetrachloroethene)

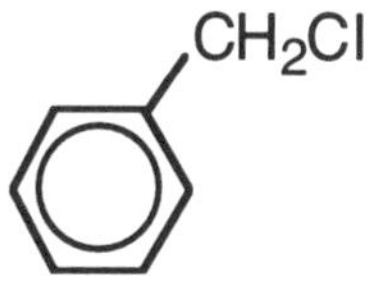

benzyl chloride (BC)
mw 126.6 $g \cdot mol^{-1}$

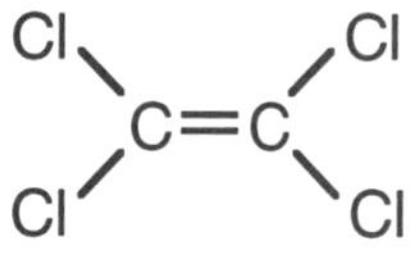

tetrachloroethene (PER)
mw 165.8 $g \cdot mol^{-1}$

Both chemicals undergo air-water exchange. Benzyl chloride also hydrolyzes to benzyl alcohol with a first-order rate constant, k_r, of 0.05 h^{-1} at 25°C (*Table 12.1, Fig. 12.1*). The concentrations of BC and PER in air are negligible. Calculate for the two flow regimes described in the first problem the concentrations of BC and PER in the River G, 10 km downstream of the waste water input. Assume a water temperature of 25°C. Note that the rate of hydrolysis of BC is pH-independent.

Answer

Provided that an immediate and complete mixing of the chemicals into the river water can be assumed, the initial concentrations can be calculated from the relation C(0) = input per unit time divided by discharge. (Note: Later the immediate mixing hypothesis will be checked by using explicit estimates for the relevant mixing processes.) The following initial concentrations are obtained:

	Input	Initial concentrations ($nmol \cdot L^{-1}$)	
	($mol \cdot s^{-1}$)	I: Q = 0.8 $m^3 \cdot s^{-1}$	II: Q = 10 $m^3 \cdot s^{-1}$
BC	6.9×10^{-5}	87	6.9
PER	2.2×10^{-5}	28	2.2

Air-water exchange of PER is liquid-film controlled (*Table 10.3*). The molecular diffusion coefficient of PER in water at 25°C is $D_w = 0.92 \times 10^{-5}$ $cm^2 \cdot s^{-1}$ (*Table 10.3*). From *Eq. 10-31* the liquid film air-water exchange velocity of PER, v_w(PER), would be calculated as 4.3×10^{-3} $cm \cdot s^{-1}$ for Situation I (Q = 0.8 $m^3 \cdot s^{-1}$), and 3.0×10^{-3} $cm \cdot s^{-1}$ for Situation

II ($Q = 10\ m^3 \cdot s^{-1}$). Since *Figure 10.8* shows v_w(PER) to be about 75% larger than the value calculated from *Eq. 10-31,* one estimates v_w(PER), i.e., v_{tot}(PER), to be $7.5 \times 10^{-3}\ cm \cdot s^{-1}$ and $5.3 \times 10^{-3}\ cm \cdot s^{-1}$ for the two respective situations.

The only significant elimination process of PER from the river is by air-water exchange. Since C_a and thus C_{eq} are zero, a simplified version of Eq. O-21 with $x = 10\ km = 10^4\ m$ can be used. Thus, the calculated concentration of tetrachloroethene at x = 10 km is

Q ($m^3 \cdot s^{-1}$)	$\varepsilon_g \equiv \frac{v_{tot} \cdot w}{Q}$ (m^{-1})	$\frac{C(10km)}{C(0)} = e^{-\varepsilon_g \cdot 10\ km}$ (-)	C(10km) ($nmol \cdot L^{-1}$)
I: 0.8	9.4×10^{-4}	8.3×10^{-5}	0.0023
II: 10	5.3×10^{-5}	0.59	1.3

Note that w = 10 m was used for both situations.

Apparently, the loss of PER to the atmosphere is significantly reduced during the large discharge event. In spite of the larger dilution of the constant mass input during the peak discharge, the concentration remains larger 10 km downstream from the input location. Three factors are involved in this effect: (1) the decrease of v_{tot}, (2) the increase of depth h, and (3) the increase of flow velocity $\bar{u}$. Note that h and $\bar{u}$, combined into the factor w/Q, act in opposite direction: Given a discharge Q and river bed width w, the flowing time, i.e., the time available for surface exchange, and the mean depth are inversely related. That explains why neither h nor $\bar{u}$ explicitly appear in Eq. O-21.

In the case of BC, two elimination processes have to be considered, air-water exchange and chemical reaction. Combining Eqs. O-7 and O-17 and setting $C_{eq} = 0$ yields

$$\frac{dC}{dx} = -\frac{k_r}{\bar{u}} C - \frac{v_{tot} \cdot w}{Q} \cdot C = -(\varepsilon_r + \varepsilon_g) C \qquad \text{(O-22)}$$

where $\varepsilon_r = k_r/\bar{u}$ and $\varepsilon_g = v_{tot} \cdot w/Q$ are the inverse length scales for elimination by chemical reaction and air-water exchange, respectively. Note that in Eq. O-22, ε_r and ε_g (dimension: length^{-1}) play the same role as the pseudo-first-order rate constants k_r and k_g (dimension: time^{-1})

which were used to describe box models (see *Section 15.3*). Their relative sizes are proportional to their importance as elimination pathways. For constant parameters, the solution of Eq. O-22 is (see *Table 15.3*)

$$C(x) = C(0)\, e^{-(\varepsilon_r+\varepsilon_g)x} \qquad \text{(O-23)}$$

According to the *CRC Handbook of Chemistry and Physics* the Henry's Law constant K_H of BC at 25°C is 0.35 atm·L·mol^{-1}; thus $K_H' \approx 0.015$; it can also be estimated from water solubility and vapor pressure (0.5 atm·L·mol^{-1}, see Problem F-1). In any case, air-water exchange of BC is mostly liquid film controlled. Although the air film may add a few percent to the overall transfer resistance, it will be disregarded. Using *Eqs. 9-25* and *10-33* with β=0.57, one gets

$$v_{tot}(BC) = v_{tot}(PER)\left(\frac{D_w(BC)}{D_w(PER)}\right)^{0.57}$$

$$\approx v_{tot}(PER)\left(\frac{mw(PER)}{mw(BC)}\right)^{0.5 \times 0.57} = 1.08\, v_{tot}(PER)$$

Thus, for BC the relative concentration decrease due to gas exchange is slightly larger than for PER (see above). In addition, hydrolysis (expressed as first-order reaction with $k_r = 0.05\ h^{-1} = 1.39x10^{-5}\ s^{-1}$) also reduces the BC concentration. The following table gives the concentration of benzyl chloride at x = 10 km:

	ε_g (m^{-1})	$e^{-\varepsilon_g \cdot 10\,km}$	ε_r (m^{-1})	$e^{-\varepsilon_r \cdot 10\,km}$	C(10km) (nM)
I	$1.01x10^{-3}$	$4.1x10^{-5}$	$3.47x10^{-5}$	0.71	$2.5x10^{-3}$
II	$5.72x10^{-5}$	0.56	$1.39x10^{-5}$	0.87	3.4

Hence for Situation I the chemical reaction has a much smaller influence on the decrease of the BC concentration than gas exchange. During the large discharge, both elimination processes are reduced leading to a concentration at x = 10 km that - in spite of the dilution effect at the source - is much greater than during low discharge. Note that for Situation II the relative importance of hydrolysis as an elimination process is larger than for Situation II.

Vertical Mixing by Turbulence

Currents in rivers and streams are turbulent (Fig. O.3). Turbulent mixing can be described by the Fickian laws (*Eqs. 9-2* and *9-15*) and by empirical turbulent diffusion coefficients E_i, where i stands for x,y,z *(Section 9.4)*. The principal source of turbulence is the friction between the water and the river bed. It can be expected that increasing roughness of the river leads to increasing turbulence, much in the same way as a large roughness causes the mean flow $\bar{u}$ to become slow (see the effect on Eq. O-4 if the friction coefficient f increases).
The size of turbulence in rivers can be scaled by the so-called *shear velocity* u*, which is given by (e.g., *Fischer et al., 1979*):

$$u^* = (g \cdot S_o \cdot R)^{1/2} \quad (m \cdot s^{-1}) \tag{O-24}$$

For wide rivers the approximation R ~ h can be used again yielding

$$u^* = (g \cdot S_o \cdot h)^{1/2}, \qquad \text{wide river} \tag{O-24a}$$

The ratio between mean flow velocity and shear velocity can be calculated from Eqs. O-4 and O-24:

$$\alpha^* = \frac{\bar{u}}{u^*} = \left(\frac{8}{f}\right)^{1/2} \tag{O-25}$$

Thus, using the typical range of f = 0.02 (smooth river bed) to f = 0.1 (rough river bed) the following range for α^* is obtained:

$$\alpha^* = 20 \text{ (smooth)} \ldots 9 \text{ (rough)} \tag{O-26}$$

Eqs. O-25 and O-26 are rather convenient for estimating u* either from measured or calculated $\bar{u}$-values.

The coefficient of vertical turbulent diffusivity, E_z, is depth dependent. It vanishes at the two boundaries, h = 0 and h = h_o (Fig. O.3), and reaches its maximum value at mid depth,

$$E_z(h) = \kappa \cdot u^* \cdot h \cdot \left(1 - \frac{h}{h_o}\right) \tag{O-27}$$

where κ = 0.41 is the *von Kàrmàn Constant.*

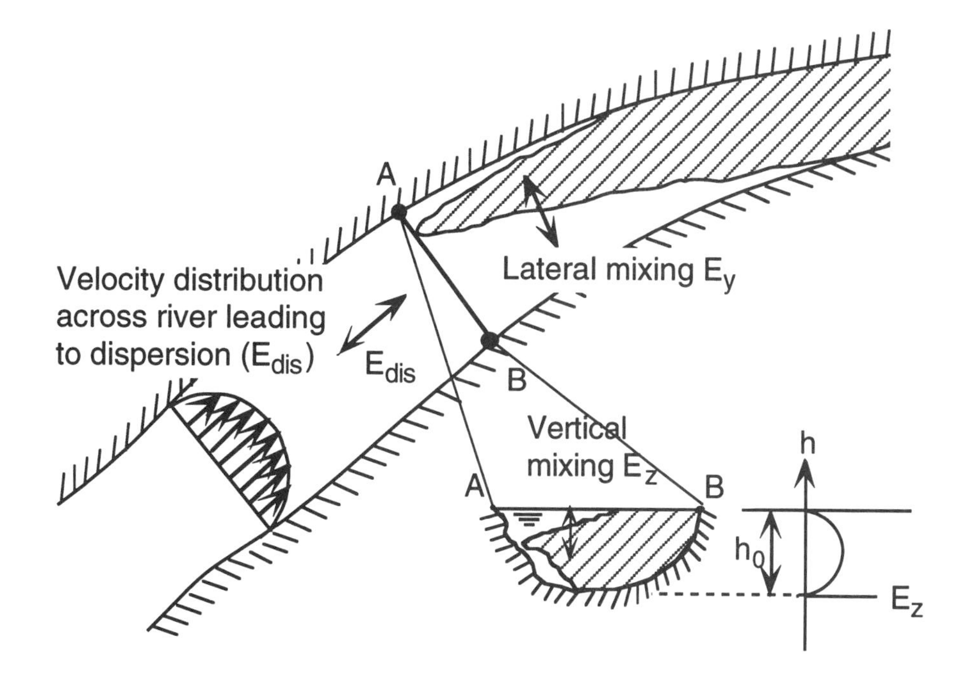

Figure O.3: Mixing processes in a river. E_y and E_z are the turbulent diffusion coefficents in the lateral and vertical direction, respectively, h_o is the maximum depth. Longitudinal dispersion, E_{dis}, results from the variation of velocity in a given cross-section of the river. A pollutant punctually added to the river in cross-section A-B mixes vertically and laterally into the whole river cross-section.

The mean value of E_z averaged over the total water depth is

$$\bar{E}_z = \frac{\kappa}{6} u^* \cdot h_o = 0.07\, u^* \cdot h_o = \frac{0.07}{\alpha^*} \bar{u} \cdot h_o$$

$$= (0.0035[\text{smooth}] \ldots 0.008[\text{rough}])\, \bar{u} \cdot h_o \qquad \text{(O-28)}$$

Note: The expression ($u^* \cdot h_o$) is another example of a case where diffusion is expressed as the product of the relevant velocity times the related diffusion distance (see *Section 9.1*, e.g., *Eq. 9-8*).

The time needed to completely mix the water column over the total depth h_o is of the order (*Eq. 9-31*)

$$\tau_z^{mix} = \frac{h_o^2}{2\,\bar{E}_z} = \frac{\alpha^*}{0.14} \cdot \frac{h_o}{\bar{u}} = (140[\text{smooth}]...64[\text{rough}]) \cdot \frac{h_o}{\bar{u}} \quad \text{(O-29)}$$

which can also be expressed as a distance of flow along the river to achieve vertical mixing:

$$x_z^{mix} = \tau_z^{mix} \cdot \bar{u} = (140[\text{smooth}]...64[\text{rough}] \cdot h_o \quad \text{(O-30)}$$

Problem
Calculate the time and distance of vertical mixing in the River G for the Situations I and II.

Answer
From Eqs. O-25 and O-28 with f = 0.04 for Situation I and f = 0.026 for Situation II one gets

$\alpha_I^* = 14$	$u_I^* = 0.028 \text{ m·s}^{-1}$	$\bar{E}_z = 4.0 \times 10^{-4} \text{ m}^2\text{·s}^{-1}$
$\alpha_{II}^* = 18$	$u_{II}^* = 0.057 \text{ m·s}^{-1}$	$\bar{E}_z = 4.0 \times 10^{-3} \text{ m}^2\text{·s}^{-1}$

Note: u* can also be calculated from Eq. O-24.

The shape of the river bed suggests that it is reasonable to approximate the maximum depth h_o by the mean depth h calculated earlier. Thus, from Eqs. O-29 and O-30

	h_o(m)	$\bar{u}$(m·s^{-1})	τ_z^{mix}(s)	x_z^{mix} (m)
I	0.20	0.40	50	20
II	1.0	1.0	125	125

Note: Incomplete vertical mixing causes the concentration at the water surface to deviate from the mean concentration and can thus affect the computation of air-water exchange (see Eqs. O-13 to O-21). Yet, for the example of benzyl chloride and tetrachloroethene in the River G, the distance to achieve vertical mixing, x_z^{mix}, is negligible relative to the total flow distance of 10 km.

Lateral Mixing by Turbulence

Like E_z, the lateral turbulent diffusion coefficient, E_y, can be characterized by u^*:

$$E_y = \theta \cdot u^* \cdot h_o \tag{O-31}$$

where θ is an empirical parameter. In straight channels with constant cross-section, θ is between 0.1 and 0.2. In natural rivers and streams it lies typically between 0.4 and 0.8. *Fischer et al. (1979)* suggest using

$$\theta = 0.6 \tag{O-32}$$

as the best estimate. In meandering rivers with variable cross section, θ can increase by a factor of 2 to 3 (i.e., $\theta \approx 1$ to 2).

From Eqs. O-31 and O-25 with $\theta = 0.6$ one obtains

$$E_y = \frac{0.6}{\alpha^*}\bar{u} \cdot h_o = (0.03[\text{smooth}]...0.07[\text{rough}]) \cdot \bar{u} \cdot h_o \tag{O-33}$$

By analogy to vertical mixing, the time and distance characteristic of lateral mixing can be expressed

$$\tau_y^{mix} = \frac{w^2}{2E_y} = \frac{\alpha^* \cdot w^2}{1.2 \cdot \bar{u} \cdot h_o} = (17[\text{smooth}]...7.5[\text{rough}] \cdot \frac{w^2}{\bar{u} \cdot h_o} \tag{O-34}$$

$$x_y^{mix} = \bar{u} \cdot \tau_y^{mix} = \frac{\alpha^* \cdot w^2}{1.2 \cdot h_o} = (17[\text{smooth}]...7.5[\text{rough}]) \cdot \frac{w^2}{h_o} \tag{O-35}$$

Problem
Calculate the characteristic time and distance of lateral mixing for the River G for the flow Situations I and II.

Answer
From Eqs. O-34 and O-35 with $\alpha_I^* = 14$ for Situation I, $\alpha_{II}^* = 18$ for Situation II, and w = 10m for both situations, one obtains

	h_o(m)	$\bar{u}$($m \cdot s^{-1}$)	E_y($m^2 \cdot s^{-1}$)	τ_y^{mix}(s)	x_y^{mix} (m)
I	0.20	0.40	3.4×10^{-3}	15000 (4.1 h)	5900
II	1.0	1.0	3.4×10^{-2}	1500 (0.4 h)	1500

Note: In this example the time and the distance of lateral mixing are significantly larger than the time and the distance of vertical mixing, although E_y is usually greater than E_z. However, the width of a river is usually much bigger than its depth, thus $\tau_y^{mix} > \tau_z^{mix}$.

Longitudinal Mixing by Dispersion

Longitudinal dispersion occurs whenever fluids are transported along a predominant direction of flow. Dispersion results from the different current velocities associated with the various stream lines passing through a given river cross-section. Since friction at the bottom of the river acts as the main "brake" that balances the gravitational forces, currents are strongly depth-dependent, and the vertical velocity gradient, the so-called vertical shear, is large. However, with respect to longitudinal dispersion the vertical shear is less effective than the velocity gradient across the river bed (lateral shear), since mixing between stream lines at different depths usually takes much less time than mixing from one side of the river bed to the other (see above). This is especially true for river bends and for cross-sections with asymmetric depth contours (Fig. O.4). Since the stream lines transport water at different mean velocities, they usually also carry different concentrations.

The techniques developed to describe transport by turbulent diffusion can also be applied to dispersion. As derived in *Eq. 9-42*, the mean advective mass flux at a given cross-section can be written as

$$\bar{F} = \bar{u} \cdot \bar{C} + \overline{u' \cdot C'} \qquad \text{(O-36)}$$

Here $\bar{u}$ is the mean current velocity across the river as defined in Eq. O-3, $\bar{C}$ is the mean concentration, and u' and C' are the *spatial* deviations from the mean (*Note: Eq. 9-42* dealt with the *temporal* deviations from the mean!)

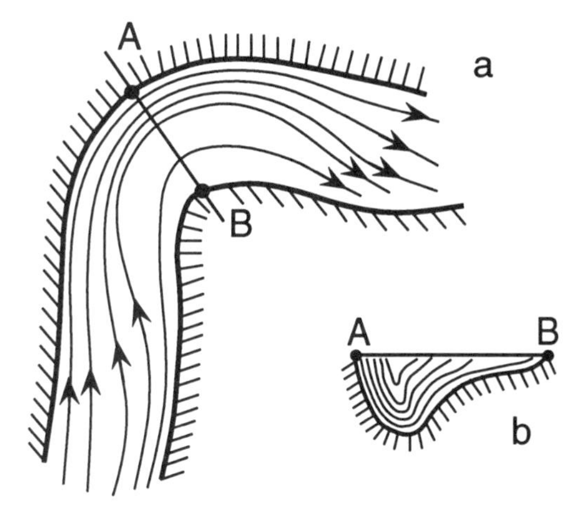

Figure O-4 (a) Stream lines at a bend are asymmetrically distributed across the river. (b) Contour lines of equal velocity in a cross-section indicate that flow velocities strongly vary laterally as well as from the water surface to the bottom. Such variations are responsible for longitudinal dispersion.

Thus,

$$u(x,y,z) = \bar{u}(x) + u'(y,z) \quad \text{(O-37)}$$

$$C(x,y,z) = \bar{C}(x) + C'(y,z) \quad \text{(O-37a)}$$

Note that the instantaneous time dependence of u and C that results from turbulent fluctuations is averaged out in the quantities defined by Eqs. O-37. However, over time scales that are relevant to assessing the behavior of a chemical, all these quantities may also depend on time.

The mass flux relative to a cross-section moving at mean velocity $\bar{u}$ along the river is

$$\bar{F}_{moving} = \bar{F} - \bar{u} \cdot \bar{C} = \overline{u' \cdot C'} \quad \text{(O-38)}$$

The mass balance equation for a water parcel which is bounded by two moving cross-sections is given by (see *Section 9.2*, especially *Fig. 9.4*)

$$\left(\frac{\partial \bar{C}}{\partial t}\right)_{moving} = -\frac{\partial \bar{F}_{moving}}{\partial x} = -\frac{\partial}{\partial x}\overline{(u'C')} \quad \text{(O-39)}$$

By analogy to the case of turbulent diffusion (*Eq. 9-44*), the dispersion coefficient E_{dis} (dimension $L^2 \cdot T^{-1}$) is defined by

$$\overline{u'C'} = - E_{dis} \frac{\partial \bar{C}}{\partial x} \tag{O-40}$$

Thus,

$$\left(\frac{\partial \bar{C}}{\partial t}\right)_{moving} = \frac{\partial}{\partial x}\left(E_{dis} \frac{\partial \bar{C}}{\partial x}\right) \tag{O-41}$$

and for E_{dis} = constant

$$\left(\frac{\partial \bar{C}}{\partial t}\right)_{moving} = E_{dis} \frac{\partial^2 \bar{C}}{\partial x^2} \tag{O-41a}$$

It is not possible to calculate E_{dis} for a natural river from first principles alone. However, starting from experiments with uniform channel flow, *Fischer et al. (1979)* have developed concepts to relate E_{dis} to other basic parameters of the river flow such as u^*, E_z, E_y, which were introduced to characterize turbulent mixing in the river. There are two qualitative arguments for the way E_{dis} should depend on other river parameters:

First, E_{dis} should increase with the maximum velocity difference across the river. Since the velocity of the stream lines that are close to the boundaries (i.e., near the river bank or the river bed) is practically zero and the maximum velocity is on the order of the mean velocity $\bar{u}$, one would expect to find E_{dis} growing with $\bar{u}$. As it turns out, E_{dis} grows in proportion to $\bar{u}^2$.

Second, imagine the various stream lines to be railway tracks running in parallel from A to B. On some tracks there are only the superfast trains, on others are ones of intermediate speed, and on some move the very slow freight trains. People travelling in these trains can jump at random from one track to the other, thus changing the speed of their journey. At time t=0 several people on different tracks start their trip at A. If they change tracks at a very high rate ("infinitely frequently"), they all end up travelling at the average speed of all the trains and will arrive at B nearly simultaneously. If, however, they never change trains, their time of arrival at B varies most strongly since those in the ultrafast trains arrive rather quickly, but the ones in the very slow trains much later.

As explained above, since the width of a river is usually much bigger than its depth, mixing between stream lines at different depths takes much less time than mixing across the river. Since the time needed to move from one extreme side of the river to the other increases as the square of the width, w^2 (Eq. O-34), one expects E_{dis} to vary as w^2/E_y.

These considerations can be combined into an empirical equation:

$$E_{dis} = \text{const} \cdot \theta \cdot (\bar{u})^2 \frac{w^2}{E_y} \qquad \text{(O-42)}$$

Substitution of E_y from Eq. O-31 yields:

$$E_{dis} = \text{const.} \cdot (\bar{u})^2 \frac{w^2}{h_o \cdot u^*} \qquad \text{(O-42a)}$$

According to *Fischer et al. (1979)*, the best choice for the constant factor appearing in Eq. (O-42) is 0.011. Adopting the optimum θ-value of 0.6 (see Eq. O-32) yields

$$E_{dis} = 0.0066\, (\bar{u})^2 \frac{w^2}{E_y} = 0.011\, (\bar{u})^2 \frac{w^2}{h_o \cdot u^*} \qquad \text{(O-43)}$$

The effect of longitudinal dispersion on the concentration of a chemical in a river can be demonstrated by the following example. Assume that at some location and time (for which x and t are arbitrarily set to 0), a total amount M of some chemical is spilled into the river such that it is immediately mixed uniformly across the constant cross-section A. The discharge Q along x shall be constant. The flow is assumed to be stationary (i.e., not changing with time).

To solve Eq. O-41a in a coordinate system which moves along the river, one can use *Eq. 9-19*:

$$\bar{C}(\xi,t) = \frac{M/A}{2(\pi \cdot E_{dis}\, t)^{1/2}} \exp\left(- \frac{\xi^2}{4E_{dis}\, t} \right) \qquad \text{(O-44)}$$

where ξ is the distance from x_m, the center of the concentration distribution moving according to the simple relation

$$x_m(t) = \bar{u} \cdot t \qquad \text{(O-45)}$$

As defined in Eq. O-37a, $\bar{C}$ is constant over the whole area A. Thus, the expression (O-44) obeys mass conservation at any time t:

$$\iiint_{River} \bar{C} \cdot dV = A \int_{-\infty}^{\infty} C(\xi,t) \cdot d\xi = M \qquad \text{(O-46)}$$

(see *Table 9.2* and *Eq. 9-21*).

In order to transform Eq. O-44 into absolute coordinates along the river, ξ is expressed as

$$\xi = x - x_m = x - \bar{u} \cdot t \qquad \text{(O-47)}$$

Thus,

$$\bar{C}(x,t) = \frac{M/A}{2(\pi \cdot E_{dis} \cdot t)^{1/2}} \exp\left(-\frac{(x - \bar{u} \cdot t)^2}{4E_{dis} \cdot t}\right) \qquad \text{(O-48)}$$

In order to understand better the meaning of this expression, imagine that it would be possible to measure the concentration $\bar{C}$ along the river *at a given time t.* Then, according to Eq. O-48, one would expect to find a Normal (or Gaussian) concentration distribution with center at $x_m(t) = \bar{u} \cdot t$ and standard deviation $\sigma(t) = (2\, E_{dis} \cdot t)^{1/2}$ (see *Table 9.2*). The maximum concentration (reached at $x = \bar{u}t$) would be

$$\bar{C}_{max}(t) = \frac{M/A}{2(\pi \cdot E_{dis} \cdot t)^{1/2}} \quad , \qquad \text{(O-49)}$$

that is, it should decreases with time as $t^{-1/2}$.

Note that the Gaussian distribution for a fixed time (Eq. O-48) *does not cause a Gaussian concentration variation in time at a fixed location x.* Imagine that the concentration cloud passes by some location x. Because of the finite longitudinal extension of the cloud (as described by σ), the passage of the cloud (or at least of most of it, since mathematically speaking the cloud has infinite size) takes some time during which σ^2 is still growing. Thus, the rising concentration curve of the cloud is always steeper than the falling curve.

Yet, this is not the only, and usually not even the most important reason why the concentration measured at a fixed location is asymmetric in time.

So far the analysis of dispersion has been limited to a situation in which the chemical is already completely mixed vertically and laterally so that all the streamlines are "occupied". In many cases chemicals enter the river from outfalls (see Fig. O-3). Remember that vertical mixing usually occurs over a short distance whereas lateral mixing may need more time (or distance). As discussed before it is mostly the lateral mixing (or rather its slowness!) which allows longitudinal dispersion.

As observations in rivers show there exists an initial phase during which the cloud is skewed. The concentration at a fixed location as a function of time has a rising slope and a long falling tail. An example is shown in Fig. O-5. During the initial phase the standard deviation σ grows more slowly than $t^{1/2}$. The time needed for the transition from the initial state of dispersion, the so-called advective period, to the Gaussian dispersion can be related to the transverse mixing time τ_y^{mix} (Eq. O-34). According to *Fischer et al. (1979)* the concentration is skewed for times

$$t < t_{skewed} = 0.8\ \tau_y^{mix} = 0.4\ \frac{w^2}{E_y} \qquad \text{(O-50)}$$

The standard deviation σ grows as $t^{1/2}$ for times

$$t > t_\sigma = 0.4\ \tau_y^{mix} = 0.2\ \frac{w^2}{E_y} \qquad \text{(O-51)}$$

A first step towards the full two-dimensional (longitudinal/lateral) modeling of dispersion in rivers is given, for instance, by the "enhanced one-dimensional model" developed by *Reichert and Wanner (1991).*

Until now it has been assumed that the chemical is added to the whole cross-section of the river during a very short time period. This leads to a narrow and sharp initial concentration peak along the river (a δ-function, as the mathematicians would call it) that is then gradually transformed into a normal distribution with growing standard deviation σ. Often the compound is added continuously or – even if it originates from an accidental spill – during a finite time Δt. Two extreme scenarios are considered. In the first scenario, the input is instantaneously "switched on" to a constant input rate μ (expressed as mass per unit time) and "turned off" after time Δt (Fig. O.6a). Even if the advective period of dispersion can be disregarded (remember, the substance is added to the whole river cross-section), dispersion would first only smooth out the

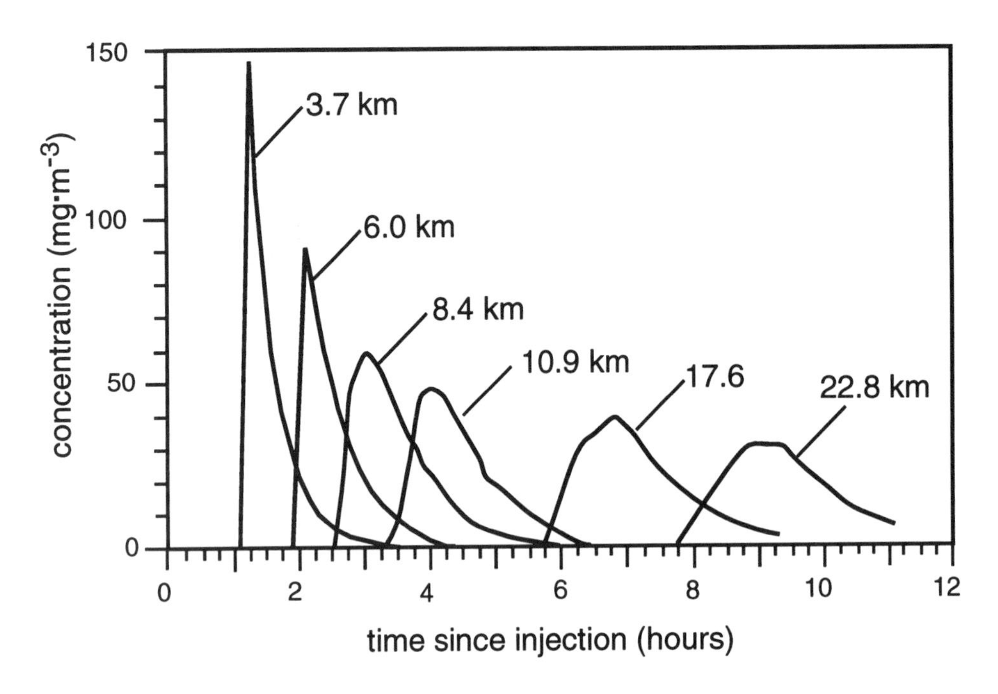

Figure O.5 Temporal evolution of a concentration cloud along a river. The curves show cross-sectional averaged dye concentrations measured at six sites in the Waikato River (New Zealand) below an instantaneous transverse line source. From *Rutherford (1994)*.

sharp concentration edges at the onset and end of the input event while the concentration in the middle part of the event, $C_{in} = \mu/Q$, would not be altered. The time, t_{ini}, needed until dispersion from both sides would meet at the center of the cloud can be estimated by using *Eq. 9-9* :

$$t_{ini} = \frac{(\Delta x/2)^2}{2\,E_{dis}} = \frac{(\Delta t \cdot \bar{u})^2}{8\,E_{dis}} \tag{O-52}$$

At time $t = t_{ini}$, the cloud can be approximated by a Gaussian distribution with maximum concentration $C_{in} = \mu/Q$, which, for times $t > t_{ini}$, decays according to the expression given by Eq. O-48. In fact, by inserting $t = t_{ini}$ and the total mass input of the event, $M = C_{in} \cdot Q \cdot \Delta t = C_{in} \cdot A \cdot \bar{u} \cdot \Delta t$, into the expression that describes the maximum concentration of a Gaussian distribution, Eq. O-49, one gets

$$\bar{C}_{max}(t) = \frac{C_{in}}{(\pi/2)^{1/2}} = 0.8\ C_{in} \quad \text{(O-53)}$$

Thus, Eq. O-48 remains approximatively valid for times $t > t_{ini}$, where t is the time when half of the input has reached the river, and M/A is replaced by $C_{in} \cdot \bar{u} \cdot \Delta t$. For the situation shown in Fig. O.6a :

$$\bar{C}(x,t) = \frac{C_{in} \cdot \bar{u} \cdot \Delta t}{2(\pi \cdot E_{dis} t)^{1/2}} \exp\left(- \frac{(x - \bar{u} \cdot t)^2}{4E_{dis} \cdot t} \right), \text{ for } t > t_{ini} \quad \text{(O-54)}$$

The second input scenario, simpler from a mathematical viewpoint but less probable to occur, is a generalization of the instantaneous (δ-)input. It is assumed that the temporal variation of the input is Gaussian and leads to an initial longitudinal variation of the concentration cloud with standard deviation

$$\sigma_o = \frac{\Delta t \cdot \bar{u}}{4} \quad \text{(O-55)}$$

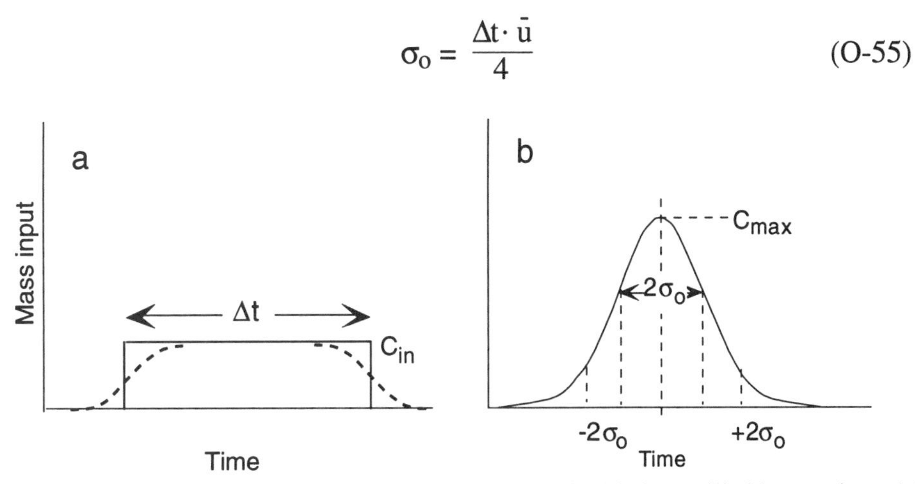

Figure O.6 Two extreme input scenarios for chemical being spilled into a river. (a) Constant input at rate μ (mass per unit time) during time Δt leading to a "rectangular" input concentration $C_{in} = \mu/Q$. The interrupted line shows how initially dispersion only acts on the edges and leaves the concentration in the middle of the cloud unchanged. (b) Gaussian input scenario: The time integral between $t = -2\sigma_o$ and $t = 2\sigma_o$ comprises 95% of the total input M (see *Table 9.2*). The maximum concentration is then given by Eq. O-58. Dispersion causes the variance to increase according to Eq. O-56.

Δt is the time during which about 95% of the substance is added to the river (see Fig. O.6b). *Eq. 9-48* then describes how the variance σ^2 grows with time:

$$\frac{d\sigma^2}{dt} = 2\,E_{dis} \qquad (9\text{-}48)$$

It has the solution

$$\sigma^2(t) = \sigma_o^2 + 2E_{dis}\cdot t \qquad (O\text{-}56)$$

Then the generalized version of Eq. O-48 *for the scenario of a Gaussian input event* is

$$\bar{C}(x,t) = \frac{M/A}{(2\pi)^{1/2}\,(\sigma_o^2 + 2E_{dis}\cdot t)^{1/2}}\exp\left(-\frac{(x-\bar{u}\cdot t)^2}{2\sigma_o^2 + 4E_{dis}\cdot t}\right) \qquad (O\text{-}57)$$

which is again a Gaussian distribution with time-dependent variance. The initial maximum concentration in the river is

$$\bar{C}_{max}(0) = \frac{M/A}{(2\pi)^{1/2}\,\sigma_o} \qquad (O\text{-}58)$$

Problem

From an adjacent corn field a total amount of 200 kg of the herbicide *atrazine* is accidentally spilled into the River G at a constant rate within a time span of 30 minutes. Estimate for both river discharge situations (see above) the maximum total concentration of atrazine attained (1) at the input location and (2) 10 km further downstream. Assume a conservative behavior of atrazine in the river. How much time does the atrazine cloud need to pass by the downstream location?

Cl

N N

CH_3

CH_3-CH_2-HN N NH—CH

CH_3

atrazine

mw 215.7 $g\cdot mol^{-1}$

Note that the process of dissolution of atrazine in the water takes some time. To make the situation not too complicated, it is assumed that the particles containing undissolved atrazine are small enough to be kept in suspension. Thus, especially in the river section just downstream of the spill, part of the total atrazine concentration that is calculated by solving the above problem may refer to the yet undissolved atrazine.

Answer

The following table summarizes the relevant hydraulic date of the River G for both discharge situations as derived above:

	I	II
Discharge Q ($m^3 \cdot s^{-1}$)	0.8	10
Mean flow velocity $\bar{u}$ ($m \cdot s^{-1}$)	0.40	1.0
Depth $h \sim h_o$ (m)	0.20	1.0
Width w, approx. (m)	10	10
Area A, approx. (m^2)	2	10
Lateral diffusivity E_y ($m^2 \cdot s^{-1}$)	3.4×10^{-3}	3.4×10^{-2}
Time of lateral mixing τ_y^{mix} (s)	15000	1500
Distance of lateral mixing x_y^{mix} (m)	5900	1500

The longitudinal dispersion coefficient E_{dis} is calculated from Eq. O-43:

I $\quad E_{dis} = 31\ m^2 \cdot s^{-1}$

II $\quad E_{dis} = 19\ m^2 \cdot s^{-1}$

If the atrazine entered the river locally on the side, the advective period would be relevant for the initial spreading of the atrazine. This period lasts $0.4 \times \tau_y^{mix}$ (Eq. O-51), i.e., about 1.6 h (or 2.4 km) for Situation I and 10 minutes (600 m) for Situation II. Often it is not exactly known how and in what form a pollutant has entered a river. Therefore the maximum concentration attained at the input location can only be estimated. For the following calculations, immediate mixing across the river is assumed.

The input rate μ is equal to 2×10^5 g / 1800 s = 111 $g \cdot s^{-1}$ = 0.52 $mol \cdot s^{-1}$ Thus, the initial concentrations $C_{in} = \mu/Q$ are 6.5×10^{-4} and 5.2×10^{-5}

mol·L^{-1}, for Situations I and II, respectively. Next, the time needed until C_{in} starts to decrease due to longitudinal dispersion, t_{ini}, is calculated from Eq. O-52. The results are summarized in the following table:

	I	II
Input M (mol)	930	930
t_{ini} (s)	2.1×10^3	21×10^3
$x_{ini} = \bar{u} \cdot t_{ini}$ (m)	840	21×10^3
C_{in} (mol·L^{-1})	(6.5×10^{-4})	5.2×10^{-5}

Recall that for Situation I, C_{in} is larger than the aqueous solubility of atrazine, which is $C_w^{sat} = 1.5 \times 10^{-4}$ mol·L^{-1} at 25°C (Appendix). Thus, dispersion has to "pull the cloud apart" before all the atrazine can go into solution.

In the case of Situation I, the maximum concentration at x = 10 km can be calculated from Eq. O-54 by taking the time needed for the cloud travelling at mean velocity, $\bar{u}$, to reach the downstream location. The relative time t is chosen as being 0 when half of the atrazine has entered the river. Thus $\bar{C}(x,t)$ has to be evaluated for $x = \bar{u} \cdot t$. In contrast, for Situation II the critical time t_{ini} is greater than the flow time; thus, dispersion has not yet become effective for diluting the atrazine concentration in the central part of the cloud where the concentration is still about C_{in}. Thus,

	I	II
Flow time t (s)	2.5×10^4	1.0×10^4
$(2E_{dis}t)^{1/2}$ (m)	1.2×10^3	not relevant
$\bar{C}(x, t=x/\bar{u})$ (mol·L^{-1})	1.5×10^{-4}	$\approx 5.2 \times 10^{-5}$

It is interesting to see that although dilution by dispersion has not yet become effective in case II, the concentration at x=10km is smaller than for case I due to the greater initial dilution at the location of the spill. Thus, in this specific example large flow rates favor large dilution but impede the effect of dispersion.

Recall that for all calculations it has been assumed that within the relevant time scales of a few hours atrazine does not undergo any degradation or other removal processes.

Sedimentation

When water containing various compounds flows over the river bed, an exchange of the chemicals between the water and the sediments occurs. In principle, the relevant processes are the same as those described for lakes (*Section 15.4*), i.e., settling and resuspension of particles to which the chemicals are sorbed, molecular diffusion between flowing water and pore water, as well as direct sorption/desorption between the water and the sediment particles. Yet, in rivers the temporal and spatial variation of these processes, as well as their relative importance, are fairly different from typical lake conditions. Therefore, the mathematical description that was introduced in *Section 15.4*, is not necessarily adequate for rivers.

Take, for instance, the case of particle settling. From a formal point of view this process can be envisioned as the vertical sinking of particles through the water column with the average velocity v_S. If the mean depth of the river, that is the cross-section divided by the width (Table O.1), is h, then the removal rate of suspended particles from the water is

$$k_S = \frac{v_S}{h} \qquad (15\text{-}51)$$

However, in rivers the settling velocity v_S has little to do with gravitional settling (e.g., Stoke's law *Eq. 15-45b*), but is mainly controlled by the turbulence of the flowing water. Consequently, v_S may strongly vary along the river from virtually zero in zones of high turbulence to several meters per day in still waters. Furthermore, in most rivers the net sedimentation rate, if averaged over several months or even years, is zero or at least very small. This means that over a long enough time period particle settling and particle resuspension balance out at practically all locations of the river, although for a given time period either settling or resuspension may be very large.

Thus, depending on the time interval of interest, the role of the sediment-water interaction is either to be described as a first-order (one-way) removal reaction (rate constant k_S) or by a (two-way) exchange reaction (rate constant k_{ex}). The former is relevant for a compound which is added to the river for a short time period only (e.g., by an accidental spill), the latter refers to the case of a compound which enters the river continously such that, depending on the actual compound concentration in the flowing water, the sediment particles (with sorbed concentrations reflecting the mean loading history of the compound) can either act as a source or a sink for the flowing water.

Section 15.4 serves as a guideline for the development of simple mathematical models for both situations. In particular, an instantaneous and reversible linear equilibrium relationship is assumed between the dissolved concentration of the compound, C_w (mol·m^{-3}), and the compound sorbed on solids, C_s (mol·kg^{-1}):

$$K_d = \frac{C_s}{C_w} \quad (m^3 \cdot kg^{-1}) \qquad (11\text{-}2)$$

where K_d is the solid-water distribution ratio. *Note that if K_d is expressed in units of (L·kg^{-1}), the numerical value of K_d is 10^3 times larger than for units of ($m^3 \cdot kg^{-1}$).* If the solid's organic matter content is controlling the sorption process, K_d can be expressed by the relation

$$K_d = f_{om} \cdot K_{om} \qquad (11\text{-}16)$$

where K_{om} is the partition constant between particulate organic matter and solution phase and f_{om} is the relative weight fraction of natural organic matter on the suspended particles. The total (dissolved and sorbed) compound concentration C_t can be expressed as the sum of dissolved and particulate concentration, C_w and C_p, respectively: $C_t = C_w + C_p$, where all concentrations are in units of mol·m^{-3}. Combining *Eqs. 15-41* and *15-42* yields

$$C_w = f_w \cdot C_t \quad \text{and} \quad C_p = (1-f_w) \cdot C_t \approx r_{sw} \cdot C_s \qquad (O\text{-}59)$$

where r_{sw} is the solid-to-solution phase ratio (kg·m^{-3}), which for rivers and lakes is about equal to the concentration of suspended solids. The fraction of the compound in solution, f_w, at sorptive equilibrium is given by

$$f_w = \frac{1}{1 + r_{sw} \cdot K_d} \qquad (11\text{-}5)$$

or

$$f_w = \frac{1}{1 + f_{om} \cdot r_{sw} \cdot K_{om}} \qquad \textit{(15-40)}$$

For the case of an *episodic addition of a compound to the river* water, the process of removal to the sediments can now be described by the simple linear model:

$$\left(\frac{\partial C_t}{\partial t}\right)_{sed} = - k_s \cdot (1 - f_w) \cdot C_t \qquad (O\text{-}60)$$

which, in analogy to Eq. O-8, can be transformed into an expression that describes the variation of C_t along the river:

$$\left(\frac{dC_t}{dx}\right)_{sed} = - \frac{k_s}{\bar{u}} (1 - f_w) \cdot C_t = - \varepsilon_s \cdot C_t \qquad (O\text{-}61)$$

where $\varepsilon_s = \frac{k_s}{\bar{u}} \cdot (1 - f_w)$. In some cases (deep river with small flow velocity), the sedimentation rate k_s may be calculated from an expression like *Eq. 15-51*, but usually it is just an empirical parameter to be determined from measurements.

Sediment-Water Interaction

The elimination process described by Eqs. O-60 or O-61 acts only on the compound that is sorbed on particles, hence on the factor $(1 - f_w)$. However, even a compound that is totally dissolved, is partially lost to the sediments by diffusion into the pore water while the water is flowing along the river (*Wanner et al., 1989*). If the time the cloud needs to pass by a fixed location is Δt, then the penetration depth due to diffusion, d_s, of the compound into the sediments is about (see *Eq. 9-9*)

$$d_s \approx (2 \cdot D_{eff} \cdot \Delta t)^{1/2} \qquad (O\text{-}62)$$

where D_{eff} is the so-called *effective* molecular diffusion coefficient of the compound in the porous sediment bed. As qualitatively explained in *Fig. 11.4* and discussed in more detail in Chapter P, due to sorption of the chemical on the sediment particles, diffusion is reduced by the factor f_{wsed} compared to the molecular diffusion coefficient in the free water,

D_w, that is, $D_{eff} = f_{wsed} \cdot D_w$, where f_{wsed} is the fraction of the compound in the pore water that is dissolved (see above *Eqs. 11-5* or *15-40*). *Note that* f_{wsed} *is usually much smaller than* f_w*, since the solid-to-solution phase ratio* r_{sw} *in the sediment is by several order of magnitudes larger than in the open water.*

A further reduction of D_{eff} is due to the effect of *tortuosity* in the sediment pore space (see *Eq. 11-155*). Tortuosity refers to the ratio between the length of a straight line connecting two points in the sediments and the real path length along the pores. Note that in the literature dealing with porous media, tortuosity χ is defined as a number greater than one (see Chapter P). The factor f used in *Eq. 11-155* is equal to χ^{-1}.

The total mass per unit area that diffuses into the sediments is roughly given by $C_w \cdot d_s \cdot \phi / f_{wsed} = f_w \cdot C_t \cdot d_s \cdot \phi / f_{wsed}$ (ϕ is porosity of the sediment, see *Eq. 11-6*). Thus, the loss rate of the compound per unit time and water volume can be written as

$$\left(\frac{\partial C_t}{\partial t}\right)_{diff} = - k_{diff} \cdot f_w \cdot C_t \tag{O-63}$$

with

$$k_{diff} = \frac{\phi}{h}\left(\frac{2 \cdot D_w}{f_{wsed} \cdot \Delta t}\right)^{1/2} \tag{O-64}$$

Note that if the passing cloud is approximated by a Gaussian distribution, the passing time can be expressed by $\Delta t = 4\sigma(t) / \bar{u}$ (see Eq. O-55). σ increases with time due to dispersion (Eq. O-56), and so does Δt, thus k_{diff} becomes smaller along the river. By analogy to Eq. O-61, the variation of C_t along the river due to diffusion into the sediments is given by

$$\left(\frac{dC_t}{dx}\right)_{diff} = - \frac{k_{diff}}{\bar{u}} \cdot f_w \cdot C_t = - \varepsilon_{diff} \cdot C_t \tag{O-65}$$

where $\varepsilon_{diff} = \frac{k_{diff}}{\bar{u}} \cdot f_w$. Note that the above calculation disregards the possible movement of the river bed.

In contrast, for the case of a *continuous loading* the (two-way) exchange processes can be described by a linear model of the form as shown in *Fig. 15.10* for the sediment-water exchange in a lake. By analogy to the last term on the righthand side of *Eq. 15-65*, the following expression can be adopted:

$$\left(\frac{\partial C_t}{\partial t}\right)_{ex} = -\, k_{ex} \cdot \left(f_w \cdot C_t - \frac{C_{ss}}{K_{ds}}\right) \qquad \text{(O-66)}$$

where C_{ss} and K_{ds} correspond to the parameters in *Eq. 11-2*, that is, C_{ss} ($mol \cdot kg^{-1}$) is the concentration on the sediment particles and K_{ds} ($kg \cdot mol^{-1}$) is the solid-water distribution ratio for sediment particles that may be different from K_d of the suspended particles. k_{ex} is the (empirical) exchange rate constant which includes all possible processes such as sediment resuspension followed by equilibration of the particles with the water, direct exchange at the sediment surface, or flow of water through the pore space of the sediments accompanied by sorption/desorption. According to recent experiments (*Eylers, 1994*) the latter process seems to be of major importance for the fast exchange of sorbing chemicals between river water and (especially sandy) sediments (see below).

The corresponding differential equation for the river coordinate x is

$$\left(\frac{dC_t}{dx}\right)_{ex} = -\, \frac{k_{ex}}{\bar{u}} \left(f_w \cdot C_t - \frac{C_{ss}}{K_{ds}}\right) \qquad \text{(O-67)}$$

Note that in the above equations the variables that refer to the sediments may vary along x. In fact, in order to build a consistent river model, in addition to the continuous concentration $C_t(x,t)$, one would also have to model continuously the compound in the sediment, $C_{ss}(x,t)$. This is beyond the intention of this chapter. The interested reader is referred to *Section 15.4*, especially to *Eq. 15-64*.

Problem

What fraction of the *atrazine* that was spilled into the River G (see above) is lost to the sediments between the input location and the station 10 km downstream if the flow regime corresponds to Situation II ? The concentration of suspended solids is 50 $mg_s \cdot L^{-1}$ with an organic matter content of $f_{om} = 0.4\ kg_{om} \cdot kg_s^{-1}$. Take a mean settling velocity $v_s = 1\ m \cdot d^{-1}$. Assume that all the atrazine is immediately dissolved in the water at the input location. The sediments of the river bed have a porosity $\phi = 0.8$; the density of the solids is $\rho_s = 2.5\ kg_s \cdot L^{-1}$ with an organic matter content $f_{om,sed} = 0.01\ kg_{om} \cdot kg_s^{-1}$.

Answer

Using the octanol-water partition constant of atrazine (log K_{ow} = 2.56, see *Appendix*) and the linear free-energy relationship shown in *Fig. 11.10* yields an estimate of the organic matter-water partition coefficient of atrazine: K_{om} = 170 $L \cdot kg_{om}^{-1}$. Application of *Eq. 15-40* with $r_{sw} \cdot f_{om} \cdot K_{om} = 3.4 \cdot 10^{-3}$ yields $(1-f_w) = 3.4 \cdot 10^{-3}$. Inserting v_s and $h \sim h_o$ = 1.0 m into *Eq. 15-51* yields $k_s = 1.0\ d^{-1} = 1.16 \times 10^{-5}\ s^{-1}$. The elimination rate due to particle settling per unit flow distance, ε_s, is then (see Eq. O-61)

$$\varepsilon_s = \frac{k_s}{\bar{u}}(1-f_w) = \frac{1.16 \text{x} 10^{-5} s^{-1}}{1.0\ m \cdot s^{-1}} \text{ x } 3.4 \text{ x } 10^{-3} = 3.9 \text{ x } 10^{-8}\ m^{-1}$$

To evaluate the loss of atrazine due to diffusion into the sediment pore space, *Eq. 15-40* is evaluated for the sediments. *Note that throughout these calculations it is assumed that all the atrazine is dissolved.* The solid-to-water phase ratio for the sediments, $r_{w,sed}$, is calculated from *Eq. 11-8*:

$$r_{sw,sed} = \rho_s \frac{1-\phi}{\phi} = 2.5\ kg \cdot L^{-1} \frac{0.2}{0.8} = 0.63\ kg_s \cdot L^{-1}$$

Thus, with $r_{w,sed} \cdot f_{om,sed} \cdot K_{om} = 1.06$, it follows $f_{w,sed} = 0.48$. When calculating the exposure time Δt of the sediments to the passing atrazine cloud (see Eq. O-64), it should be remembered that for Situation II the cloud still has part of its "rectangular" shape (Fig. O.6a) when passing by the station at x = 10 km. Thus, Δt can be approximated by the duration of the spill (30 minutes).

The molecular diffusion coefficient of atrazine in water is estimated from its molecular mass (215.7 $g \cdot mol^{-1}$) using *Eq. 9-30* and with O_2 as the reference substance (as in *Table 10.3*): D_w(atrazine) = $8.1 \text{x} 10^{-6}\ cm^2 \cdot s^{-1}$ = $8.1 \text{x} 10^{-10}\ m^2 \cdot s^{-1}$. Thus, from Eq. O-64

$$k_{diff} = \frac{0.8}{1.0\ m}\left(\frac{2 \text{x} 8.1 \text{x} 10^{-10}\ m^2 \cdot s^{-1}}{0.48 \text{x} 1.8 \text{x} 10^3\ s}\right)^{1/2} = 1.1 \text{x} 10^{-6}\ s^{-1},$$

and from Eq. O-65

$$\varepsilon_{diff} = \frac{1.1 \text{x} 10^{-6}\ s^{-1}}{1.0\ m \cdot s^{-1}} \text{ x } 0.9966 = 1.1 \text{x} 10^{-6}\ m^{-1}$$

Note that the influence of sorption of atrazine on the sediment particles while diffusing into the pores, as expressed by the factor $(f_{w,sed})^{1/2}$ in the denominator of Eq. O-64, does not influence the result very much. The following table summarizes the results:

	Particle settling	Diffusion into sediment
$\varepsilon(m^{-1})$	3.9×10^{-8}	1.1×10^{-6}
reduction at $\Delta x = 10$ km: $e^{-\varepsilon \cdot \Delta x}$	0.9996	0.989

Thus, neither particle settling nor diffusion into the sediments play a significant role for the elimination of atrazine from the river. However, the flow of water through the pore space of the sediments accompanied by sorption/desorption may be a more important mechanism. According to *Eylers (1994)*, this process can be quantified by the following exchange velocity v_{sed}:

$$v_{sed} = \frac{0.42}{g} \frac{K_q \bar{u}^2}{\lambda} \left(\frac{H}{h}\right)^{3/8} \qquad \text{(O-68)}$$

Here K_q is the hydraulic conductivity of the sediments (units: $m \cdot s^{-1}$), a quantity that will be discussed more extensively in Chapter P on porous media (see Eq. P-5), λ is the wavelength (horizontal distance) of the ripples or dunes at the bottom of the river bed, and H is the height of the dunes. The flow-through mechanism is especially important for sandy river beds where ripples and dunes often occur. The removal rate for the total concentration in the river then depends on the mean depth h of the river and on the fraction of the substance that is retained on the sediment particles. The latter is expressed by $(1 - f_{w,sed})$ where $f_{w,sed}$ is the fraction of the compound in the pore water that is dissolved. Thus,

$$k_{sed} = \frac{v_{sed}}{h} (1 - f_{w,sed}) \qquad \text{(O-69)}$$

In order to estimate the order of magnitude of v_{sed} for the River G, the following assumption are made: $K_q = 10^{-3}\ m \cdot s^{-1}$, $\lambda \approx 3 \cdot h = 3$ m, $H \approx h/10 = 0.1$ m. Eq. O-68 yields $v_{sed} = 6.0 \times 10^{-6}\ m \cdot s^{-1}$. Thus, with $f_{w,sed} = 0.48$, $k_{sed} = 3.1 \times 10^{-6}\ s^{-1}$, which is of the same order of magnitude as k_{diff}. Yet, a larger sediment conductivity K_q or a smaller dune wavelength λ can make k_{sed} much larger.

All Things Considered: The Reduction of Total Load and Maximum Concentration

As shown above, the concentration of a compound that is transported in a river is affected by various mixing and elimination processes. Their relative importance for reducing the riverborne mass flow and the maximum concentration depends on the characteristics of the river as well as on the compound under consideration. This section gives a summary of the relevant rate constants by emphasizing the simplest descriptions. In order to compare their relative importance, all processes will be approximated by first-order rate constants, either in time (k-rates, units s^{-1}) or in space along the river (ε-rates, units m^{-1}).

A distinction has to be made between processes that (a) affect both the total mass and the concentration in the river, and (b) affect only the concentration but leave the mass in the river unchanged. Processes which move the compound from the open river water to another environmental compartment (atmosphere, sediment) are also listed among (a). The following processes belong to the first category:

(1) *Chemical and biological reactions*, such as hydrolysis, biodegradation, photolysis and other chemical transformations. The relevant expressions are given by Eqs. O-6 and O-8.

(2) *Air-water exchange*, see Eqs. O-14, O-17, and O-19.

(3) *Removal to the sediments by particle settling*, see Eqs. O-60, O-61

(4) *Diffusion into the pore space of the sediments*, see Eqs. O-64, O-65.

The following processes belong to the second category (no reduction of total mass in the river water):

(5) *Longitudinal dispersion*

(6) *Dilution by merging rivers and by infiltrating groundwater*

Vertical and lateral mixing are not considered here. It is assumed that a given mass flux of a compound is mixed instantaneously into the corresponding volume flux of water. Of course, close to the input of waste water, vertical and lateral mixing is the most important mechanism reducing the maximum concentration in the river. Also note that processes (3), (4), and (5) are only effective in reducing the concentration of the compound for the case of an episodic input. If a compound is continuously added to the river, the sediments reach an equilibrium with the river water and thus do not act as a sink anymore. Similarly, dispersion does not reduce the maximum concentration of a compound that is permanently added to the river.

In order to compare the effects of dispersion and dilution with the other processes, they will both be discribed by pseudo-linear models. *In the following expressions the overbar used to describe the mean concentration along the river will be omitted. Instead, the subscript t will remind the reader that all equations are expressed in terms of the total concentration.*

The rate constant of *longitudinal dispersion* is defined as the relative change with time of the maximum concentration of a concentration cloud. For the *case of a δ-input*, it can be calculated from Eq. O-49 as

$$\left(\frac{\partial C_{t,max}}{\partial t}\right)_{dis} = - k_{dis} \cdot C_{t,max} \qquad \text{(O-70)}$$

with

$$k_{dis}(t) = \frac{1}{2 \cdot t} \qquad \text{(O-71)}$$

Since according to Eq. O-54, the concentration cloud of a chemical that is introduced into a river at a *constant rate during time Δt* assumes Gaussian shape for times $t > t_{ini}$ (see Eq. O-52 for definition), Eq. O-71 also holds for this case. For the *case of a Gaussian input* with initial variance σ_o^2, the maximum concentration can be derived from Eq. O-54 by setting $x = \bar{u} \cdot t$:

$$C_{t,max}(t) = C_t(x=\bar{u} \cdot t,\, t) = \frac{M/A}{(2\pi)^{1/2}\,(\sigma_o^2 + 2E_{dis} \cdot t)^{1/2}} \qquad \text{(O-72)}$$

Differentiation with respect to time t yields

$$k_{dis}(t) = \frac{E_{dis}}{\sigma_o^2 + 2E_{dis} \cdot t} \qquad \text{(O-73)}$$

The rate constant of *dilution* is calculated from the definition of concentration, C_t, as mass flux per unit time, m, divided by volume flux of water, Q:

$$C_t = \frac{m}{Q} \qquad \text{(O-74)}$$

The derivative of this equation with respect to time, where m is constant, yields

$$\left(\frac{\partial C_t}{\partial t}\right)_{dil} = -\frac{m}{Q^2} \cdot \frac{dQ}{dt} = -\frac{C_t}{Q} \cdot \frac{dQ}{dt} = -k_{dil} \cdot C_t \qquad \text{(O-75)}$$

with

$$k_{dil} = \frac{1}{Q} \cdot \frac{dQ}{dt} \qquad \text{(O-76)}$$

Problem

On November 1, 1986, a fire destroyed a storehouse at Schweizer-halle near Basel (Switzerland). During the fighting of the fire, several tons of various pesticides and other chemicals were flushed into the River Rhine (*Wanner et al., 1989*). One of the major constituents discharged into the river was *disulfoton* , an insecticide. An estimated quantity of 3.3 metric tons reached the river within a time period of about 12 hours leading to a massive killing of fish and other aquatic organisms

On their trip of about 8 days to Lobith, 700 km downstream of Schweizerhalle at the border between Germany and the Netherlands, the spilled chemicals underwent various transformations, were transferred to other environmental compartments (atmosphere, sediments), and were diluted by dispersion and merging rivers. Calculate how much these processes contribute to the reduction of disulfoton from Schweizerhalle to Lobith (a) in terms of the maximum concentration in the river, and (b) in terms of the total load. All relevant information is given in Table O.2.

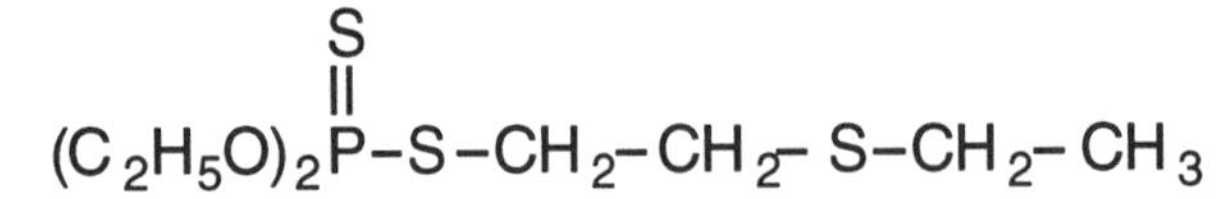

$$(C_2H_5O)_2P(=S)\text{–}S\text{–}CH_2\text{–}CH_2\text{–}S\text{–}CH_2\text{–}CH_3 \qquad \text{mw} \quad 274.4\ g \cdot mol^{-1}$$

disulfoton

Table O.2 Relevant Information on the Behavior of Disulfoton in the River Rhine

River Rhine	
Distance from Schweizerhalle to Lobith	$x_o = 700$ km
Mean flow velocity	$\bar{u} = 1\ m \cdot s^{-1}$
Mean depth	$h = 5$ m
Discharge of Rhine at Schweizerhalle	$Q_o = 750\ m^3 \cdot s^{-1}$
Discharge of Rhine at Lobith	$Q_1 = 2300\ m^3 \cdot s^{-1}$
Concentration of particulate organic material	$[POM] = 4 x 10^{-6}\ kg_{om} \cdot L^{-1}$
Settling velocity of particles	$v_s = 2\ m \cdot d^{-1}$
Coefficient of longitudinal dispersion	$E_{dis} = 2.8 x 10^3\ m^2 \cdot s^{-1}$
Air-water transfer velocity	
in air for H_2O	$v_a(H_2O) = 5 x 10^{-3}\ m \cdot s^{-1}$
in water for O_2	$v_w(O_2) = 2 x 10^{-5}\ m \cdot s^{-1}$
Water temperature (approximative)	$T_w = 10°C$
pH	7.9
Disulfoton	
Henry's Law constant at 10°C	$K_H = 1 x 10^{-3}\ atm \cdot L \cdot mol^{-1}$
Air-water transfer velocities in air (v_a)and in water (v_w) are reduced by a factor of 5 relative to H_2O and O_2, respectively	
Organic matter-water partition coefficient	$K_{om} = 750\ L \cdot kg_{om}^{-1}$
Molecular diffusion coefficient in water at 10°C	$D_w = 3.8 x 10^{-10}\ m^2 \cdot s^{-1}$
Rate constants [a] for	
abiotic hydrolysis	$k_{hyd} = 4.1 x 10^{-3}\ d^{-1}$
photochemical transformation (reaction with $^1O_2$)	$k_{photo} = 6.9 x 10^{-4}\ d^{-1}$
biological transformation (at 5 $\mu g \cdot L^{-1}$)	$k_{biol} = 0.14\ d^{-1}$
Accident	
Duration of spill (approx.)	$\Delta t = 12$ hours
Total input	$M = 3.3 x 10^3$ kg

[a] For details see *Wanner et al. (1989)*

Answer

Calculate first the fraction in particulate form, $(1\text{-}f_w)$, at sorption equilibrium by assuming that hydrophobic partitioning is the dominant sorption mechanism. By analogy to *Eq. 15-40* and using the [POM] and K_{om} values listed in Table O.2, you get

$$(1 - f_w) = \frac{[POM] \cdot K_{om}}{1 + [POM] \cdot K_{om}} \simeq 3 \times 10^{-3}$$

Hence, $f_w \sim 1$. Therefore, for all the processes affecting the dissolved species, the dissolved concentration can be set equal to the total concentration.

The initial phase of vertical and lateral mixing will not be considered here. Not enough details on the exact mode of input of the polluted water are known. Furthermore, a hydroelectric power station a few kilometers downstream of the spill caused enhanced mixing across the river cross-section, which cannot be described with the theory as presented in this chapter. In fact, the total mass of disulfoton which entered the river (see Table O.2) was indirectly calculated from measurements on samples taken at Village Neuf. This sampling station is located 14 km downstream of the spill, i.e., below the power station. At this station, a disulfoton concentration of about 100 $\mu g \cdot L^{-1}$ was measured.

The following evaluation is based on the list of transformation and mixing processes as listed above:

(1) *Chemical and biological transformations.* By far the largest contribution comes from biological transformation; hydrolysis and photochemical reactions combined contribute less than 4% to the total reaction rate constant:

$$k_r = k_{hyd} + k_{photo} + k_{bio} \approx k_{bio} = 0.14 \text{ d}^{-1}$$

Note that the biological transformation is assumed to be first-order. If river sections with higher concentrations were considered, a Michaelis-Menten approach would have to be taken (Chapter 14 and Wanner et al., 1989).

(2) *Air-water exchange.* Based on the exchange velocities adjusted for disulfoton, $v_a = 1 \times 10^{-3}$ $m \cdot s^{-1}$ and $v_w = 4 \times 10^{-6}$ $m \cdot s^{-1}$, and the

nondimensional Henry's Law constant at 10°C, $K_H' = 4.3 \times 10^{-5}$, one concludes that the exchange is air-side controlled, and thus $v_{tot} = v_a \cdot K_H' = 4.3 \times 10^{-8}$. Hence,

$$k_{gas} = k_g \cdot f_w = \frac{v_{tot}}{h} \cdot f_w = 8.6 \times 10^{-9}\ s^{-1} = 7.4 \times 10^{-4}\ d^{-1}$$

(3) *Particle sedimentation*. The removal rate of particles, $k_s = 2\ m \cdot d^{-1} / 5\ m = 0.4\ d^{-1}$, is fairly large, but only a minor fraction of disulfoton is sorbed to the particles:

$$k_{sed} = k_s \cdot (1 - f_w) = 1.2 \times 10^{-3}\ d^{-1}$$

(4) *Diffusion into the sediments*. There is not enough information on the sediments of the Rhine that would allow an accurate calculation of $f_{w,sed}$. To get an order of magnitude estimate, the data of the River G are used, that is $\phi = 0.8$, $\rho_s = 2.5\ kg_s \cdot L^{-1}$; thus $r_{sw,sed} = 0.63\ kg_s \cdot L^{-1}$. Taking $f_{om} = 0.01\ kg_{om} \cdot kg_s^{-1}$ and the K_{om} of disulfoton listed in Table O.2 yields $f_{w,sed} = 0.17$. Initially, the duration of the spill is 0.5 days. In order to calculate its duration at Lobith, t_{ini}, the time until longitudinal dispersion becomes effective, has to be calculated. Application of Eq. O-52, using the coefficients listed in Table O.2, yields $t_{ini} = 23$ hours. Thus, for $t > t_{ini}$ the size of the cloud can be calculated from Eq. O-56 with $\sigma_o = 0$. For $t = 8$ days, the duration of the cloud has increased to 0.72 days only; thus the average duration is roughly $\Delta t = 0.6$ days. Inserting all this information into Eq. O-64 yields (note $D_w = 3.8 \times 10^{-10}\ m^2 \cdot s^{-1} = 3.3 \times 10^{-5}\ m^2 \cdot d^{-1}$)

$$k_{diff} = \frac{0.8}{5\ m} \left(\frac{2 \times 3.3 \times 10^{-5}\ m^2 \cdot d^{-1}}{0.17 \times 0.6\ d} \right)^{1/2} = 4.1 \times 10^{-3}\ d^{-1}$$

Although not considered in the original analysis by *Wanner et al. (1989)*, it is instructive to compare the size of k_{diff} to the possible effect of the flow of water through the sediments as described by Eq. O-68. Due to the lack of adequate data, the following parameters are to be considered as rough estimates: $K_q = 10^{-3}\ m \cdot s^{-1}$, $\lambda \approx 3 \cdot h = 15$ m, $H \approx h/10 = 0.5$ m. Eq. O-68 yields $v_{sed} = 1.2 \times 10^{-6}\ m \cdot s^{-1}$. Thus, from Eq. O-69 with $f_{wsed} = 0.17$,

$$k_{sed} = 0.2 \times 10^{-6}\,s^{-1} = 1.7 \times 10^{-2}\,d^{-1}$$

which is four times larger than k_{diff}. Due to the speculative character of this number, it will not be included in the summary table below; but this result suggests that further characterization of the Rhine sediments might be warranted.

The following processes only reduce the maximum concentration, not the mass flux of disulfoton.

(5) *Longitudinal dispersion.* According to Eq. O-71 the rate, k_{dis}, at which the maximum concentration is reduced, decreases with flow time t. As a first approach, Eq. O-66 is evaluated for t = 4 days, i.e., for half of the total travel time from Schweizerhalle to Lobith. Thus,

$$\bar{k}_{dis} \sim k_{dis}(t = 4d) = \frac{1}{2x4\ d} = 0.125\ d^{-1}$$

Yet, since $k_{dis}(t)$ is not a linear function of the flow time t, a better way to evaluate $\bar{k}_{dis}$ is to calculate the ratio of the maximum concentrations for t = 8 d and t = 0 d, respectively. The first value, calculated from Eq. O-54 with $x=\bar{u}\cdot t$, is 27.5 μg/L, the second value is simply

$$C_{max}\ (t{=}0) = \frac{M}{Q\cdot\Delta t} = \frac{3.3x10^{6}\ g}{750\ m^{3}\cdot s^{-1}\ x\ 4.32x10^{4}\ s} = 0.1 g\cdot m^{-3} = 100\ \mu g/L$$

Thus,

$$\frac{C_{max}\ (t{=}8d)}{C_{max}\ (t{=}0d)} = 0.277$$

According to the solution for Eq. O-70, this ratio has to be equal to $e^{-k_{dis} x 8days}$, thus,

$$\bar{k}_{dis} = -\frac{1}{8d}\ln(0.275) = 0.16\ d^{-1}$$

This value will be used for the following comparisons with all the other processes.

(6) *Dilution by merging rivers.* The rate constant is given by Eq. O-76. Obviously, the discharge Q(t) increases stepwise along the Rhine,

not continuously. However, in order to calculate a mean dilution rate, k_{dil}, Eq. O-76 can be solved for constant k_{dil}:

$$Q(t) = Q(0)e^{k_{dil} \cdot t}$$

From Table O.2 one has Q(t)/Q(0) = 2300/750 = 3.1. Thus,

$$k = \frac{1}{t} \ln [Q(t)/Q(0)] = \frac{1}{8d} \ln(3.1) = 0.14 \text{ d}^{-1}$$

The results are summarized in the following table:

	Rate k (d^{-1})	% of load red.	% of reduction of max. conc.
Transformation, transfer			
Hydrolysis (k_{hyd})	4.1×10^{-3}	2.7	0.91
Photochemical transf. (k_{photo})	6.9×10^{-4}	0.5	0.15
Biological transf. (k_{bio})	1.4×10^{-1}	92.8	31.1
Air-water exchange (k_{gas})	7.4×10^{-4}	0.5	0.16
Particle sedimentation (k_{sed})	1.2×10^{-3}	0.8	0.27
Diffusion into sed. (k_{diff})	4.1×10^{-3}	2.7	0.91
Total removal (k_{rem})	0.151	100	33.5
Transport			
Dispersion (k_{dis})	0.16	-	35.3
Dilution (k_{dil})	0.14	-	31.1
Total of all processes (k_{tot})	0.45	-	100
Remaining total load at Lobith[a]	29.9%		
Remaining max. conc. at Lobith[a]	2.1%		

[a] The relative remaining mass flux is calculated from $e^{-k_{rem} \cdot t}$ (t=8days), the relative remaining maximum concentration from $e^{-k_{tot} \cdot t}$.

Roughly two thirds of the disulfoton has disappeared from the water between Schweizerhalle and Lobith. About 90% of this loss was due to biological transformation. The drop of the maximum concentration was mainly due to dispersion (~34%), dilution (~31%), and biodegradation (31%).

Measurements taken at various locations between Schweizerhalle and Lobith are in good agreement with these calculations (see *Wanner et al., 1989,* for more details).

CASE STUDIES

● *Case Study O-1 Chloroform in the Mississippi River*

On August 19, 1973, a barge carrying three tanks of trichloromethane (chloroform) was damaged on the Mississippi River at Baton Rouge, Louisiana (*Neely et al., 1976*). Two tanks containing a total of 7.8×10^5 kg of chloroform were damaged and their contents spilled to the river.

$CHCl_3$ (structure: C bonded to three Cl and one H)

trichloromethane (chloroform)

mw	$119.4\ g \cdot mol^{-1}$
T_m	–63.5°C
T_b	61.7°C
C_w^{sat} (25°C)	$6.5 \times 10^{-2}\ mol \cdot L^{-1}$
K_H (25°C)	$4.0\ L \cdot atm \cdot mol^{-1}$
K_{ow} (25°C)	85
ρ_{liquid} (20°C)	$1.489\ g \cdot cm^{-3}$

Some data on the hydrology of the Mississippi River at the time of the accident:

Q	$= 7.5 \times 10^3\ m \cdot s^{-1}$	discharge
$\bar{u}$	$= 0.57\ m \cdot s^{-1}$	mean flow velocity
w	= 1'200 m	width
h	= 11 m	mean depth
h_o	= 20 m	maximum depth (estimated)
S_o	$= 2 \times 10^{-5}$	slope of river bed

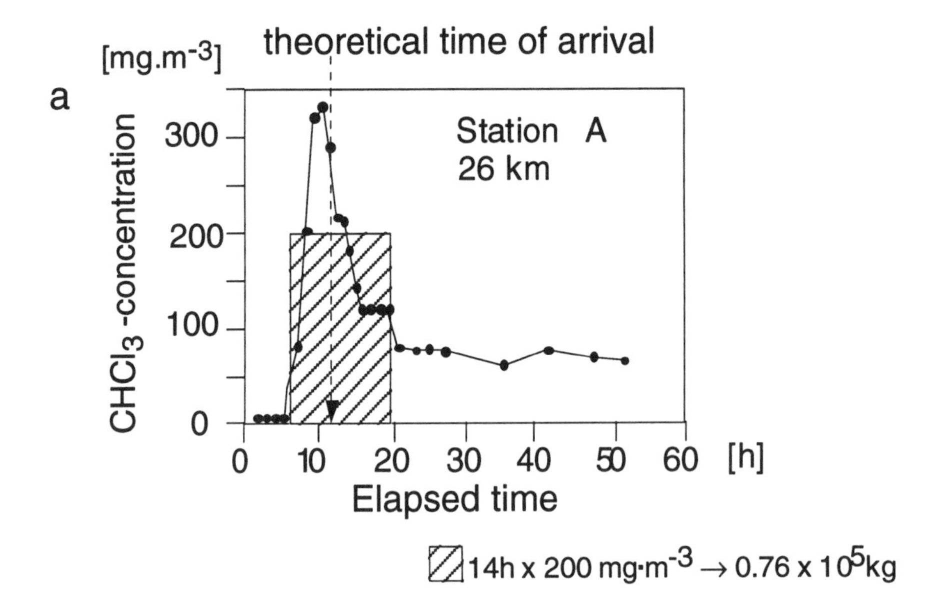

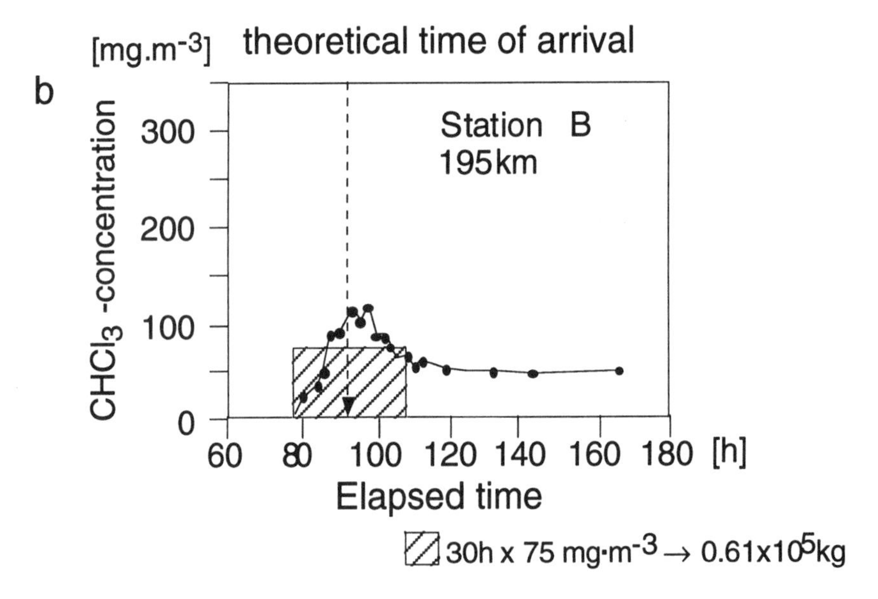

Figure O-7 Concentration of chloroform as a function of time measured at (a) Station A (26 km downstream of spill), and (b) Station B (195 km downstream of spill at New Orleans). The rectangular areas serve to estimate the total mass flux originating from the rupture of the first tank. Data from *Neely et al. (1976).*

Further information about the accident:

(1) The first tank (3.9×10^5 kg) ruptured at 2:40 p.m. (*Note:* This time is chosen as reference time, $t = 0$). All the chloroform was lost within a very short time.
(2) The second tank (3.9×10^5 kg) began to leak at 10:00 p.m. ($t = 7.3$ h); its contents were lost over a 45-minute period.
(3) Water samples were taken at 26 km (station A) and 195 km (station B, New Orleans) from the point of the accident (Fig. O-7)
(4) It was found that at the time when chloroform first was detected at station A, the chloroform was evenly distributed across the river.

Problem (a)
Which processes determine the fate of the chloroform on its journey along the Mississippi River? What kind of information (about the chloroform or the river) is needed in order to quantify these processes? Assess separatly the change (1) of the total mass and (2) of the maximum concentration of chloroform along the river.

Hints and Help
Air-water exchange is an obvious candidate that deserves closer inspection. It may be helpful to estimate the molecular diffusivity of chloroform in water (D_w) and then use *Fig. 10-8*. As an alternative take a mean wind speed of 1 $m \cdot s^{-1}$ to calculate the total transfer velocity. *Neely et al. (1976)* give $v_{tot} = 2.2 \times 10^{-3}$ $cm \cdot s^{-1}$ as a realistic mass transfer velocity of chloroform during the accident. In case that this value should greatly differ from your own estimate, continue your calculation with the value from *Neely et al.* .

Problem (b)
Consider the rupture of the first tank as an individual event during which all of the chloroform was spilled into the river during a very short time period (say within less than 5 minutes). Calculate the time of arrival of the chloroform peak as well as the maximum concentration at stations A and B. Assume that all the chloroform is immediately dissolved in the water (is this at all possible?). Compare your result with Fig. O-7 and draw your conclusions.

Hints and Help

Discuss time and distance for vertical and lateral mixing. Note that a friction factor α^* and the shear velocity u* can be calculated from the available river data.

The distance for lateral mixing may require some special thoughts. From the reported observations one knows that it must be smaller than 26 km (remember that complete lateral mixing was found at station A). Even if you assume that the accident occurred in the middle of the river and thus the distance of mixing to the river banks is only w/2 (600 m), the lateral turbulent diffusivity, E_y, that is formally calculated from the available data may still be too small. In this case, for all further calculations use a value for E_y that just guarantees complete lateral mixing over the lateral distance w/2 between the spill and station A (26 km downstream).

By comparing the calculated and observed time of arrival of the chloroform peak at station A, you will also find that during the first stretch (from the point of the accident to station A) the mean flow velocity was greater than 0.57 $m \cdot s^{-1}$. Apparently, the chloroform was first flowing in the center of the river where the real flow velocity is larger. Use the observed $\bar{u}$-value for the mixing and dispersion calculations between the point of input and station A, but the values listed above beyond station A.

Problem (c)

From the calculations performed under (b), it becomes evident that the peak concentration of chloroform measured at both stations is significantly smaller than the predicted values. A minor part of this discrepancy can be explained by loss to the atmosphere (calculate the corresponding relative concentration reduction at both stations). Furthermore, the tailing of the concentration lasts much longer than predicted by a simple dispersion model. *Neely et al.(1976)* explain this by assuming that there was a pool of pure chloroform laying on the bottom of the river at the place of the accident from which chloroform slowly dissolved into the river water and drifted downstream.

The essence of the Neely model are the following elements:

(1) 20% of the first tank (i.e., 20% of 3.9 x 10^5 kg) is immediately released to the river water. The cloud drifts downstream and is broadened by dispersion. In addition, air-water exchange leads to some reduction of the chloroform concentration.

(2) The remaining portion of the first tank (i.e., 80% of 3.9×10^5 kg) rests on the bottom of the river from which it is introduced by a first order reaction into the flowing water (rate constant $k_b = 3 \times 10^{-3}\ h^{-1}$). Dispersion of this "cloud" can be disregarded, but not air-water exchange.

(3) All of the second tank (3.9×10^5 kg) is added to the bottom reservoir from which it leaks out at the same rate starting at time $t_2 = 7.6$ h . Treat this cloud as described under (2).

Calculate the concentrations of chloroform as a function of time for stations A and B and compare the results with Fig. O-7.

Hints and Help
Since all the relevant reactions are linear, you can calculate each contribution separately and then simply add them up. Air-water exchange can be accounted for by applying the appropriate reduction factor (which is of course different for station A and B) to the results.

● *Case Study O-2 Gas Exchange Experiments with Halogenated Compounds Exhibiting Very Different Henry's Law Constants*

In order to study the air-water exchange at river cascades, *Cirpka et al. (1993)* performed an experiment in the River Glatt near Zürich (Switzerland). On January 28, 1992, five halogenated compounds (Table O.3) with very different Henry's law constants were injected into the river at a constant rate during 2.5 hours about 1 km upstream of the first of four cascades to guarantee vertical and transverse mixing. The concentration of each compound was measured just upstream and downstream of each cascade.

For the following considerations one of the river sections in between the cascades, Section # 2, is chosen. The relevant data for this section are summarized in Table O.4.

Table O.3 Characteristic Data of Chemical Tracers Used for the Gas Exchange Experiment: Nondimensional Henry's Law Constant (K_H'), Molecular Diffusion Coefficient in Air (D_a) and Water (D_w), Octanol-Water Partition Coefficient (K_{ow}), All Valid for 4°C. Data from *Cirpka et al. (1993)*

Compound	K_H' (-)	D_w ($cm^2 \cdot s^{-1}$)	D_a ($cm^2 \cdot s^{-1}$)	K_{ow} (-)
Sulfur hexafluoride (SF_6)	82.4	$6.23 \cdot 10^{-6}$	0.187	< 50
1,1,2-Trichlorotrifluoroethane (R113)	8.58	$4.41 \cdot 10^{-6}$	0.127	100
Trichloroethene (C_2HCl_3)	0.137	$5.26 \cdot 10^{-6}$	0.160	240
Trichloromethane ($CHCl_3$)	0.0564	$5.51 \cdot 10^{-6}$	0.178	93.3
Tribromomethane ($CHBr_3$)	0.00717	$5.04 \cdot 10^{-6}$	0.165	251

Table O.4 Geometric, Hydraulic, and Chemical Data for Section #2 of the River Glatt (from *Cirpka et al., 1993*)

River discharge	Q	3.3 $m^3 \cdot s^{-1}$
Water depth	$h_o \sim h$	0.48 m
Mean flow velocity	$\bar{u}$	0.45 $m \cdot s^{-1}$
Distance from injection site		
Beginning of Section # 2	x_o	1440 m
End of Section # 2	x_1	2661 m
Difference in altitude	Δz_o	8.56 m
Width	w	16.2 m
Water temperature	T_w	0°C
Air temperature	T_a	4°C

The following concentrations were measured at the upstream and downstream boundaries of Section # 2:

	C_{up} (upstream)	C_{down} (downstream)
SF_6 ($ng \cdot L^{-1}$)	13.8 ± 0.6	7.25 ± 0.40
R113 ($\mu g \cdot L^{-1}$)	1.03 ± 0.016	0.57 ± 0.015
C_2HCl_3 ($\mu g \cdot L^{-1}$)	2.36 ± 0.010	1.68 ± 0.025
$CHCl_3$ ($\mu g \cdot L^{-1}$)	2.92 ± 0.033	2.26 ± 0.060
$CHBr_3$ ($\mu g \cdot L^{-1}$)	4.18 ± 0.046	3.59 ± 0.073

The concentrations in air, C_a, of all these compounds are very small and can be disregarded in the following considerations.

Problem (a)

Check whether the assumption of vertical and lateral homogeneity from the point of tracer injection to the first cascade (1 km downstream) is justified. At the start of the injection, a short pulse of uranin (sodium-fluorescein, a fluorescent dye) was added to the river in order to measure the travel time of the water. At each station, the samples were taken 1.5 hours after the uranin peak had passed by. Based on this information justify why longitudinal dispersion can be disregarded in the evaluation of the experiment.

Hints and Help

The friction factor, f, calculated from Eqs. O-4 or O-4a may turn out to be extremely large, which would indicate a very rough river bed. In fact, in the River Glatt in between the cascades there are small drops every 50 m. Use this fact to explain the rather extreme value of f.

Problem (b)

Calculate the "excess ratio" $R = C_{up}/C_{down}$ for all five compounds (1) from the concentration measured above and below the river section (R is then called R_{meas}) and (2) from the linear air-water exchange model, Eq. O-21 (R is then called R_g). Compare R_{meas} with R_g.

Hints and Help

For the evaluation of Eq. O-21 you need to know the total transfer velocity v_{tot} for all compounds. Use *Eq. 10-31* or *Fig. 10.8* for the water-side transfer velocity v_w and *Eq. 10-28* with a wind speed of $u_{10} \sim 1\ m \cdot s^{-1}$ for the air-side velocity v_a, and combine these parameters with the compound-specific information given in Table O.3.

Problem (c)

As it will turn out, all the R_{meas}-values are systematically larger than the corresponding R_g-values. This means that the mechanism of air-water exchange is significantly underestimated by the theory represented by Eq. O-21. Consider two effects that could explain the observed discrepancy.

(1) The drops in the river bed as well as surface waves significantly increase the effective area of gas exchange per unit river length relative to the geometric value, $w \cdot x$.

(2) There exists an additional mechanism of air-water exchange which is based on the injection of air bubbles into the water and subsequent gas exchange between bubbles and water. As shown by *Cirpka et al. (1993)*, this effect can be quantified by the excess ratio

$$R_b = 1 + A\, K_H' \,[1 - \exp(-B/K_H')] \qquad \text{(O-77)}$$

where A and B are two parameters to be determined by observation. R_b increases monotonically from $R_b = 1$ (if $K_H' = 0$) to $R_b = 1 + A \cdot B$ (if $K_H' \rightarrow \infty$).

Assume that for $CHBr_3$, which has the smallest K_H', the discrepancy between R_{meas} and R_g is completely due to the first effect. Calculate the modified exchange area $(w \cdot x)_{eff}$ and the modified $R_{g,eff}$ for all other compounds. Calculate the remaining difference between R_{meas} and $R_{g,eff}$ that may still exist for the compounds (except for $CHBr_3$), and express it as

$$R_b = \frac{R_{meas}}{R_{g,eff}}$$

Convince yourself that R_b increases with K_H' and estimate the fitting parameters A and B of Eq. O-77.

Hint and Help

As shown by *Cirpka et al. (1993)* in greater detail, the process of bubble injection is important for rivers with steps and cascades. Note that in this process, the influence of the Henry's Law constant on the overall air-water exchange flux is extended to much greater K_H'-values than it is the case for the surface exchange two-film model. As demonstrated by *Fig. 10.9*, for the latter v_{tot} becomes independent of K_H if $K_H > \sim 1$ L·atm·mol^{-1} (corresponding to $K_H' > 0.04$ for T = 4°C). In contrast, evaluation of the River Glatt case study along the guidelines described above shows that for the process of bubble-induced gas exchange the influence of K_H' is extended to much larger values. However, for small K_H'-values the bubble process is not effective. Develop a qualitative argument which supports this findings.

For further information on air-water exchange at river cascades read *Cirpka et al. (1993)*.

● *Case Study O-3 Using a Volatile Organic Compound as Tracer for Quantifying Oxygen Demand in a Polluted River*

Problem (a)

A colleague who has to survey the water quality in River A tells you that he needs to quantify the oxygen consumption in the river downstream of a sewage discharge. He shows you the following measurements of dissolved oxygen as well as concentration data for tetrachloroethene (PER) that is introduced into the river with a constant input by the sewage plant.

Distance from the sewage outlet (km)	O_2-concentration (mg·L^{-1})	PER-concentration ng·L^{-1})
2.0	8.9	810
7.0	9.1	690
14.0	9.0	570
19.0	9.1	490

He recognizes that along the whole river section considered, O_2 seems to be roughly constant, but about 2.2 to 2.4 mg·L^{-1} below the O_2-saturation (equilibrium) concentration of 11.3 mg·L^{-1} (the temperature of the river water is 10 °C). He also tells you that the PER concentration in air is virtually zero. However, your colleague is unable to quantify the oxygen consumption, J ($mgO_2 \cdot L^{-1} \cdot d^{-1}$), in the well-mixed river. Realizing that in the river section considered, there are no cascades present that could complicate calculations of gas exchange rates (see Case Study O-2), and using the information given below for river A, you sit down, and, an hour later, you give your friend the answer. How large is J?

River A: (rectangular cross-section)

River discharge	Q	$20\ m^3 \cdot s^{-1}$
Width	w	15 m
Slope	S_o	0.002
Friction factor	f	0.1

Assumption: No other major inlets along the river section considered.

Problem (b)

Your colleague also tells you about a flood prevention project in which it is planned to double the width of river A to 30 m, and to lower the friction factor to f = 0.025. He wants to know how the concentration of O_2 and PER would be affected in the river when assuming that everything else (including the PER input and the oxygen consumption rate) would remain the same. Try to answer this question. Semi-quantitative arguments are sufficient!

P. ORGANIC COMPOUNDS IN POROUS MEDIA

TRANSPORT, MIXING, AND REACTIONS

Introduction

Among the three major aqueous environments that are discussed in this book (lakes, rivers, porous media), the porous media represent the biggest challenge to those who are looking for simple but powerful principles for assessing the behavior of organic contaminants. To fall back on the simplifications that were successfully employed for lakes and rivers seems to be rather naive, at least at first sight. Regarding the complex physical structure of, e.g., most soils and aquifers, it is difficult to imagine that concepts like the completely mixed box model, the one-phase (pure liquid) model, or the one-dimensional diffusion/advection approach could lead to meaningful results. Every porous system is an individual case for which general principles seem to have little value.

And yet, without the courage for the "terrible simplification" it is difficult to extract, from the information gained for one particular system, the general principles that can be used for other situations. The aim of this chapter is to demonstrate that simplifications are necessary not only because of the limits of our understanding, but also because they help to shift these limits, and to gain new insights into the basic mechanisms that determine the fate of organic compounds in porous media.

Deliberately, this chapter remains simple – too simple, as some of the readers who are familiar with the subject may think. However, the advantage of the simple approach will be to demonstrate, step by step, how the various processes contribute to the "final" story and also to reveal the similarities to other environmental systems treated in this book. The chapter does not try to compete with the vast specialized literature on the physics and chemistry of porous media.

Simplicity requires choice; to choose wisely, the options should be known among which the choice is possible. The following characteristics make the porous media special (see also Fig. P-1):

(1) There are always *at least two phases* involved, that is, *water* and *solids*, that are of comparable importance for the fate of organic compounds in porous media. A third phase (*air*) becomes important in the so-called unsaturated zone ; recent studies show that a fourth phase, that is *colloids*, are important for the transport of chemicals in porous media. Furthermore, if extreme situations are analyzed

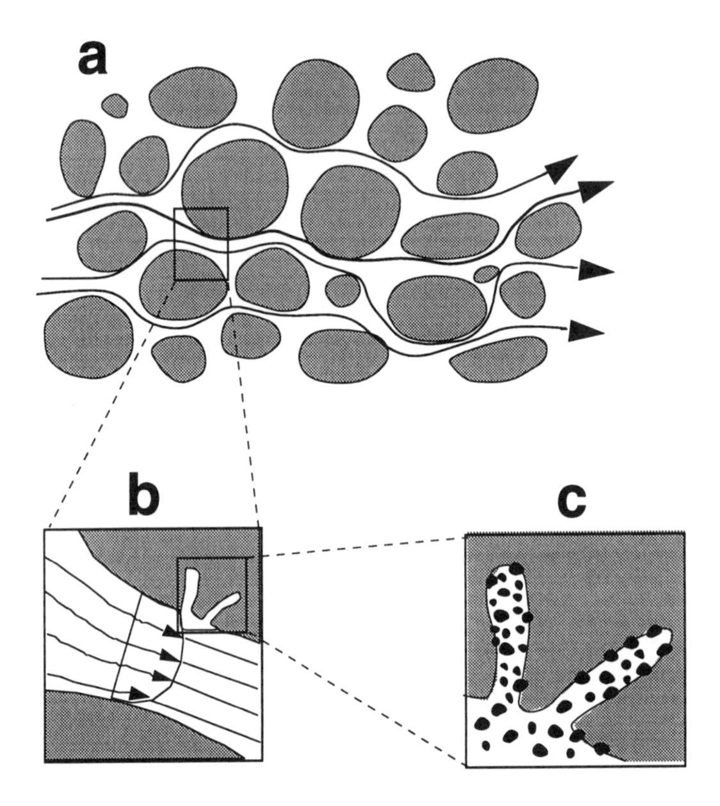

Figure P.1 Heterogeneity is one of the main properties of porous media; it characterizes not only the scales shown in the figure, but occurs also on larger scales up to the size of the whole porous system. Three important mechanisms of transport and mixing in porous media are (a) Interpore dispersion caused by mixing of pore channels; (b) intrapore dispersion caused by nonuniform velocity distribution and mixing in individual channels; (c) dispersion and retardation of solute transport caused by molecular diffusion between open and dead-end pores as well as between the water and the surface of the solids.

such as chemical dump-sites, also phases like "nonaqueous phase liquids" (NAPLs) have to be considered.

(2) Compared to rivers and lakes, *transport* in porous media is generally *slow, three-dimensional* and*spatially variable* due to heterogeneities in the medium. The velocity of transport differs by orders of magnitude between the phases of air, water, colloids, and solids. Due to the small size of the pores, transport is seldom turbulent. Molecular diffusion and dispersion along the flow are the main producers of "randomness" in the mass flux of chemical compounds.

(3) Heterogeneities exist on all spatial scales, from the micropores, the larger pores and particles to the macrostructure of the aquifer (Fig. P.1). Therefore, the description of transport in porous media strongly depends on the scale of interest. For instance, in natural aquifers, dispersion is much larger than in artificial porous media that are often used to study the behavior of chemicals. The reason is that in natural systems there are large scale heterogeneities that cannot be included in laboratory systems since they are usually too small. Therefore, quantification of dispersion as deduced from measurements in laboratory systems cannot be extrapolated to field situations.

(4) Field information on a specific porous medium remains scarce compared to its complexity. In fact, it is much simpler to take water samples in a river or in a lake than drilling holes into a ground-water system that are spaced closely enough in order to depict the spatial variation of the system. Ironically, in many cases it is easier to invent complex models than to collect the field data that would be needed to validate the model for a given system. Therefore, mathematical models of aquifers are important additional tools to explore the potential behavior of aquifers under different conditions.

To summarize, porous media share certain properties with lakes, others with rivers. They are three-dimensional like lakes, but nonturbulent. They have much larger solid-to-solution phase ratios, r_{sw}, than lakes; typical values lie between 1 and 10 $kg \cdot L^{-1}$, compared to about 1 x 10^{-6} $kg \cdot L^{-1}$ for lakes. Porous media usually have a predominant direction of flow like rivers, but the flow is slower and often not confined by well-defined vertical and lateral boundaries.

The structure of this chapter is similar to the one of Chapter O on rivers.

Most of the sections will be restricted to the two major phases of the *saturated zone, water and solids*. All the equations will be written for *one dimension* only, that is, for the x-axis, which is chosen in the direction of the mean flow. This explains and (hopefully) justifies the choice of a rather special groundwater system as the main playground to exemplify the following processes:

(a) Transport by the mean flow
(b) Longitudinal dispersion and diffusion
(c) Sorption / desorption between water and solids
(d) Chemical and biological transformation
(e) Transport on colloids

The relevant parameters are summarized in Table P.1.

Transport by the Mean Flow

In order to describe the flow of water through a porous medium, some basic geometric characteristics have to be discussed. In the saturated zone a given bulk volume V_{tot} (e.g., 1 m^3) consists of the volume V_w filled with water and the volume $V_s = V_{tot} - V_w$ occupied by solids. The porosity ϕ is defined by (see *Section 11.2*)

$$\phi = \frac{V_w}{V_{tot}} = \frac{V_w}{V_w + V_s} \qquad (11\text{-}6)$$

Most of the pore volume is connected so that water can flow through. However, there are pores that are completely cut off from the main pore volume or that are dead ends that are barely participating in the flow of the water (Fig. P-1). In the following, the parameter ϕ refers always to the so-called *effective porosity*, i.e., to the pore volume that is available for the bulk flow.

The flow rate of water through the porous medium per unit total (bulk) area perpendicular to the direction of flow, the so-called *specific discharge q*, is related to the *effective* mean flow velocity in the pores along the x-axis, $\bar{u}$, by

$$q = \phi \cdot \bar{u} \quad (m^3 \cdot s^{-1} \cdot m^{-2} = m \cdot s^{-1}) \qquad (P\text{-}1)$$

Table P.1 Definition of Characteristic Parameters for Porous Media

			Typical value in GWS
ϕ	(Effective) porosity ($\phi < 1$)	-	0.31
ρ_s	Density of solid material	$kg \cdot m^{-3}$	2'500
χ	Tortuosity ($\chi > 1$)	-	1.5
K_q	Hydraulic conductivity	$m \cdot s^{-1}$	$1 x 10^{-3}$
κ	Permeability	m^2	$1 x 10^{-10}$
f_{om}	Weight fraction of organic matter in solid phase of aquifer	-	0.0015
h_{wt}	Height of water table (relative to some fixed depth, e.g., above sea level)	m	
x	Distance along the main flow	m	
S_{wt}	$= -\frac{dh_{wt}}{dx}$ Slope of the water table	-	
q	Flow of water per unit bulk area (specific discharge)	$m \cdot s^{-1}$	
$\bar{u}$	$= q/\phi$ effective mean flow velocity in pores along the x-axis	$m \cdot s^{-1}$	
ν	Kinematic viscosity of water $\nu\,(5°C) = 1.52 x 10^{-6}$, $\nu\,(15°C) = 1.14 x 10^{-6}$	$m^2 \cdot s^{-1}$	
Mixing parameters			
E_{dis}	Coefficient of longitudinal dispersion	$m^2 \cdot s^{-1}$	$(4 - 40) x 10^{-5}$
α_L	Dispersivity [a]	m	3

[a] α_L depends strongly on the length scale L over which the transport of chemicals is considered

Note that q has the units of a velocity and is always smaller than the effective velocity in the pores, $\bar{u}$, since $\phi < 1$. Both specific discharge q and effective mean velocity, $\bar{u}$, are vectors, that is, quantities defined by the three scalar components along the coordinate axes x, y, z. Vector variables lead quickly to rather complicated partial differential equations that can usually not be solved analytically. For simplicity, the following discussion is restricted to one-dimensional flow. It is assumed that the x-axis of the (local) coordinate system is always lying in the direction of the mean flow. Since the mean flow does usually not follow a straight line, the x-axis is not necessarily a straight line either.

The concentration C_w is the mass of a dissolved chemical per unit (pore) water volume. Its temporal change at a fixed location x due to the effect of the mean flow along the x-axis (advection) can be calculated for a small rectangular test volume located between x and (x + Δx) and with area a_x perpendicular to the x-axis (see *Fig. 9.4*):

$$\phi \cdot a_x \cdot \Delta x \left(\frac{\partial C_w}{\partial t}\right)_{ad} = a_x \cdot q(x) \cdot C(x) - a_x \cdot q(x+\Delta x) \cdot C(x+\Delta x) \qquad \text{(P-2)}$$

Division by $\phi \cdot a_x \cdot \Delta x$, taking the limit $\Delta x \rightarrow 0$, and using Eq. P-1 yields

$$\left(\frac{\partial C_w}{\partial t}\right)_{ad} = \frac{1}{\phi}\frac{\partial}{\partial x}\left(-q \cdot C_w\right) = \frac{1}{\phi}\frac{\partial}{\partial x}\left(-\phi \cdot \bar{u} \cdot C_w\right) \qquad \text{(P-3)}$$

Note: The continuity equation for three-dimensional flow of an incompressible liquid requires $\frac{\partial q_x}{\partial x} + \frac{\partial q_y}{\partial y} + \frac{\partial q_z}{\partial z} = 0$, thus in one dimension q = const. can be assumed. Then Eq. P-3 reduces to

$$\left(\frac{\partial C_w}{\partial t}\right)_{ad} = -\frac{q}{\phi} \cdot \frac{\partial C_w}{\partial x} = -\bar{u}\frac{\partial C_w}{\partial x} \qquad \text{(P-4)}$$

This expression, the analogy of Eq. O-2 for rivers, describes the fastest and most important mode of transport in groundwater. Thus, a central task of the hydrologist is to develop a theory that allows him to predict the effective velocity $\bar{u}$ (or the specific flow rate q). Like the Darcy-Weisbach equation for rivers (Eq. O-4), there is an important equation for groundwater flow, the well-known *Darcy's Law*. In its original version formulated by Darcy in 1856, the equation describes the one-dimensional flow through a vertical filter column. The characteristic properties of the column (i.e., of the aquifer) are described by the so-called *hydraulic conductivity,* K_q (units: $m \cdot s^{-1}$). Based on Darcy's Law, *Dupuit* derived an approximate equation for quasi-horizontal flow:

$$q = -K_q \frac{dh_{wt}}{dx} \quad (m \cdot s^{-1}) \qquad \text{(P-5)}$$

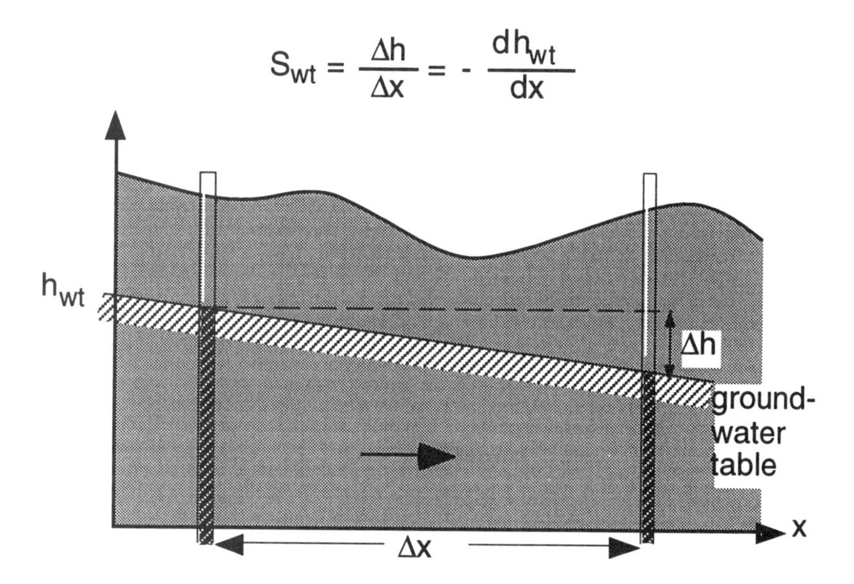

Figure P.2 The hydraulic gradient, defined as the slope of the (unconfined) groundwater table, $S_{wt} = \Delta h/\Delta x$, is a measure for the horizontal pressure gradient that drives the flow through a porous medium from high to the low pressure.

where dh_{wt}/dx is the slope of the height of the (unconfined) groundwater table along the x-axis, i.e., the so-called *hydraulic gradient* (Fig. P.2). For the groundwater hydrologist, this slope is the common way to quantify horizontal pressure gradients, since it can easily be determined by comparing the water levels in adjacent wells located along the direction of the mean flow.

Commonly, the hydraulic conductivity is designed by K; K_q is used here in order to distinguish it from the various distribution ratios (K_d, K_{om}, etc.). K_q depends not only on the structure of the porous medium (i.e., porosity, tortuosity, and grain size distribution), but also on the property of the liquid that is flowing through the pores (in fact, on its kinematic viscosity ν), and finally on gravity (described by the acceleration of gravity, g).

The *permeability* κ, related to K_q by

$$\kappa = \frac{\nu}{g} K_q \quad (m^2), \qquad \text{(P-6)}$$

is independent of gravity and of the viscosity of the liquid and thus yields a better characterization of the physical structure of the aquifer than K_q. Water at T = 20°C has a kinematic viscosity $\nu = 10^{-6}\ m^2 \cdot s^{-1}$. With $g \sim 10\ m \cdot s^{-2}$, κ and K are numerically related by $\kappa\ (m^2) \sim 10^{-7} (m \cdot s) \cdot K_q\ (m \cdot s^{-1})$.

As mentioned above, κ depends only on the geometry and structure of the porous medium. Several investigations demonstrate that for the case of perfect spheres of uniform size, κ increases as the square of the particle radius, r :

$$\kappa = a \cdot r^2 \qquad \text{(P-7)}$$

where a is an empirical constant. Figure P.3 illustrates how the r^2-dependence of κ can be understood based on the law of *Hagen-Poiseuille* stating that the (laminar) flow through a tube with radius r is proportional to r^4, provided that all other parameters (pressure, viscosity of liquid, length of tube) are kept constant.

In fact, Eq. P-7 can also be used for aquifers that consist of particles of different size, if r is defined as the mean particle radius. Then the parameter a depends on the particle size distribution. It increases with porosity ϕ that, in turn, is linked to the so-called sorting coefficient, So (see caption of Fig. P.4). Small So values indicate greater uniformity of the particles, large So values indicate a greater variance of the particle size, that is, a denser packing (small particles fill the space in between the larger ones). Therefore, porosity ϕ and permeability κ decrease with increasing sorting coefficient So. More sophisticated theories include additional characteristics of the particle size distribution (e.g., see *Freeze and Cherry, 1979*).

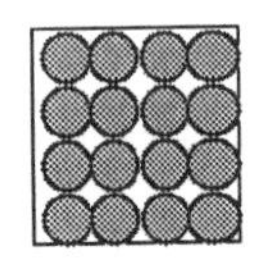

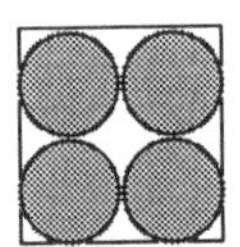

(1) Pore size = const. r^2
(2) Flow per pore $Q(r)$ = const. r^4 (Hagen-Poiseuille)
(3) Number of pores per unit area $N(r)$ = const. r^{-2}
(4) Total flow per unit area $q = Q(r) \cdot N(r)$ = const. r^2

Figure P.3 A simplified groundwater matrix consisting of spherical grains of equal radius r serves to demonstrate why the hydraulic conductivity (i.e., the flow per unit area for constant $\Delta h / \Delta x$ is proportional to r^2.

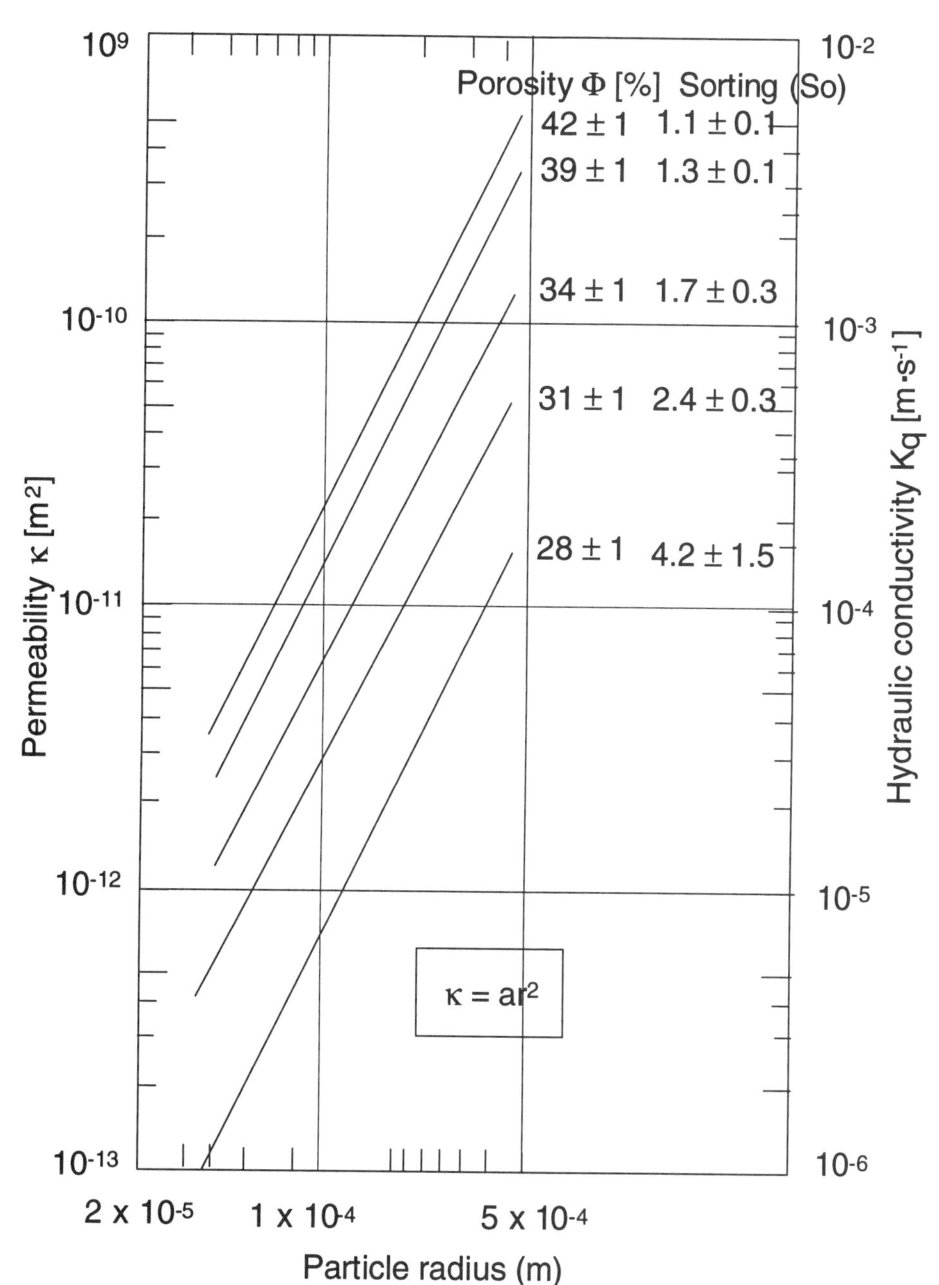

Figure P.4 Permeability κ as a function of (mean) particle radius r for different aquifer porosity ϕ, which, in turn, depends on the sorting coefficient $So=(r_{75}/r_{25})^{1/2}$, where r_{25} and r_{75} characterize the particle radii larger than, respectively, 25% and 75% of the radii of all the aquifer particles. The hydraulic conductivity K_q (right scale) refers to water at 20°C. Redrawn from *Lerman (1979)*.

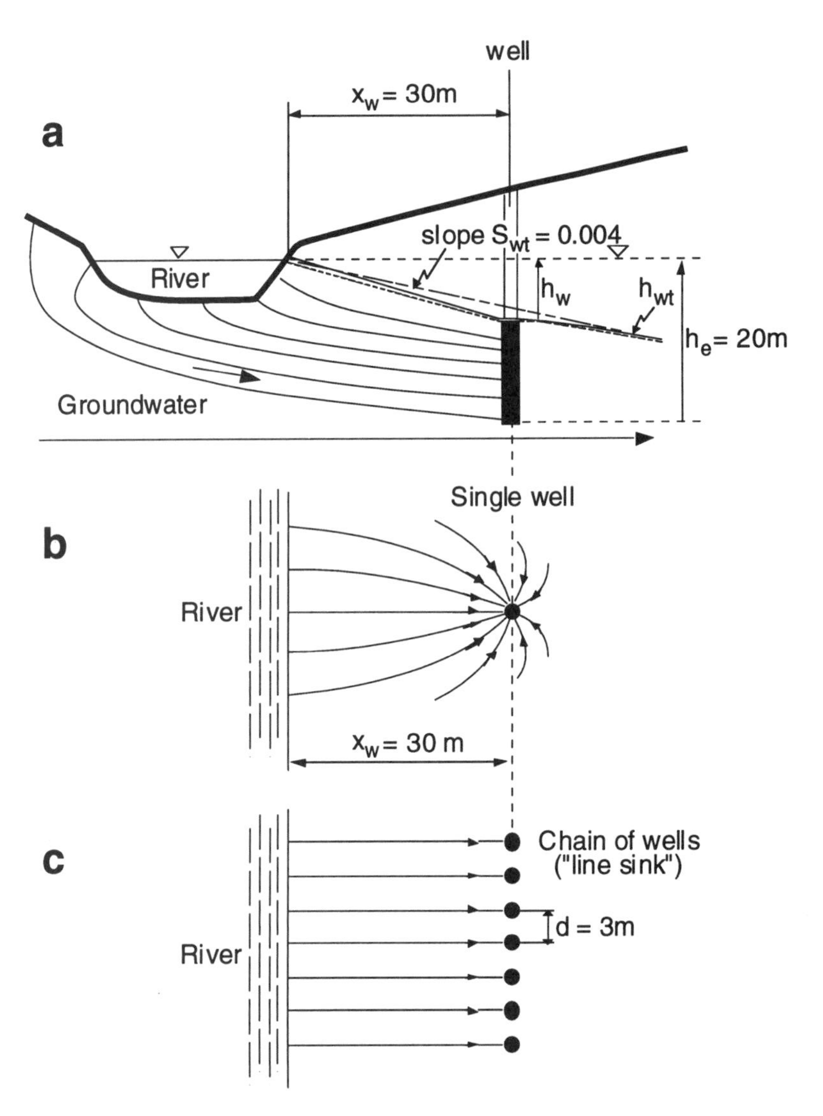

Figure P.5 (a) The stream lines in the Groundwater System S (GWS) are crossing the River R at a right angle. The slope of the water table in the GWS is S_{wt} = 0.004. The aquifer is assumed to be homogeneous with porosity $\phi = 0.31$ and permeability $\kappa = 1 \times 10^{-10}$ m^2. Water is infiltrating from the river into the aquifer. (b) Pumping of water from a single well located at distance x_w = 30 m from the river causes a local distortion of the stream lines in the groundwater flow. (c) Pumping from a number of wells, located along the river at distance x_w = 30 m, each d = 3 m apart from its neighbors, keeps the flow field quasi linear.

In this chapter, all phenomena and processes that are introduced, step by step, will be exemplified for a specific aquifer, the *Groundwater System S* (GWS). It is assumed that the flow in the GWS is one-dimensional, consisting of parallel stream lines crossing the *River R* (see Chapter O) at a right angle (Fig. P.5a). Water infiltrates from the River R into the aquifer. Groundwater systems that are fed by a river are often used for drinking water supply by wells located in the vicinity of the river. If the height of the water table is locally reduced by pumping water from a single well, the groundwater flow is distorted. Only part of the pumped water originates from the river; due to the local perturbation of the hydraulic gradient some water flows from the opposite side into the well (see Fig. P.5b). Finally, the well is also fed by water that crosses the river bed at greater depth and is not directly influenced by the river water.

An adequate quantitative description of such a situation requires a two- or even three-dimensional approach. Today, a great variety of numerical models are available that allow to solve such models almost routinely. However, from a didactic point of view numerical models are less suitable as illustrative examples than equations that can still be solved analytically. Therefore, an alternative approach is chosen. In order to keep the flow field quasi one-dimensional, the single well is replaced by a dense array of wells located along the river at a fixed distance x_w (Fig. P.5c). Ultimately, the set of wells can be looked at as a *line sink*. This is certainly not the usual method to exploit aquifers! Nonetheless, from a qualitative point of view a single well has very similar properties to the line sink.

Problem

Consider the *River R* from which water (at T=20°C) is infiltrating through a saturated zone into the *Groundwater System S* (GWS). The flow in the GWS crosses the river at a right angle. The slope of the (undisturbed) water table is $S_{wt} = - dh_{wt}/dx = 0.004$ (Fig. P.5a). The aquifer is assumed to be homogeneous with porosity $\phi = 0.31$. Its permeability κ is 1×10^{-10} m^2. Calculate the specific discharge of the groundwater system.

Answer

Inserting the permeability $\kappa = 1 \times 10^{-10}$ m^2 and $g/\nu \sim 10^7$ m^{-1}s^{-1} into Eq. P-6 yields the hydraulic conductivity $K_q = (g/\nu)\kappa = 1 \times 10^{-3}$ m·s^{-1}. Thus, the specific discharge calculated from Eq. P-5 is

$$q = -1 \times 10^{-3} \text{ m·s}^{-1} (-0.004) = 4 \times 10^{-6} \text{ m·s}^{-1}$$

Note that according to the orientation of the x-axis (see Fig. P.5a), the gradient dh_{wt}/dx is negative; thus, q is positive indicating a flow along the positive x-axis. In the following discussion this situation is called the *Natural Flow Regime*.

Note that a real aquifer is usually not homogeneous. Often the permeability at the transition from the river into the aquifer is reduced due to a process called colmatation. As a result, the slope of the water table is significantly larger at the point of infiltration than farther away from the river. In order to remain simple, such effects are disregarded.

Problem
At the distance x_w = 30 m from the River R, parallel to the river bank, there is an array of wells regularly spaced with distance d = 3m in between the wells (Fig. P.5c). The water level in the wells lies 60 cm below the water surface of the river (h_w = 0.6m). The lower end of the wells is located 20 m below the water surface in the river (h_e = 20 m). Estimate how much water is pumped from each well.

Hints and Help
The geometry of the flow modified by the wells is a superposition of linear and radial flow and thus not ideal for an analytical discussion. As explained above, a one-dimensional approximation is justified, nonetheless, provided that the distance in between the wells is small. As a first approximation it can then be assumed that each well captures the water from the groundwater that flows through a section with width equal to the spacing of the wells, d. Furthermore, the mean slope of the water table that is drawing water into the well can be approximated by h_w/x_w.

Answer
The area from which water is captured is $a_x = (h_e - h_w) \cdot d$ = 19.4m x 3m = 58 m^2.

According to the approximate one-dimensional model described above, the flow per unit area to the wells is (Eq. P-5b)

$$q = -1 \times 10^{-3}\ m \cdot s^{-1} \left(-\frac{0.6m}{30m} \right) = 2 \times 10^{-5}\ m \cdot s^{-1}$$

Thus, the pump rate in each well, Q_{well}, is

$$Q_{well} = q \cdot a_x = 2x10^{-5}\ m \cdot s^{-1} \times 58\ m^2 = 1.16 \times 10^{-3}\ m^3 \cdot s^{-1}$$
$$\approx 1.2\ L \cdot s^{-1}$$

In the following discussion this situation is called the *Pump Regime I*.

Problem
The pump rate of the individual pumps is increased to Q_{well} = 2.5 $L \cdot s^{-1}$. How big is h_w, once a new steady state is reached?

Answer
Intuitively, it is clear that the water level in the well drops until the slope of the surface of the water table, h_w/x_w, becomes large enough in order to produce a flow into the well that equals Q_{well}. This flow can be expressed by

$$Q_{well} = K_q \frac{h_w}{x_w} \cdot d \cdot (h_e - h_w) = \gamma \cdot h_w \cdot (h_e - h_w)$$

with $\gamma = K_q \cdot d / x_w = 1 \times 10^{-4}\ m \cdot s^{-1}$. Solving the quadratic equation for h_w yields

$$h_w = \frac{1}{2}(h_e \pm [h_e^2 - 4 \cdot Q_{well}/\gamma]^{1/2}$$

Inserting h_e = 20m, Q_{well} = 0.0025 $m^3 \cdot s^{-1}$, yields the two solutions h_w = 1.3 m and h_w = 18.7 m. The first value is the one that the system will attain if the pump rate is continuously increased to 2.5 $L \cdot s^{-1}$. This situation will be called *Pump Regime II.* The second solution, although formally correct, is physically unrealistic; it yields the same Q_{well}, in spite of the much steeper slope of the water table, since the thickness of the layer reaching the well is smaller. However, the physical conditions underlying the validity of Eq. P-5 are no longer fulfilled since at such large hydraulic gradients the flow breaks off and becomes unsaturated. The following table summarizes both regimes:

Pump Regime I	Q_{well} = 1.2 $L \cdot s^{-1}$, h_w = 0.6 m
Pump Regime II	Q_{well} = 2.5 $L \cdot s^{-1}$, h_w = 1.3 m

Problem

A cloud of a nonreactive (conservative) pollutant is transported along the river. How much time does it take for the pollutant to reach the wells? Discuss all three regimes, i.e., the Natural Regime as well as the Pump Regimes I and II.

Answer

According to Eq. P-1, the effective flow velocity along the x-axis is

$$\bar{u} = \frac{q}{\phi} = K_q \frac{h_w}{x_w} \cdot \frac{1}{\phi}$$

Since the travel time of a pollutant to the wells is $t_o = x_w/\bar{u}$, the following results are obtained:

Table P.2 Specific Discharge q, Effective Mean Flow Velocity $\bar{u}$, and Travel Time t_o for Different Flow Regimes in the Groundwater System S

Regime	q ($m \cdot s^{-1}$)[a]	$\bar{u}$ ($m \cdot d^{-1}$)[b]	t_o (days)
Natural	4×10^{-6}	1.1	27
I	2×10^{-5}	5.6	5.4
II	4.5×10^{-5}	12.5	2.4

[a] With $K_q = 1 \times 10^{-3}\ m \cdot s^{-1}$.
[b] With $\phi = 0.31$.

Note: The travel time calculated by this simple model is not quite correct. More realistic results are obtained by including longitudinal dispersion, the subject discussed in the next section.

Longitudinal Dispersion

In porous media the flow of water and the transport of solutes is complex and three-dimensional on all scales (Fig. P.1). A one-dimensional description needs an empirical correction that takes account of the three-dimensional structure of the flow. Due to the different length and irregular shape of the individual pore channels, the flow time between two (macroscopically separated) locations varies from one channel to another. As discussed for rivers (Chapter O), this causes *dispersion*, the so-called *interpore* dispersion. In addition, the nonuniform velocity distribution within individual channels is responsible for *intrapore dispersion.* Finally, molecular diffusion along the direction of the main flow also contributes to the longitudinal dispersion/diffusion process. For simplicity, transversal diffusion (as discussed for rivers) is not considered here. The discussion is limited to the one-dimensional linear case for which simple calculations without sophisticated computer programs are possible.

In analogy to the description of dispersion in rivers, the dispersive flux relative to the flowing water, F_{dis}, can be described by an equation of the First Fickian Law type (see Eq. O-40):

$$F_{dis} = - \phi \cdot E_{dis} \frac{\partial C_w}{\partial x} \quad \text{(P-8)}$$

where E_{dis} is the coefficient of longitudinal dispersion and C_w is the (mean) concentration in the pore water. This flux leads to an additional term in the transport Equation P-4:

$$\left(\frac{\partial C_w}{\partial t}\right)_{trans} = \left(\frac{\partial C_w}{\partial t}\right)_{ad} + \left(\frac{\partial C_w}{\partial t}\right)_{dis} = - \bar{u} \frac{\partial C_w}{\partial x} + \frac{\partial}{\partial x}\left(E_{dis} \frac{\partial C_w}{\partial x}\right) \quad \text{(P-9)}$$

which for E_{dis} = const. becomes

$$\left(\frac{\partial C_w}{\partial t}\right)_{trans} = - \bar{u} \frac{\partial C_w}{\partial x} + E_{dis} \frac{\partial^2 C_w}{\partial x^2} \quad \text{(P-10)}$$

E_{dis} combines the effects of the different processes causing dispersion. For the case of inter-/intrapore dispersion it is usually written as *(Freeze and Cherry, 1979)*:

$$E_{dis} = \frac{D_w}{\chi} + \alpha_L \bar{u} \quad \text{(P-11)}$$

D_w is the molecular diffusion coefficient of the chemical in water (Chapter G), χ is tortuosity, and α_L is the *(longitudinal) dispersivity* (dimension: L). Tortuosity χ, a number greater than 1, describes the ratio between the real length along the pores and the length of a straight line connecting the two macroscopically separated endpoints. (*Note that the factor f appearing in Eq. 11-155 corresponds to* χ^{-1}.) Typical values of the dispersivity α_L for field systems with flow distances of up to about 100 m lie between 1 and 100 m. Since α_L depends strongly on the scale of the aquifer, test columns used in the laboratory have much smaller α_L values, which are between 0.001 and 0.01 m. Such values should not be extrapolated to the field. *Brusseau (1993)* summarizes the present knowledge on the dependence of α_L on the characteristic structure of the aquifer.

Problem

The dispersivity of the GWS (System S) is $\alpha_L = 3$ m, tortuosity is $\chi = 1.5$. Calculate the dispersion coefficient, E_{dis}, for *2,4-dinitrophenol* at 15°C for the three flow regimes (see Table P.2). The pH of the groundwater is 7.5.

OH

NO_2

NO_2

2,4-dinitrophenol

mw 184.1 $g \cdot mol^{-1}$

Answer

The pK_a of 2,4-dinitrophenol (2,4-DNP) is 3.94 (*Schwarzenbach et al., 1988*). Thus, at pH = 7.5, > 99.99% of the chemical is present in its anionic form, and sorption can, therefore, be neglected (see Chapter J). When further assuming that 2,4-DNP does not undergo any transformation, the compound can be assumed to exhibit a conservative behavior in the aquifer.

The molecular diffusion coefficient of 2,4-dinitrophenol is estimated from $D_w(O_2)$ using *Eq. 9-25*. At 25°C, $D_w(O_2) = 2.1 \times 10^{-5}\ cm^2 \cdot s^{-1}$ (see footnote c of *Table 10.3*). At 15°C, $D_w(O_2)$ is slightly smaller: $2.0 \times 10^{-5}\ cm^2 \cdot s^{-1}$ (Note: Problem G-1 helps in understanding the temperature dependence of molecular diffusivity.) Thus,

$$D_w(\text{2,4-DNP}, 15°C) = \left(\frac{mw(O_2)}{mw\ (\text{phenol})}\right)^{1/2} D_w(O_2, 15°C)$$

$$= \left(\frac{32}{184}\right)^{1/2} \times 2.0 \times 10^{-5}\ cm^2 \cdot s^{-1} = 8.3 \times 10^{-6}\ cm^2 \cdot s^{-1} = 8.3 \times 10^{-10}\ m^2 \cdot s^{-1}$$

From Eq. P-11 for the Natural Regime ($\bar{u} = 1.1\ m \cdot d^{-1} = 1.3 \times 10^{-5} m \cdot s^{-1}$) one obtains

$$E_{dis}\text{ (phenol, 15°C)} = \frac{8.3\text{x}10^{-10}\ m^2{\cdot}s^{-1}}{1.5} + 3m \text{ x } 1.3\text{x}10^{-5}\ ms^{-1}$$
$$= 5.5 \text{ x } 10^{-10}\ m^2{\cdot}s^{-1} + 3.9 \text{ x } 10^{-5}\ m^2{\cdot}s^{-1} \approx 3.9 \text{ x } 10^{-5}\ m^2{\cdot}s^{-1}.$$

It appears that for the present case the influence of the molecular diffusivity on E_{dis} is negligible. In fact, in most natural groundwater flows Eq. P-11 reduces to

$$E_{dis} = \alpha_L \cdot \bar{u} \qquad \text{(P-11a)}$$

The following table gives the results for all regimes:

Regime	$\bar{u}$ ($m{\cdot}s^{-1}$)	E_{dis} ($m^2{\cdot}s^{-1}$)
Natural Regime	$1.3 \text{ x } 10^{-5}$	$3.9 \text{ x } 10^{-5}$
Pump Regime I	$6.5 \text{ x } 10^{-5}$	$2.0 \text{ x } 10^{-4}$
Pump Regime II	$14.5 \text{ x } 10^{-5}$	$4.4 \text{ x } 10^{-4}$

Advection versus Dispersion

The relative importance of advective versus dispersive transport for a given distance x_o can be expressed by the nondimensional *Peclet Number* (see *Table 15.12*):

$$Pe = \frac{x_o \cdot \bar{u}}{E_{dis}} \qquad \text{(P-12)}$$

With Eq. P-11a this reduces to

$$Pe = \frac{x_o}{\alpha_L} \qquad \text{(P-13)}$$

The ratio between transport distance and dispersivity determines whether transport is dominated by dispersion ($x_o \ll \alpha_L$, $Pe \ll 1$) or by advection ($x_o \gg \alpha_L$, $Pe \gg 1$). Since the scale of the heterogeneities grows with the size of the system, α_L increases with x_o, as well. Thus, Pe does not necessarily increase linearly with distance x_o.

Problem
Assess the relative importance of dispersion versus advection in the GWS (Fig. P.5) for 2,4-dinitrophenol infiltrating from the river and detected in one of the wells. Discuss all three flow regimes.

Answer
Note that if E_{dis} is approximated by Eq. P-11a, the Peclet Number, Pe, does not depend on the flow regime. From Eq. P-13 with $x_o = x_w = 30$ m:

$$Pe = \frac{30 \text{ m}}{3 \text{ m}} = 10 > 1$$

Thus, for the transport from the river to the well, advection is more important than dispersion. In order to assess the implications of this conclusion for, e.g., the dynamic relationship between the concentration in the river and in the well, the solution of the transport equation P-10 for a time-dependent input concentration at $x = 0$, $C_{in}(t)$, will be discussed below. Note that the input into the aquifer is determined by the concentration variation in the river from which the water is infiltrating. In the river typical time scales of change are of the order of minutes (in case of an accidental spill) to days, but seasonal variations also exist for some chemicals. In contrast, transport within the aquifer is slower than most riverine concentration variations. The question arises how the river dynamics is transmitted to the groundwater. Three different cases are discussed:

(1) the pulse input;
(2) the step input;
(3) the fluctuating (sinusoidal) input with period τ.

Although the partial differential equation P-10 is linear and looks rather simple, explicit analytical solutions can only be derived for special cases. They are characterized by the size of certain nondimensional numbers that completely determine the shape of the solutions in space and time. A reference distance x_o and a reference time t_o are chosen that are linked by

$$t_o = \frac{x_o}{\bar{u}} \qquad \text{(P-14)}$$

t_o is the time needed for the water in the aquifer to flow with the effective mean velocity $\bar{u}$ from the point of infiltration ($x = 0$) to x_o. The following nondimensional coordinates are introduced:

$$\xi = \frac{x}{x_o} \quad , \quad \theta = \frac{t}{t_o} = \frac{\bar{u}}{x_o} \cdot t \qquad \text{(P-15)}$$

By using the transformation rules

$$\frac{\partial}{\partial t} = \frac{\partial}{\partial \theta} \cdot \frac{d\theta}{dt} = \frac{1}{t_o} \frac{\partial}{\partial \theta} \qquad \text{(P-16)}$$

$$\frac{\partial}{\partial x} = \frac{\partial}{\partial \xi} \frac{d\xi}{dx} = \frac{1}{x_o} \frac{\partial}{\partial \xi} \qquad \text{(P-16a)}$$

Eq. P-10 can be transformed into

$$\left(\frac{\partial C_w}{\partial \theta}\right)_{trans} = - \frac{\partial C_w}{\partial \xi} + \frac{1}{Pe} \frac{\partial^2 C_w}{\partial \xi^2} \qquad \text{(P-17)}$$

where the Peclet Number was defined in Eq. P-12. Given the geometry of the system, the solutions of Eq. P-17, if expressed in terms of the relative distance, ξ, and the relative time, θ, are thus completely determined by the Peclet Number.

(1) The pulse input
Consider the case of a pollution cloud in the river passing by the infiltration location during the time Δt, which shall be very short compared to the time t_o needed for the groundwater to travel from the river to the wells. During the event, the concentration in the river is C_{in}; before and after the event, C is ≈ 0.

Note: As discussed in Chapter O, a pollution cloud caused by an accidental spill and travelling along a river often has the shape of a normal distribution (see Fig. O.6b). In order to keep the following considerations simple, it is assumed that the variance of the cloud in the river is still small at the time when it passes by the GWS. Otherwise, the following considerations would have to be modified in a similar way as explained in Eqs. O-56 and O-57.

The total mass input by infiltration into the aquifer per unit area perpendicular to the flow is (see Eq. P-1)

$$m_{in} = \Delta t \cdot q \cdot C_{in} = \Delta t \cdot \bar{u} \cdot \phi \cdot C_{in} \ \ (\text{mol} \cdot \text{m}^{-2}) \tag{P-18}$$

The fate of the pollutant moving in the aquifer along the stream lines is determined by the advection-dispersion equation P-10. For Pe » 1, i.e., for locations x » $E_{dis} / \bar{u}$, the concentration cloud can be envisaged to originate from an infinitely short input at x=0 of total mass m_{in} (a so-called δ-input) that by dispersion is turned into a normal distribution function along the x-axis with growing standard deviation (see *Table 9.2* and *Eq. 9-19*). Since the arrival of the main pollution cloud at some distance x is determined by the effective flow velocity $\bar{u}$, the symmetry point (or maximum value) of the normal distribution moves as $\bar{u} \cdot t$. Thus, the solution is identical to the one derived for dispersion in rivers (Eq. O-48)

$$C(x,t) = \frac{m_{in} / \phi}{2(\pi \cdot E_{dis} \cdot t)^{1/2}} \exp\left(-\frac{(x - \bar{u} \cdot t)^2}{4E_{dis} \cdot t}\right) \tag{P-19}$$

The integral over x of this expression is constant (mass conservation), and its time-dependent standard deviation is (*Eq. 9-20*):

$$\sigma(t) = (2E_{dis} \cdot t)^{1/2} \tag{P-20}$$

Its maximum moves along x at the effective mean flow velocity, $\bar{u}$ (Eq. O-45):

$$x_m = \bar{u} \cdot t \tag{P-21}$$

Obviously, everything that has been said in Chapter O on dispersion in rivers also applies to the one-dimensional flow in aquifers.

Remember that from the nondimensional version of the advection-dispersion equation, Eq. P-17, the Peclet Number Pe was identified as the only parameter that determines the shape of the concentration distribution in the aquifer. By introducing relative coordinates for space (ξ) and time (θ) as defined in Eq. P-15, Eq. P-19 takes the form

$$C(x,t) = C(\xi,\theta) = \frac{\hat{C}}{(2\pi)^{1/2}\, s} \exp\left(-\frac{(\xi - \theta)^2}{2s^2}\right) \tag{P-22}$$

where

$$s = \left(\frac{2\theta}{Pe}\right)^{1/2} \quad \text{(nondimensional standard deviation)} \tag{P-23}$$

$$\hat{C} = \frac{m_{in}}{x_o \cdot \phi} = \frac{\Delta t \cdot \bar{u}}{x_o} \cdot C_{in} \qquad \text{(P-24)}$$

Note that Eq. P-22 describes a normal distribution with maximum at $\xi = \theta$ (or $x = \bar{u} \cdot t$):

$$C_{max} = \frac{\hat{C}}{(2\pi)^{1/2} s} = \frac{1}{2}\left(\frac{Pe}{\pi \cdot \theta}\right)^{1/2} \cdot \hat{C} \qquad \text{(P-25)}$$

or

$$\frac{C_{max}}{C_{in}} = \frac{1}{2}\left(\frac{\Delta t \cdot \bar{u} \cdot \phi}{x_o}\right) \mathrm{x} \left(\frac{Pe}{\pi \cdot \theta}\right)^{1/2} \qquad \text{(P-26)}$$

Problem

A pollution cloud of *2,4-dinitrophenol* in the River R (duration $\Delta t = 1$ hour, concentration $C_{in} = 50\ \mathrm{ng \cdot L^{-1}}$) is passing by the Groundwater System S. Calculate the maximum concentration reached at the wells for the three regimes. Compare these values to the maximum concentrations reached 3 m away from the river.

Answer

It is convenient to choose $x_o = x_w = 30$ m as the point of reference. (*Note however that the following calculation does not depend on this choice!*) Thus, Eq. P-22 has to be solved for $\xi = 1$, where $x = x_o$. The maximum concentration is approximately reached when $\theta = \xi = 1$. (*Note:* It is true that $C(x,t)$ as a function of x for a *fixed time t* has its maximum at $x_m = \bar{u} \cdot t$. In contrast, the maximum of $C(x,t)$ as a function of t for a *fixed location x* is only approximately reached at time $t_m = x / \bar{u}$, but for $Pe \geq 3$ the error involved is small and will be disregarded.) Since the Peclet Number $Pe = 10$ is independent of the pump regime, C_{max} depends on the pump regime only via the dependence of $\hat{C}$ on $\bar{u}$ (Eq. P-24). For instance, for the Natural Regime with $\bar{u} = 1.3 \times 10^{-5}\ \mathrm{m \cdot s^{-1}}$:

$$\hat{C} = \frac{3600\ \mathrm{s} \times 1.3 \times 10^{-5}\ \mathrm{m \cdot s^{-1}}}{30\ \mathrm{m}} \times 50\ \mathrm{ng \cdot L^{-1}} = 7.8 \times 10^{-2}\ \mathrm{ng \cdot L^{-1}}$$

The nondimensional standard deviation s follows from Eq. P-23

$$s_{30} = \left(\frac{2}{10}\right)^{1/2} = 0.45 \qquad \text{(at } x = 30 \text{ m)}$$

At x = 3 m, which is 10% along the stretch from the river to the wells, the nondimensional coordinates for the transition of the maximum concentration are $\xi = \theta = 0.1$. Thus,

$$s_3 = \left(\frac{2 \times 0.1}{10}\right)^{1/2} = 0.14 \quad (\text{at } x = 3 \text{ m})$$

Finally, the maximum concentrations are calculated from Eq. P-25 as summarized below:

Regime	$\hat{C}$ (ng·L^{-1})	C_{max} (ng·L^{-1})[a] x = 30 m	C_{max} (ng·L^{-1})[a] x = 3 m
Natural	0.078	0.069	0.22
I	0.39	0.35	1.10
II	0.90	0.80	2.5

[a] With C_{in} = 50 ng·L^{-1}

The maximum concentration is found at x = 3 m for the Pump Regime II, it reaches 5% of the concentration in the river water (50 ng·L^{-1}). This value is an upper limit, since dilution of the infiltrating river water with uncontaminated groundwater is disregarded. In fact, according to these calculations C decreases only because of dispersion; the total mass of 2,4-dinitrophenol does, however, not change. Remember that according to the approximation made for the dispersion coefficient E_{dis} (Eq. P-11a), the Peclet Number does not depend on $\bar{u}$. Therefore, the influence of the pump regime on the concentration at the well is *not* caused by a change of the transport conditions in the aquifer (e.g., flow time t_o), but simply by the fact that more dinitrophenol enters the groundwater during the passage of the pollution cloud if $\bar{u}$ is large.

Note: The fact that x_w = 30 m was chosen as the reference scale to transform Eq. P-19 into a nondimensional form (Eq. P-22) may cause some confusion. The reader may have been tempted to change the reference point in order to calculate the concentration at x = 3 m. Yet, once x_o is chosen, the concentration can be calculated for any distance by using the relative scales, ξ and θ. In fact, any other choice would have been appropriate. Particularly, one could also use x_o = 3 m. Then, the Peclet Number would decrease by a factor of 10 (Pe = 1, see Eq. P-13). Also, the relative time for the passing of the concentration maximum at x = 3 m would now be θ = 1. Combining these changes into Eq. P-23 increases s by a factor of 10 (θ increases by 10, Pe decreases by 10, and *the square root leaves a factor of 10 in the nominator). This change is compensated for by a corresponding tenfold increase of $\hat{C}$ (Eq. P-24) caused by the change of x_o from 30 m to 3 m. Thus, C(x,t) of Eq. P-22 and C_{max} (Eq. P-25) remain unchanged, as obviously should be the case.*

(2) The step input

Next, the situation is considered in which at time t = 0 the concentration of a chemical in the infiltrating river water suddenly changes from C_o to C_1 and then remains at C_1 (Fig. P.6a). Since in this context only concentration changes are relevant, it can be assumed that the initial

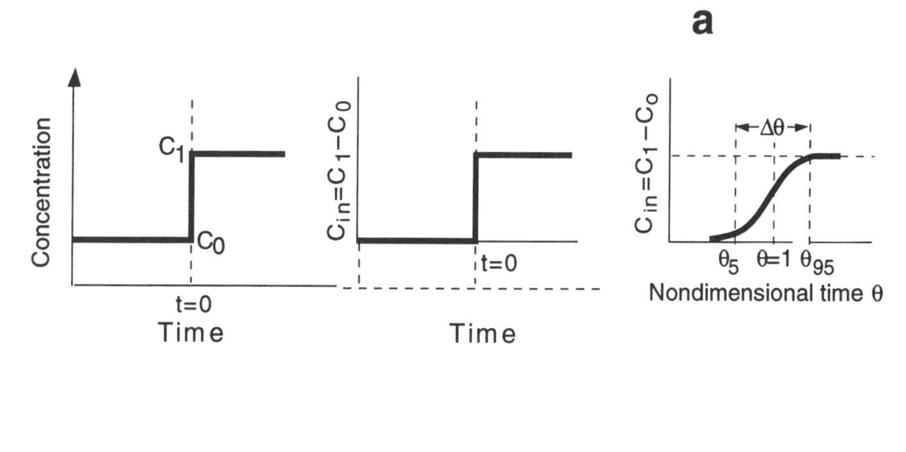

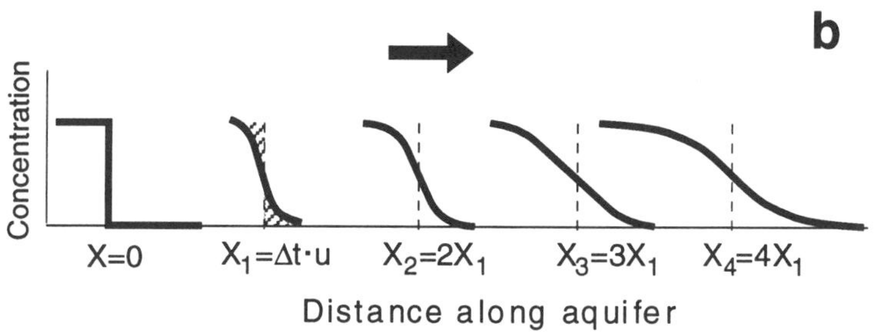

Figure P.6 The step input: (a) By introducing a normalized concentration, $C = C_o$, the step input can be described as a sudden change in the river concentration from 0 for $t < 0$ to C_{in} for $t \geq 0$. (b) When the step function moves along the aquifer, the front becomes smoother. The slope of the front can be described by the complementary error function erfc(y) relative to the moving coordinate $x = \bar{u}t$ (Eqs. P-31 or P-32). θ_5 and θ_{95} are the nondimensional times when C in the aquifer reaches 5% and 95% of C_{in}, respectively.

concentration C_o is 0. This situation is called a step input.

As for the pulse input, the evolving concentration C(x,t) can be envisioned as the result of two simultaneous processes: (1) advective transport along x at effective flow velocity $\bar{u}$ and (2) smoothing by dispersion relative to the moving front, $x = \bar{u} \cdot t$. Imagine a very long tube extended along the x-axis and filled with water. A wall at x = 0 divides the tube into two sections. To the left (x < 0) the water contains a chemical at concentration C_{in}; to the right (x > 0) the concentration is zero. At time t = 0, the wall is suddenly removed, and the chemical starts to diffuse into the right section (diffusion coefficient D). Because of the special symmetry of the initial situation, the concentration along the x-axis at any time should be such that the deficit on the left side exactly mirrors the addition on the right side (Fig. 6b). Particularly, the concentration curve at x = 0 should always remain at $C_{in}/2$. In fact, the solution to the diffusion problem without advection is given by (e.g., *Crank, 1975*)

$$C(x,t) = \frac{C_{in}}{2} \operatorname{erfc}\left(\frac{x}{2(Dt)^{1/2}}\right) \tag{P-27}$$

The numerical values of the *complementary error function*, erfc(y), defined by

$$\operatorname{erfc}(y) = 1 - \frac{2}{(\pi)^{1/2}} \int_0^y e^{-\eta^2} d\eta \tag{P-28}$$

can be found in most mathematical handbooks (e.g., *Abramowitz and Stegun, 1965*). It drops monotonously from erfc(- ∞) = 2 to erfc(0) = 1 and erfc(∞) = 0, but most of the variation occurs between y = ± 1. The relationship

$$\operatorname{erfc}(-y) = 2 - \operatorname{erfc}(y) \tag{P-29}$$

is responsible for the mentioned symmetry at y = 0.

In the same way as the additional effect of advection was taken into account for the pulse input by modifying *Eq. 9-19* into Eq. O-48 or P-19, the solution (Eq. P-27) is changed into

$$C(x,t) = \frac{C_{in}}{2} \operatorname{erfc}\left(\frac{x - \bar{u} \cdot t}{2(E_{dis} \cdot t)^{1/2}}\right) \tag{P-30}$$

to obtain the solution of the advection-dispersion equation for the step input function. Note that D has been substituted by the dispersion coefficient E_{dis}.

In order to discuss the properties of this solution it again is helpful to transform Eq. P-30 into a nondimensional form by choosing a reference distance x_o and applying Eq. P-15:

$$\frac{C(x,t)}{C_{in}} = \frac{1}{2} \operatorname{erfc}\left(\frac{\xi - \theta}{2(\theta/Pe)^{1/2}}\right) \qquad \text{(P-31)}$$

The temporal change of C(x,t) at a fixed location, say at $x = x_o$, where $\xi=1$, can be qualitatively described by distinguishing three phases:

(1) The actual time t is still much smaller than t_o, the time needed for the front to reach x_o by travelling at the effective flow velocity $\bar{u}$; then the argument of the error function

$$y = \frac{x - \bar{u}\cdot t}{2(E_{dis} t)^{1/2}} = \frac{\xi - \theta}{2(\theta/Pe)^{1/2}} \qquad \text{(P-32)}$$

is still much larger than 1; i.e., $C(x,t) \sim 0$.

(2) The central part of the front passes by x_o, i.e., $y \sim 0$ and $\operatorname{erfc}(y) \sim 1$, i.e. $C(x,t) \sim C_{in}/2$.

(3) The actual time t is much larger than t_o, i.e., the main portion of the front has passed x_o. Then $y \rightarrow -\infty$, i.e., $\operatorname{erfc}(y) \sim 2$ and $C \sim C_{in}$.

To quantify the time interval during which the front is moving past x_o, the two times t_1 and t_2 are chosen when C(x,t) reaches 5% and 95% of C_{in}, respectively (Fig. P.6a). Thus, one is looking for the critical values y_5 and y_{95} for which

$$\operatorname{erfc}(y_5) = 0.1, \quad \operatorname{erfc}(y_{95}) = 1.9 \qquad \text{(P-33)}$$

Note that because of Eq. (P-29), $y_5 = -y_{95}$. From tabulated values of erfc(y) one finds $y_5 = 1.16$ thus $y_{95} = -1.16$. Solving Eq. P-32 for θ with $y = \pm 1.16$ and $\xi = 1$ yields the corresponding (nondimensional) times, θ_5 and θ_{95}, which mark the passing of the concentration front. The duration of the front passage, $\Delta\theta = \theta_{95} - \theta_5$, increases with the Peclet Number, Pe. Some values are tabulated in Table P.3.

Table P.3 Nondimensional Time Interval of the Passage of a Step Front at $\xi = 1$ for Different Peclet Numbers Pe. θ_5 And θ_{95} Are the Nondimensional Times when the Concentration Reaches 5% and 95% of C_{in}, Respectively [a]

Pe	θ_5	θ_{95}	$\Delta\theta = \theta_{95} - \theta_5$
1 [b]	0.14	7.2	7
3	0.28	3.5	3.2
10	0.49	2.1	1.6
30	0.657	1.522	0.87
100	0.793	1.261	0.47
300	0.874	1.143	0.269
1000	0.929	1.076	0.147
10000	0.977	1.024	0.047

[a] θ_5 and θ_{95} are calculated from

$$\theta_{5/95} = \frac{1}{2}[2 + \frac{5.382}{Pe} \pm \{(2 + \frac{5.382}{Pe})^2 - 4\}^{1/2}]$$

[b] Note that for the case of fixed boundary conditions at x = 0, Eq. P-31 is only an approximate solution of the differential equation (P-10). The approximation is not appropriate if Pe $\lesssim$ 1.

Problem

On April 5, at noon, the 2,4-dinitrophenol concentration in the River R at the GWS suddenly increases from 0 to 50 $ng \cdot L^{-1}$ and then remains constant. At what time does the concentration in the wells of the GWS reach 25 $ng \cdot L^{-1}$ and 47.5 $ng \cdot L^{-1}$, respectively? Calculate the time for all three flow regimes. When are these concentrations reached 3 m from the river if no water is pumped from the wells (Natural Regime)?

Answer

Again, $x_o = 30$ m is chosen as the point of reference. The concentration 25 $ng \cdot L^{-1}$ (this is half of C_{in}) is reached at $\theta = 1$, i.e., at time $t_{50} = t_o = x_o/\bar{u}$. The corresponding times for all regimes are summarized in Table P.2.

Since Pe = 10 is constant for all flow regimes, θ_{95} can be directly taken from Table P.3: $\theta_{95} = 2.1$; thus $t_{95} = 2.1 \cdot t_o = 2.1 \cdot x_o/\bar{u}$. The results are summarized below.

Regime	Time since April 5		Absolute time	
	t_{50} (days)	t_{95} (days)	t_{50}	t_{95}
Natural	27	57	May 2	June 1st
I	5.4	11.3	April 10, at night	April 16
II	2.4	5.0	April 7, at night	April 10

There are two ways to calculate θ_{50} and θ_{95} for the location x = 3 m. One way is to solve Eq. P-32 for ξ = 3 m/30 m = 0.1 and y = ± 1.16. An alternative possibility is to choose x_o = 3 m as the new reference point. Then Eq. P-13 yields Pe = 3 m/3 m = 1 and Table P.2 yields θ_{95} = 7.2. Since the mean flow time t_o = 3 m/1.1 m·d^{-1} = 2.7 d, it follows

$$t_{95} = 7.2 \times 2.7 \text{ d} = 20 \text{ days} \quad \text{(April 25)}$$

(3) The sinusoidal input with period τ

Often the concentration of a chemical substance in a river shows periodic fluctuations with typical periods of, e.g., a day (diurnal fluctuations), a week, or even a year (annual fluctuations). The fluctuations propagate into the aquifer by diffusion and advection. In addition to the Peclet Number, the nondimensional *Attenuation Number*

$$\Omega = \frac{\omega x_o}{\bar{u}} \qquad \text{(P-34)}$$

determines how far the fluctuations are felt within the aquifer and how the phase shifts relative to the phase of the "driving" river concentration. The angular frequency ω is related to the period τ by

$$\omega = \frac{2\pi}{\tau} \qquad \text{(P-35)}$$

If the concentration in the river varies as

$$C_{in}(t) = C_o + C_1 \sin(\omega t) \qquad \text{(P-36)}$$

then the solution of Eq. P-10 can be written in the form

$$C(x,t) = C_o + e^{-\alpha x} \cdot C_1 \cdot \sin(\omega t - \beta x) \qquad \text{(P-37)}$$

The coefficients α and β, which depend on the size of $\bar{u}$, E_{dis}, and ω, describe the attenuation of the fluctuations in the aquifer and the corresponding phase shift. Note the similarity of the situation with the case of a one-box model that is driven by a periodic external force as discussed in *Chapter 15* (see *Eqs. 15-23* and *15-25*).

The nondimensional version of Eq. P-37 is

$$\frac{C(\xi,\theta)}{C_o} = 1 + \frac{C_1}{C_o} e^{-\hat{\alpha}\xi} \sin(\Omega\theta - \hat{\beta}\xi) \qquad \text{(P-38)}$$

where the nondimensional space and time coordinates, ξ and θ, are defined in Eq. P-15, and $\hat{\alpha} = \alpha x_o$, $\hat{\beta} = \beta x_o$. The nondimensional attenuation coefficients $\hat{\alpha}$ and $\hat{\beta}$ are (*Roberts and Valocchi, 1981*)

$$\hat{\alpha} = \frac{Pe}{2} \left\{ \left[\frac{1}{2} + \frac{1}{2} \left(1 + \left[\frac{4\Omega}{Pe} \right]^2 \right)^{1/2} \right]^{1/2} - 1 \right\} \qquad \text{(P-39)}$$

$$\hat{\beta} = \frac{Pe}{2} \left\{ \frac{1}{2} \left[1 + \left(\frac{4\Omega}{Pe} \right)^2 \right]^{1/2} - \frac{1}{2} \right\}^{1/2} \qquad \text{(P-40)}$$

Figure P.7 shows how $\hat{\alpha}$ and $\hat{\beta}$ vary with Pe and Ω. If Pe and Ω have very different values, then Eqs. P-39 and P-40 can be approximated by

$$\Omega \ll Pe \; : \; \hat{\alpha} = \frac{\Omega^2}{Pe} \quad , \quad \hat{\beta} = \Omega \qquad \text{(P-41)}$$

$$\Omega \gg Pe \; : \; \hat{\alpha} = \hat{\beta} = \left(\frac{Pe \cdot \Omega}{2} \right)^{1/2} \qquad \text{(P-42)}$$

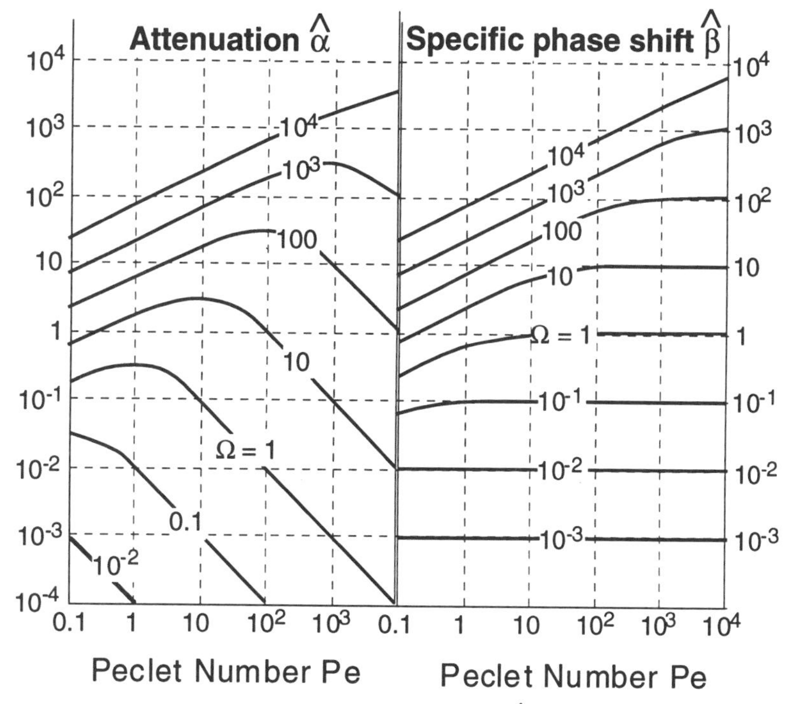

Figure P.7 Nondimensional attenuation coefficient $\hat{\alpha}$ and specific phase shift $\hat{\beta}$ for sinusoidal input into a saturated groundwater system as a function of the Peclet Number Pe. The numbers attributed to the different lines give the nondimensional Attenuation Number Ω. See text for definitions.

Problem

Continuous measurements of *2,4-dinitrophenol* in the River R passing by the GWS show a superposition of diurnal and annual concentration variations. Calculate how much of this variation can still be observed at one of the wells and determine the phase shift in days between the oscillations in the river and in the well. Include into the analysis also a monthly oscillation (30 days). Make all calculations for the three flow regimes.

Answer

As mentioned before, the Peclet Number does not depend on the flow regime (Pe = 10). In contrast, the Attenuation Number Ω is proportional to $\bar{u}$ (Eq. P-33). The point of reference is chosen as x_o = 30 m. The angular frequencies for the diurnal, monthly, and annual variations are, respectively: $\omega_1 = \frac{2\pi}{1\text{d}} = 6.28\ \text{d}^{-1}$, $\omega_{30} = 0.209\ \text{d}^{-1}$, $\omega_{365} = 0.0172\ \text{d}^{-1}$.

As an example, the Attenuation Number Ω is calculated from Eq. P-33 for the *Natural Regime and the diurnal input variation:*

$$\Omega = \frac{30\ \text{m x } 6.28\ \text{d}^{-1}}{1.1\ \text{m}\cdot\text{d}^{-1}} = 171$$

Table P.4 summarizes the characteristic numbers for all cases.

Table P.4 Attenuation Numbers for a Nonsorbing Substance in the GWS between the River R and the Wells, Calculated for Three Typical Periods (Diurnal, Monthly, Annual).

Regime	Pe	Attenuation Number Ω		
		Diurnal	Monthly	Annual
Natural	10	171	5.7	0.47
Pump Regime I	10	34	1.1	0.092
Pump Regime II	10	15	0.50	0.041

Next, $\hat{\alpha}$ and $\hat{\beta}$ are calculated from Eqs. P-39 and P-40. For instance, the *daily variation in the Pump Regime II* yields the following coefficients (use Table P.4: Pe = 10, Ω = 15, i.e., $\left(\frac{4\Omega}{\text{Pe}}\right)^2 = 36$)

$$\hat{\alpha} = \frac{10}{2} \{[0.5 + 0.5\,(1 + 36)^{1/2}]^{1/2} - 1\} = 5\,\{[3.54]^{1/2} - 1\} = 4.4$$

$$\hat{\beta} = \frac{10}{2} \{\frac{1}{2}(37)^{1/2} - \frac{1}{2}\}^{1/2} = 8.0$$

Since $x_o = x_w = 10$ m was chosen as the reference point, Eq. P-37 has to be evaluated at $\xi = 1$. Thus, the sinusoidal variation at the well is attenuated by the factor

$$\exp(-\hat{\alpha}\cdot\xi) = \exp(-4.4 \times 1) = 0.012 \quad , \quad \tau = 1 \text{ day, Regime II}$$

i.e., by 1.2% of the full amplitude C_1. The phase shift is

$$\phi = -\hat{\beta} \cdot \xi = -8.0 \times 1 = -8.0 \quad , \quad \tau = 1 \text{ day, Regime II}$$

This shift is more than one full period (note that the sinus has a period of $2\pi = 6.28$). In fact, the shift expressed in number of periods is

$$\Delta = -\frac{8.0}{2\pi} = -1.27 \quad , \quad \tau = 1 \text{ day, Regime II}$$

Table P.5 Attenuation and Retardation of Sinusoidal Tracer Input from the River R Calculated for the Wells of the Groundwater System S.

		Natural Regime	Pump Regime I	Pump Regime II
Attenuation, $e^{-\hat{\alpha} \cdot \xi}$ (at $\xi=1$)				
Period [a]	day	$2.4\text{x}10^{-11}$	$2.0\text{x}10^{-4}$	$1.2\text{x}10^{-2}$
	month	0.20	0.89	0.98
	year	0.98	1.00	1.00
Phase shift, $\Delta = \frac{\hat{\beta} \cdot \xi}{2\pi}$ (at $\xi=1$)				
Period [a]	day	4.6	2.0	1.27
	month	0.69	0.17	0.079
	year	0.074	0.015	$6.5\text{x}10^{-3}$
Time lag of oscillation, $\Delta\tau$ (days)				
Period [a]	day	4.6	2.0	1.27
	month	21	5.1	2.4
	year	27	5.3	2.4
For comparison advective flow time, t_o (days)		27	5.4	2.4

[a] Corresponds to $\omega_1 = 6.28 \text{ d}^{-1}$, $\omega_{30} = 0.209 \text{ d}^{-1}$, $\omega_{365} = 0.0172 \text{ d}^{-1}$.

Table P.5 summarizes the relevant parameters for all periods and pump regimes.

Since the period is one day, the variation at the well lags behind by 1.27 days = 30 hours. Note that this is much shorter than the mean advective travelling time from the river to the well, t_o (2.4 days for Pump Regime II).

The following features can be extracted from these results:

(1) The attenuation strongly decreases with increasing length of the characteristic period τ. Particularly, for periods τ that are smaller than the mean advective flow time t_o, the signal of the concentration variation is practically absent at x_o.

(2) The attenuation decreases from the Natural Regime to Pump Regime II. This is consistent with (1), since the mean advective flow time t_o is smallest for Pump Regime II. A short advective flow time increases the change of the oscillations to reach the well.

(3) The time lag of the oscillation increases with increasing period τ and approaches (but does not surpass) the advective flow time t_o.

Effect of Sorption

As mentioned before, porous media have much larger solid-to-solution phase ratios (r_{sw}) than surface waters (lakes and rivers). Therefore, even the transport of a chemical with moderate to small solid-water distribution ratios (K_d) may be influenced by sorption processes. The basic mathematical tools which are needed to quantify the effect of sorption on transport are described in *Chapters 11* and *15*.

Although the case of finite sorption/desorption dynamics has recently earned a lot of research interest (see, e.g., *Brusseau, 1992*), the following discussion will be restricted to the case of a *linear and instantaneous local equilibrium* between the compound sorbed on solids, C_s ($mol \cdot kg^{-1}$), and the dissolved compound concentration, C_w ($mol \cdot m^{-3}$):

$$K_d = \frac{C_s}{C_w} \quad (m^3 \cdot kg^{-1}) \qquad (11\text{-}2)$$

For porous media it is convenient to choose C_w as the key variable since it is this concentration that is determined in filtered water samples taken at a well. The dynamic equation for C_w is derived from Eq. P-10 by introducing an additional term, ψ, that describes the sorption/desorption process:

$$\frac{\partial C_w}{\partial t} = - \bar{u}\frac{\partial C_w}{\partial x} + E_{dis}\frac{\partial^2 C_w}{\partial x^2} + \psi \qquad \text{(P-43)}$$

ψ, the amount of chemical added to the pore water per unit pore volume and time, is linked to the corresponding change of the sorbed concentration, C_s:

$$\psi = \left(\frac{\partial C_w}{\partial t}\right)_{sorption} = - r_{sw}\left(\frac{\partial C_s}{\partial t}\right)_{sorption} \qquad \text{(P-44)}$$

where r_{sw} is the solid-to-solution phase ratio, which is related to the density of the solid phase, ρ_s, and the porosity, ϕ, by

$$r_{sw} = \rho_s \frac{1-\phi}{\phi} \quad (\text{kg}\cdot\text{m}^{-3}) \qquad \textit{(11-8)}$$

With the instantaneous equilibrium, *Eq. 11-2*, it follows

$$\psi = - K_d \cdot \rho_s \cdot \frac{1-\phi}{\phi}\left(\frac{\partial C_w}{\partial t}\right) \qquad \text{(P-45)}$$

Inserting into Eq. P-43 and rearranging the terms yields

$$R_f \frac{\partial C_w}{\partial t} = - \bar{u}\frac{\partial C_w}{\partial x} + E_{dis}\frac{\partial^2 C_w}{\partial x^2} \qquad \text{(P-46)}$$

where

$$R_f = 1 + K_d \cdot \rho_s \cdot \frac{1-\phi}{\phi} \qquad \text{(P-47)}$$

is the *retardation factor* for a sorbing chemical. Note that for a very weakly sorbing chemical, R_f is equal to 1 ($K_d \approx 0$) and larger than 1 otherwise. The inverse of R_f is equal to f_w, the fraction in dissolved form as defined by *Eq. 11-5*. It also corresponds to the transport velocity of a chemical in the aquifer relative to the transport velocity of water or of a nonsorbing compound.

If R_f does not change along x, i.e., if the aquifer matrix is homogeneous, then the transformation

$$E^*_{dis} = E_{dis}/R_f, \quad \bar{u}^* = \bar{u}/Rf \qquad \text{(P-48)}$$

changes Eq. P-46 into the advection-diffusion equation of a nonsorbing compound (Eq. P-10). Note that the Peclet Number Pe (Eq. P-12) is not altered by Eq. P-48; a sorbing species has the same Pe value as water. Thus, the solutions of Eq. P-10 derived above for three different input scenarios (peak, step, sinusoidal) are also valid for sorbing chemicals if E_{dis} and $\bar{u}$ are replaced by E^*_{dis} and $\bar{u}^*$, respectively. The consequences of this substitution for the behavior of a chemical in groundwater are discussed separately for the three cases:

(1) *Pulse input*: The nondimensional solution, Eq. P-22, remains valid if the following modified definition is used:

$$t_o \rightarrow t^*_o = R_f \frac{x_o}{\bar{u}} = R_f\, t_o \qquad \text{(P-49)}$$

Note that the relative time, $\theta = t/t_o$, *at a given location* along the stream lines is the same for all chemicals, although t and t_o depend on R_f. Thus, the growth of the nondimensional standard deviation s (Eq. P-23) along the stream lines is not affected by sorption. All the compounds have the same standard deviation along the stream lines when they pass by some fixed location x. Of course, the less sorbing substances arrive sooner at x then the more sorbing ones. In contrast, an observer measuring C_w as a function of time at a *fixed location x*, e.g., at the well, has a different view: Since the concentration clouds are passing by the point of observation at their individual velocity, $\bar{u}/R_f$, the C(t) curve appears as a broader signal for the slow (sorbing) compound than for the fast (nonsorbing) one.

Liquid chromatography can be used to picture the transport of sorbing chemicals in an aquifer. If at a given time the concentration distributions of various compounds along the column of an ideal system could instantaneously be measured, then their standard deviations, σ, would be linearly related to the distance that they have already travelled. When the concentration cloud reaches the detector at the end of the column, the concentration is determined as a function of time. Thus, the early peaks are narrower than the later ones.

Sorption also affects the absolute value of C_w, particularly $C_{w,max}$. Given the total mass input m_{in} (Eq. P-24), only the fraction m_{in}/R_f is dissolved; thus, all concentrations in the dissolved phase are reduced by the factor R_f.

(2) *Step input*: All the conclusions drawn for the pulse input can directly be transferred to the step input. The concentrations, if expressed for

the nondimensional coordinates ξ and θ, are not affected by sorption; the shape of the concentration curve along x for a *fixed time* is independent of sorption. Yet, when the front passes by a fixed location x, the time needed for the concentration to increase from, say, 5% to 95% of the maximum concentration grows with R_f. This can directly be deduced from Table P.2, where the duration of the passage of the front is quantified by the nondimensional time interval $\Delta\theta = \theta_{95} - \theta_5$. This value does not depend on sorption, but after transformation back into real time it does (see Eq. P-49):

$$\Delta t^* = t_o^* \cdot \Delta\theta = R_f \cdot t_o \cdot \Delta\theta = R_f \cdot \Delta t \qquad \text{(P-50)}$$

Of course, the absolute time for the front to reach a given location increases in proportion to R_f.

(3) *Sinusoidal input:* All properties of the sinusoidal input scenario can be expressed by the two nondimensional numbers, Pe and Ω. Whereas Pe is not affected by sorption, the Attenuation Number Ω is modified (Eq. P-33):

$$\Omega \rightarrow \Omega^* = \frac{\omega x_o}{\bar{u}/R_f} = R_f \cdot \Omega \qquad \text{(P-51)}$$

Thus, sorption affects $C(\xi, \theta)$ in the same way as increasing the angular frequency ω by R_f or reducing the period τ by R_f^{-1}. As shown in Fig. P.7, such a change causes both the attenuation $\hat{\alpha}$ and the specific phase shift $\hat{\beta}$ to grow.

Problem

Calculate the retardation factor, R_f, for tetrachloroethene (PER, see *Chapter 15.2*) in the Groundwater System S. The fraction of organic matter in the solid aquifer material is 1.5‰; the density of the solids is $\rho_s = 2.5\ g \cdot cm^{-3}$.

Answer

The K_d value of PER is determined by the fraction of organic matter in the solid aquifer material, f_{om} = 0.0015, and the corresponding distribution coefficient, K_{om}:

$$K_d = f_{om} \cdot K_{om} \qquad \textit{(11-16)}$$

K_{om} is estimated from the octanol-water partition constant, K_{ow}, using the concept of linear free energy relationship. According to *Fig. 11.10* and with log K_{ow} (PER) = 2.88 (*Appendix*),

$$\log K_{om}\ (\text{PER}) = 0.82 \cdot \log K_{ow}\ (\text{PER}) + 0.14 = 2.50,$$

thus, K_{om} (PER) = 320 $L \cdot kg^{-1}$ and K_d (PER) = 0.48 $L \cdot kg^{-1}$. (Note that all the empirical relations given for K_d and K_{om} in *Chapter 11* refer to values expressed in units of $L \cdot kg^{-1}$.) Thus, in order to be consistent the solid density ρ_s has to be expressed in units of $kg \cdot L^{-1} = g \cdot cm^{-3}$). From Eq. P-47 the retardation factor of PER is ($\phi = 0.31$)

$$R_f\ (\text{PER}) = 1 + 0.48\ L \cdot kg^{-1} \times 2.5\ kg \cdot L^{-1} \times \frac{0.69}{0.31} = 3.7 \approx 4$$

Problem

Calculate the *annual* sinusoidal variation of the PER concentration in the wells of the Groundwater System S relative to the variation in the River R. Compare this number with the relative variation of a nonsorbing chemical such as *2,4-dinitrophenol.* Determine the time lag in the well relative to the variation in the river. Use all three flow regimes.

Answer

With the approximate retardation factor of PER, $R_f \approx 4$, the Attenuation Number Ω for the annual variations becomes (see Table P.4) $\Omega^*_{NR} = 1.9$, $\Omega^*_I = 0.37$, $\Omega^*_{II} = 0.16$, respectively. Attenuation and retardation are calculated from Eq. P-38 as explained for the case of the nonsorbing solute. The results are summarized in Table P.6. They show that the attenuation is not affected for PER that sorbs only weakly. In contrast, the time lag strongly increases and in all cases reaches nearly the advective flow time. The latter is four times larger than for the nonsorbing chemical: $t^*_o = 4 \cdot t_o$.

Measurements in an aquifer at the River Aare in Switzerland by *Schwarzenbach et al. (1983)* show a time lag of about 4 months between the concentration in the river and in a well situated about 13 m from the river. The corresponding amplitude ratio is about 0.8. In the same aquifer, the annual temperature variation shows a shift of 1 month from the river to the observation well. This demonstrates the more conservative (or less "sorbing") nature of water temperature versus PER.

Table P.6 Attenuation and Retardation of an *Annual* Input Signal of Tetrachloroethene (PER) at the Wells of the Groundwater System S Compared to the Situation of a Nonsorbing Chemical (see Table P.4). R_f (PER) = 4.

	Natural Regime	Pump Regime I	Pump Regime II
Attenuation, $e^{-\hat{\alpha}\cdot\xi}$ (at ξ=1)			
Nonsorbing	0.98	1.00	1.00
PER	0.73	0.99	1.00
Phase shift, $\Delta = \frac{\hat{\beta}\cdot\xi}{2\pi}$ (at ξ=1)			
Nonsorbing	0.074	0.015	6.5×10^{-3}
PER	0.28	0.059	0.025
Time lag of oscillation, $\Delta\cdot\tau$ (days)			
Nonsorbing	27	5.4	2.4
PER	100	21	9.5
Advective flow time (days)			
Nonsorbing	27	5.4	2.4
Sorbing	110	22	9.6

Effect of Transformation Processes

As a last step, a first-order (linear) reaction is added to the advective-diffusive equation of a sorbing substance, Eq. P-46:

$$R_f \frac{\partial C_w}{\partial t} = -\bar{u}\frac{\partial C_w}{\partial x} + E_{dis}\frac{\partial^2 C_w}{\partial x^2} - k_{r,w}\cdot C_w - (R_f - 1)k_{r,s}\cdot C_w \qquad \text{(P-52)}$$

where $k_{r,w}$ and $k_{r,s}$ are the first-order reaction rates for the dissolved and sorbed phase, respectively. This equation can also be written as

$$\frac{\partial C_w}{\partial t} = -\bar{u}^*\frac{\partial C_w}{\partial x} + E_{dis}^*\frac{\partial^2 C_w}{\partial x^2} - k_r^* C_w \qquad \text{(P-53)}$$

where $\bar{u}^*$, E_{dis}^* were defined in Eq. P-48, and with the apparent first-order rate constant

$$k_r^* = \frac{k_{r,w}}{R_f} + \frac{R_f - 1}{R_f} k_{r,s} \qquad \text{(P-54)}$$

Steady state solutions of Eq. P-53 are discussed in *Table 15.12*. The solution depends on two nondimensional numbers, on the Peclet Number Pe introduced earlier (Eq. P-12) and on the *Damköhler Number*

$$Da = \frac{E_{dis}^* \cdot k_r^*}{(\bar{u}^*)^2} \qquad \text{(P-55)}$$

In nondimensional form, the steady-state version of Eq. P-53 is *(Table 15.12, Eq. d)*

$$\frac{d^2 C_w}{d\xi^2} - Pe \frac{dC_w}{d\xi} - Pe \cdot Da^2 \cdot C_w = 0 \qquad \text{(P-56)}$$

with solution (Table 15.12, Eq. e)

$$C_w(\xi) = C_w\left(\frac{x}{x_o}\right) = A_1 \cdot e^{\lambda_1 \xi} + A_2 \cdot e^{\lambda_2 \xi} \qquad \text{(P-57)}$$

The coefficients A_1 and A_2 are fixed by two boundary conditions. The eigenvalues λ_1 and λ_2 are *(Table 15.12, Eq. f)*

$$\lambda_{1,2} = \frac{Pe}{2}[1 \pm \{1 + 4 \cdot Da\}^{1/2}] \qquad \text{(P-58)}$$

Figure 15.13 shows schematically the shape of the solution $C_w(\xi)$ for different values of Pe and Da.

Note: There are several definitions of the Damköhler Number Da in the literature. Equation P-55 corresponds to the definition adopted in *Table 15.12*. Several authors distinguish between two (or more) Damköhler Numbers (see, e.g., *Domenico and Schwartz, 1990.*) The Damköhler Number Da_I defined by

$$Da_I = \frac{k_r \cdot x_o}{\bar{u}} \qquad \text{(P-55a)}$$

is the ratio of the time needed for advection (over distance x_o) versus the typical reaction time. In contrast, the Damköhler Number

$$Da_{II} = \frac{k_r \cdot x_o^2}{E} \qquad \text{(P-55b)}$$

is the ratio of the time needed for diffusion (molecular, turbulent, or dispersive) versus the reaction time. Note that the ratio of Da_{II} and Da_I is the Peclet Number Pe. The definition used in Eq. P-55 is a mixture of these numbers, i.e.,

$$Da = \frac{Da_I^2}{Da_{II}} \qquad \text{(P-55c)}$$

Problem

The concentration of benzylchloride in the River R at GWS shows a constant value of 3 $\mu g \cdot L^{-1}$. When the infiltrating river water is flowing through the GWS, benzylchloride is hydrolized and also sorbs to the particles of the aquifer. The rate of hydrolysis is assumed to be the same for the dissolved and the sorbed phase. Possible biodegradation of benzylchloride is disregarded.

The water temperature is 10°C. Assume that the rate of hydrolysis at 10°C is four times smaller than at 25°C. Calculate for the Natural Regime the concentration of benzylchloride at x = 3 m and at the wells (x = 30 m). The octanol-water partition coefficient of benzylchloride is $\log K_{ow}$ = 2.30 (*Hansch and Leo, 1979*). Use $\phi = 0.31$, $f_{om} = 0.0015\ kg_{om} \cdot kg_s^{-1}$ and $\rho_s = 2.5\ g \cdot cm^{-3}$ to characterize the solid phase of the aquifer (Table P.1).

benzylchloride

Hints and Help

One of the boundary conditions can be chosen at x = 0: $C_w = C_{River}$. Due to hydrolysis and biodegradation, benzylchloride eventually disappears completely from the groundwater. Therefore, it can be assumed that C_w = 0 for $x \rightarrow \infty$. This serves to formulate the second boundary condition.

Answer

The half-life of benzylchloride with respect to hydrolysis at 25°C is 15 hours *(Table 12.7)*. Thus, at 10°C the first-order rate constant is (see *Eq. 12-11)*

$$k_r = k_{r,w} = k_{r,s} = \frac{\ln 2}{4 \times 15 \text{ h}} = \frac{0.693}{60 \text{ h}} = 0.012 \cdot \text{h}^{-1} = 3.2 \times 10^{-6} \text{s}^{-1}$$

From *Fig. 11.10* follows

$$\log K_{om} = 0.82 \times 2.30 + 0.14 = 2.02 \ (\text{L} \cdot \text{kg}^{-1})$$

thus $K_{om} = 106 \text{ L} \cdot \text{kg}^{-1}$ and $K_d = 0.0015 \times 106 \text{ L} \cdot \text{kg}^{-1} = 0.16 \text{ L} \cdot \text{kg}^{-1}$.

The retardation factor of benzylchloride is calculated from Eq. P-47:

$$R_f = 1 + 0.16 \text{ L} \cdot \text{kg}^{-1} \times 2.5 \text{ kg} \cdot \text{L}^{-1} \times \frac{0.69}{0.31} = 1.9 \approx 2$$

The relevant coefficients of Eq. P-53 for the Natural Regime are

$$\bar{u}^* = \frac{1.3 \times 10^{-5} \text{ m} \cdot \text{s}^{-1}}{2} = 6.5 \times 10^{-6} \text{ m} \cdot \text{s}^{-1}$$

$$E_{dis}^* = \frac{3.9 \times 10^{-5} \text{ m}^2 \cdot \text{s}^{-1}}{2} = 2.0 \times 10^{-5} \text{ m} \cdot \text{s}^{-1}$$

$$k_r^* = \frac{3.2 \times 10^{-6} \text{ s}^{-1}}{2} = 1.6 \times 10^{-6} \text{ s}^{-1}$$

The Peclet Number (Eq. P-12, with $x_o = 30$ m) is affected neither by sorption nor by reaction: Pe = 10. The Damköhler Number is (Eq. P-55)

$$Da = \frac{2.0 \times 10^{-5} \times 1.6 \times 10^{-6}}{(6.5 \times 10^{-6})^2} = 0.8$$

Note that for the special case, $k_{r,w} = k_{r,s}$, Da is independent of the retardation factor R_f.

The eigenvalues are (Eq. P-58)

$$\lambda_{1,2} = \frac{10}{2} [1 \pm \{1 + 4 \times 0.8\}^{1/2}]$$

$$\lambda_1 = 15, \quad \lambda_2 = -5$$

Since $C_w(x \rightarrow \infty) = 0$, the coefficient A_1 of Eq. P-57 that belongs to the first eigenvalue λ_1 must be zero. Then, A_2 is just equal to the concentration in the infiltrating water: $A_2 = 3\ \mu g \cdot L^{-1}$. Thus,

$$C_w(\xi) = 3\ \mu g \cdot L^{-1} \cdot \exp(-5\ \xi)$$

Evaluated at $\xi = 1$ ($x_o = 30$ m) and $\xi = 0.1$ ($x = 3$ m)

$$C_w\ (x = 30\ m) = 3\ \mu g \cdot L^{-1} \times 6.7 \times 10^{-3} = 0.02\ \mu g \cdot L^{-1}$$

$$C_w\ (x = 3\ m) = 3\ \mu g \cdot L^{-1} \times 0.6 = 1.8\ \mu g \cdot L^{-1}$$

This result demonstrates that it may be worthwhile to drill wells not too close to the river even at the expenses of a reduced yield of water.

Solutions of Eq. P-53 for *nonsteady-state conditions* are difficult to obtain analytically, yet numerical procedures are straight-forward. For the case of slow reactions, more precisely for Da « 1, the solution can be approximated by multiplying the time-dependent solution for a conservative substance with the linear decay term $\exp(-k_r^* \cdot t)$. Thus, for the pulse input Eq. P-19 is modified into

$$C(x,t) \sim \frac{m_{in}\ e^{-k_r^* \cdot t}}{2(\pi E_{dis}^* t)^{1/2}} \exp\left(-\frac{(x-\bar{u}^* t)^2}{4 E_{dis}^* t}\right), \quad Da \ll 1 \qquad \text{(P-59)}$$

and correspondingly for Eq. P-30 (step input) and Eq. P-37 (sinusoidal input).

The Role of Colloids in Pollutant Transport

The influence of colloids on the transport of chemicals in groundwater has recently become an important issue in groundwater research. Colloids can be important for the transport of contaminants; they may also influence remediation efforts of contaminated aquifers. A recent review of the topic is given by *Corapcioglu and Jiang (1993)*.

It would lie far beyond the aim of this chapter to introduce the state-of-the art concepts that have been developed to quantify the influence of colloids on transport and reaction of chemicals in an aquifer. Instead, a few effects will be discussed on a purely qualitative level. In general, the presence of colloidal particles, like dissolved organic matter (DOM), enhances the transport of chemicals in groundwater. Figure P.8 gives a conceptual view of the relevant interaction mechanisms of colloids in saturated porous media. A simple model consists of just three phases, the dissolved (aqueous) phase, the colloid (carrier) phase and the solid matrix (stationary) phase. The distribution of a chemical between the phases can – as a first step – be described by an equilibrium relation as introduced in *Section 15.4* to discuss the effect of colloids on the fate of polychlorinated biphenyls (PCBs) in Lake Superior (see *Table 15.10* and Chapter J).

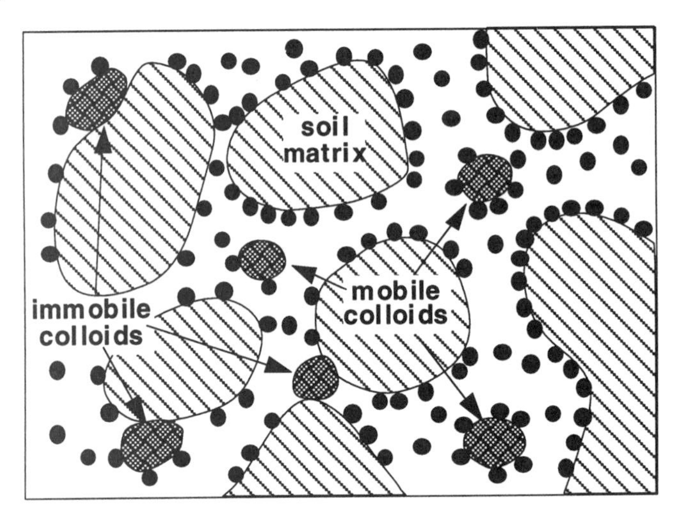

Figure P.8 Conceptual view of a saturated porous medium with colloids. The black dots represent the contaminant that is dissolved, sorbed on soil matrix particles, or sorbed on colloids. The colloids themselves can sorb on the soil matrix. Adapted from *Corapcioglu and Jiang (1993)*

For the case of the PCBs it has been shown that sorption on colloids greatly enhances the "dissolved" concentration, which is usually determined on filtered samples in which most colloids are still present. For PCBs in lakes, the colloidal fraction of the chemical may be the most important phase. Since it contributes neither to air/water exchange nor to sedimentation (only the larger particles, not the colloids, sink to the bottom), the effect of the colloids can result in an increase of the residence time of the chemical in the lake.

In contrast, transport of chemicals through groundwater systems is generally enhanced by colloids. Though colloids may become immobile due to attachment to the solid matrix, part of them are moving with the water. Thus, chemicals that are sorbed on colloids and not on the solid matrix are not (or not as strongly) retarded as one would calculate by using the retardation factor R_f.

An additional process that enhances transport is due to the typical velocity distribution within the pores of the aquifer. Since colloids mainly move in the central parts of the larger pores, their effective flow velocity may be larger than $\bar{u}$. Remember that $\bar{u}$ represents an average of all the velocities in the aquifer. Colloids have the tendency to select the faster stream lines. This phenomenon is called hydrodynamic chromatography *(Small, 1974; Stoistis et al., 1976).*

Even a simple mathematical model for transport on colloids in an aquifer must include dynamic equations for the dissolved phase and for the colloids. The second equation describes the migration, immobilization, and detachment of the colloids. More sophisticated models include dynamic equations for sorption and desorption of the chemical onto colloids and the stationary solid phase. The consequences of the latter phenomenon are shortly discussed in the last part of this chapter.

Models and Reality

The aim of this chapter was to present some of the basic mechanisms that determine the fate of organic pollutants in porous media. In order to achieve a general understanding, simple models were presented that can be analyzed by analytical techniques and that have explicit mathematical solutions.

The price that was paid for this kind of simplicity is obvious to the specialist. There are numerous reasons why real systems do not behave as described above. However, one can only understand the complex case once the simple one has been analyzed.

A recent review article by *Brusseau (1994)* gives an overview on the real behavior of reactive contaminants in heterogeneous porous media. The reader can also find a large up-dated list of references on transport in porous media. Just a few phenomena are shortly mentioned here:

(1) The presence of macropores as well as of "dead end zones" causes a real breakthrough curve to look quite differently from that given by Eq. P-27. Real breakthrough curves may rise earlier (early breakthrough) and smear out later. It may take longer to reach the full concentration since the exchange of water and chemicals is

slow between the macropore space and the space of the pores that are less well connected to the main flow.

(2) Sorption kinetics: Sorption and desorption may not always be fast compared to other processes, such as advection and dispersion. In addition, the sorption equilibrium does not necessary follow a linear relationship.

(3) Transport in unsaturated aquifers, especially in their gaseous phase, may lead to rather complicated situations.

(4) Kinetics of microbial growth: In the case of a sudden contamination of an aquifer by a chemical compound, the build-up of the appropriate population of microorganisms able to decompose the compound may be a slow and complicated process. Such situations have to be simultaneously analyzed by field observations and models, but the latter may become rather complicated.

(5) The heterogeneity of porous media with respect to their hydraulic permeability poses one of the most difficult problems. This is especially true for aquifers formed by glacial and fluvial deposits. Prediction of breakthrough curves may become impossible if a few long macropores or highly conducting layers are present in which water moves at a speed 10 or 100 times faster than the effective mean velocity. Such situations are still full of surprises, even to the specialist.

CASE STUDIES

● *Case Study P-1 Assessing the Effect of River Water Pollution on the Quality of the Bank Filtrate Used for Drinking Water Supply*

You are responsible for the safe operation of a drinking water supply system that gets its raw water from a well located close to a river. From tracer experiments you know that the effective mean flow velocity, $\bar{u}$, is 3 $m \cdot d^{-1}$ and that the distance along the streamline from the point of infiltration to the well is x_w = 18 m. The dispersivity of the aquifer for this distance of flow is α_L = 5 m. In order to be prepared for a possible pollution event in the river you are interested in the following question:

Problem (a)

In order to prevent polluted river water from reaching your drinking water system, you want to know how much time you have to turn off the pumps once a pollution cloud in the river has reached the location adjacent to the well. In your considerations you assume that the concentration of the pollutant suddenly increases from 0 to a value 10 times above the maxiumum tolerable drinking water concentration and then remains at this level. Take the worst case scenario and calculate how much time you have to turn off the pumps.

How much does this time change if you assume that the concentration in the river reaches 1000 times the maximum tolerable drinking water concentration?

Hints and Help

Use the equations derived for the step input to model the pollution scenario. Think about the appropriate choice of R_f in order to cover the worst case.

Problem (b)

One day, measurements in the river suddenly show rather high concentrations of 2,4-dinitrophenol. Further observations show that the values vary on a weekly basis and reach maximum concentrations C_{max} = 3.5 $\mu g \cdot L^{-1}$ during the week and minimum concentrations

C_{min} = 1.3 μg·L^{-1} over the weekend. Calculate the maximum concentration of 2,4-dinitrophenol to be expected at the drinking water well. The pH in the infiltrating water is 7.8.

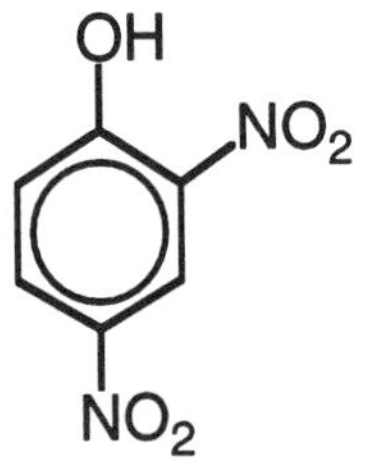

OH, NO_2, NO_2 (structure)

mw 184.1 g·mol^{-1}

2,4-dinitrophenol

Hints and Help

Assume first that the concentration in the river can be described by a sinusoidal function with a period of one week. Then remember that any deviation from the sinus mode can be accounted for by superposition of harmonic oscillations of *higher* frequencies (Fourier series). Use Fig. P.7 to prove that the attenuation of these faster modes is larger than the attenuation of the weekly mode and that therefore they do not affect the result calculated for the pure sinus input alone.

Problem (c)

You wonder how much the weekly variation of trichloroethene, another pollutant, would be reduced from the river to the well. Use the following information on the aquifer: porosity ϕ = 0.37, solid phase density ρ_s = 2.5 g·cm^{-3}, organic matter content of solid phase f_{om} = 0.01. Compare this attenuation with the case of a concentration variation at the well that is caused by a short pollution pulse of trichloroethene in the river. To what fraction is the maximum concentration reduced from the river to the pumping well?

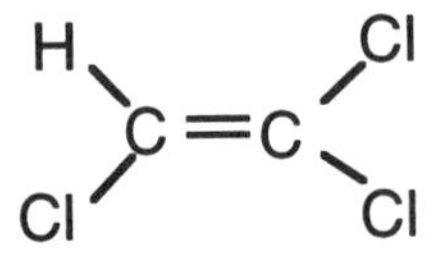

H, Cl, C=C, Cl, Cl (structure)

mw 131.4 g·mol^{-1}

trichloroethene

● *Case Study P-2 Determining the Hydraulic Properties of an Aquifer from Observed Concentration Time Series in Two Adjacent Wells*

You are interested in the hydraulic and geochemical properties of an aquifer, particularly in its dispersivity and average organic matter content. Fortunately, at your disposition you have long-term time series of water temperature and tetrachloroethene (PER) measured in two adjacent wells, 15 m apart from each other along the main flow direction of the groundwater. From earlier tracer experiments you are pretty sure that the two wells are located on the same streamline of the aquifer, i.e., that they "see" the same water passing by, although not at the same time.

Both parameters, water temperature and PER concentration, show a clear annual variation. The maximum (minimum) water temperature is registered in the first well about 18 days earlier than in the second one, and the temperature amplitudes are approximately equal in both wells. In contrast, the time lag observed for the concentration variation of PER between the two wells is 6 months. The concentration amplitude of PER in the downstream well, i.e., the difference between maximum and mean concentration, is only about 65% of the amplitude in the upstream well, but the annual mean concentrations are equal in both wells.

Estimate the dispersivity α_L of the aquifer and the retardation factor R_f of PER. Calculate the organic matter content, f_{om}, of the aquifer material by using a rough estimate for the solid phase density, $\rho_s = 2.5\ g \cdot cm^{-1}$, and for the effective porosity $\phi = 0.35$.

$Cl_2C{=}CCl_2$ mw 165.8 $g \cdot mol^{-1}$

tetrachloroethene
(PER)

Hints and Help

Assume, as a first approximation, that water temperature is a conservative, nonretarded parameter. You can then interpret the R_f value calculated for PER as a retardation factor *relative* to water temperature. You may want to estimate the retardation of water temperature. How could this be done? It may be helpful to assume, as a rough estimate, an average value for the specific heat of the solids in the aquifer, c_p, of 1 $J \cdot kg^{-1} \cdot K^{-1}$.

● *Case Study P-3* *Determining the Characteristic Transport Properties from Measurements of Breakthrough Curves in Laboratory Columns*

You work in a research laboratory, and part of your duty is to determine the sorption behavior of a series of polycyclic aromatic hydrocarbons in laboratory columns. In one series of experiments you use a column that is 1 m long and has a diameter of 5 cm. The flow of water through the column is regulated by a pump. The total through flow of water is held constant at 20 mL per hour. The column material is sterile to avoid biological transformation processes. Prior to every experiment the column is carefully flushed and then kept saturated with pure water.

Problem (a)

With the first experiment, conducted with chloride, you want to determine the hydraulic parameters of the column. At time t = 0, the chloride concentration at the input is set to $C_{in} = 100\ \text{mg}\cdot\text{L}^{-1}$ and then kept constant. Time series of chloride concentrations are measured at the outlet of the column. The results are given in the table below. Determine the porosity ϕ and the dispersion coefficient E_{dis} of the column.

Problem (b)

The second experiment is conducted with naphthalene. At t = 0 the concentration of naphthalene is set to a constant value of $C_{in} = 1\ \mu\text{mol}\cdot\text{L}^{-1}$ and the naphthalene concentration measured in the outlet as a function of time (see table below). Calculate the retardation factor R_f as well as the relative organic content, f_{om}, of the column material. Use $\rho_s = 2.5\ \text{g}\cdot\text{cm}^{-3}$ and the parameters determined for chloride.

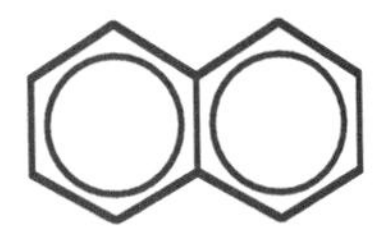

mw 128.2 $\text{g}\cdot\text{mol}^{-1}$

naphthalene

Data from chloride experiment

Time since beginning (hours)	chloride concentration ($mg \cdot L^{-1}$)	Time since beginning (hours)	chloride concentration ($mg \cdot L^{-1}$)
12	0	44	60
16	1	48	67
20	4	52	75
24	10	56	80
28	18	64	89
32	28	80	96
36	39	90	100
40	51		

Data from naphthalene experiment

Time since beginning (hours)	naphthalene concentration ($nmol \cdot L^{-1} \cdot L^{-1}$)	Time since beginning (hours)	naphthalene concentration ($nmol \cdot L^{-1} \cdot L^{-1}$)
100	0	425	660
200	30	450	620
250	100	475	660
300	200	500	710
350	380	550	790
400	490	600	850

● *Case Study P-4 Migration of Chemical Pollutants from Dump Site through Clay Liner into the Groundwater*

A clay liner of 1 m thickness protects the underlying groundwater from chemically polluted soil. The characteristics of the clay liner are given in Problem J-1. The hydrogeologists assure you that no water can flow from the polluted area into the groundwater. However, you are also concerned with transport by diffusion and decide to estimate the minimum time needed for the four compounds listed in Problem J-1 to reach the groundwater.

Hints and Help
Estimate the transport time by assuming that the chemicals diffuse from a reservoir of constant concentration into the liner that at time t = 0 is assumed to be uncontaminated. Equation P-11 may give you an idea how to calculate the effective diffusion coefficient through the liner. Select the most critical compound among the substances listed above. What is the criterion?

● *Case Study P-5 Interpreting Real Breakthrough Curves*

A field experiment was conducted at the Canadian Air Forces Base Borden, Ontario, to study the behavior of organic pollutants in a sand aquifer under natural conditions (*Mackay et al., 1986*). Figure P.9 shows the results of two experiments, the first one for tetrachloroethene, the second one for chloride. Both substances were added as short pulses to the aquifer. The curves designed as "ideal" were computed according to Eqs. P-19 or P-22; the measured data clearly deviate from the ideal curve. The "nonideal" curves were constructed by *Brusseau (1994)* with a mathematical model that includes various factors causing nonideal behavior.

Give qualitative reasons to explain the experimental data.

Hints and Help
The concentrations curves in Fig. P.9 are plotted versus *pore volumes*, V_p, instead of time. V_p sums up the amount of water (per unit area) flowing through the aquifer expressed in units of pore space (per unit area) contained in the aquifer between input and output (location of measurement). Convince yourself that this variable is equivalent to time and that for ideal flow of a nonsorbing substance the input and output curves should be identical if shifted by $\Delta V_p = 1$.

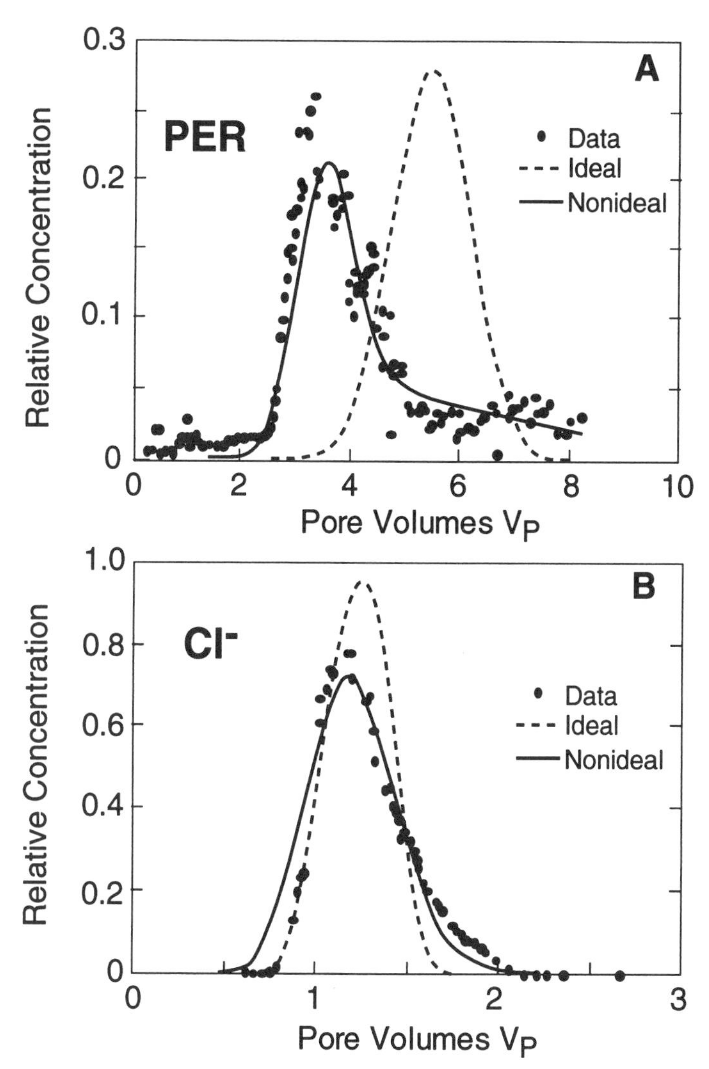

Figure P.9 Measured breakthrough curves from the Borden, Ontario, field experiment for tetrachloroethene (PER) and chloride (Cl^-). From *Brusseau (1994)*.

Personal Notes

Q. BIBLIOGRAPHY

Abramowitz, M. and A. Stegun, *Handbook of Mathematical Functions,* Dover, New York, 1965.

Achman, D.A., K.H. Hornbuckle, and S.J. Eisenreich, "Volatilization of polychlorinated biphenyls in Green Bay, Lake Michigan," *Environ. Sci. Technol.*, **27**, 75-87 (1993).

Appelo, C.A.J. and D. Postma, *Geochemistry, Groundwater and Pollution*, A.A. Balkema Publishers, Rotterdam, Netherlands, 1993.

Atkins, P.W., *Physical Chemistry*, W.H. Freeman, San Francisco CA, 1978.

Banerjee, S., "Calculation of water solubility of organic compounds with UNIFAC-derived parameters," *Environ. Sci. Technol.,* **19**, 369-370 (1985).

Bartholomew, G., W. Gene, and F.K. Pfaender, "Influence of spatial and temporal variations on organic pollutant biodegradation in an estuarine environment," *Appl. Environ. Microbiol.,* **45**, 103-109 (1983).

Bidleman, T.F., "Atmospheric processes," *Environ. Sci. Technol.*, **22**, 361-367 and 726-727 (errata) (1988).

Bruckmann, P., W. Kersten, W. Funcke, E. Balfanz, J. König, J. Theisen, M. Ball, and O. Päpke, "The occurrence of chlorinated and other organic trace compounds in urban air," *Chemosphere*, **17**, 2363-2380 (1988).

Brunner, S., E. Hornung, H. Santl, E. Wolff, O.G. Piringer, J. Altschuh, and R. Brüggemann, "Henry's law constant for polychlorinated biphenyls: Experimental determination and structure-property relationships," *Environ. Sci. Technol.*, **24**, 1751-1754 (1990).

Brusseau, M.L., "Transport of rate-limited sorbing solutes in heterogeneous porous media: Applicaton of a one-dimensional multifactor non ideality model to field data," *Water Resour. Res.*, **29**, 2485-2487 (1992).

Brusseau, M.L., "The influence of solute size, pore water velocity, and intraparticle porosity on solute dispersion and transport in soil," *Water Resour. Res.*, **29**, 1071-1080 (1993).

Brusseau, M.L., "Transport of reactive contaminants in heterogeneous porous media," *Rev. Geophys.*, **32**, 285-313 (1994).

Capel, P.D., W. Giger, P. Reichert, and O. Wanner, "Accidental input of pesticides into the Rhine River," *Environ. Sci. Technol.,* **22**, 992-996 (1988).

Chiou, C.T., "Partition coefficients of organic compounds in lipid-water systems and correlations with fish bioconcentration factors," *Environ. Sci. Technol.,* **19**, 57-62, (1985).

Chiou, C.T., P.E. Porter, and D.W. Schmedding, "Partition equilibria of nonionic organic compounds between soil organic matter and water," *Environ. Sci. Technol.,* **17**, 227-231 (1983).

Chiou, C.T., R.L. Malcolm, T.I., Brinton, and D.E. Kile, "Water solubility enhancement of some organic pollutants and pesticides by dissolved humic and fulvic acids," *Environ. Sci. Technol.,* **20**, 502-508 (1986).

Chow, V.T., *Open-Channel Hydraulics*, McGraw-Hill, New York, 1959.

Cirpka, O., P. Reichert, O. Wanner, S.R. Müller, and R.P. Schwarzenbach, "Gas exchange at river cascades: Field experiments and model calculations," *Environ. Sci. Techn.,* **27**, 2086-2097 (1993).

Clark, W.M., *Oxidation-Reduction Potentials of Organic Systems*, Williams and Wilkins, Baltimore, MD, 1960.

Cohen, B.A., L.R. Krumbolz, H.-Y. Kim, and H.F. Hemond, "*In situ* biodegradation of toluene in a contaminated stream: Part 2. Laboratory studies," *Environ. Sci. Technol.*, **29**, 117-125 (1995).

Corapcioglu, M.Y. and S. Jiang, "Colloid-facilitated groundwater contaminant transport," *Water Resour. Res.*, **29**, 2215-2226 (1993).

Crank, J., *The Mathematics of Diffusion*, 2nd Ed., Clarendon Press, Oxford, 1975.

CRC Handbook of Chemistry and Physics, 66th Ed., CRC Press, Boca Raton, FL, 1985-1986.

Deneer, J.W., T.L. Sinnige, W. Seinen, and J.L.M. Hermens, "Quantitative structure-activity relationships for the toxicity and bioconcentration factor of nitrobenzene derivatives towards the guppy (Poecilia reticulata)," *Aquat. Toxicol.,* **10**, 115-129 (1987).

Domenico, P.A. and F.W. Schwartz, *Physical and Chemical Hydrology,* Wiley, New York, 1990.

Dunnivant, F.M., D. Macalady, and R.P. Schwarzenbach, "Reduction of substituted nitrobenzenes in aqueous solutions containing natural organic matter," *Environ. Sci. Technol.* **26**, 2133-2141 (1992).

Eylers, H., "Transport of adsorbing metal ions between stream and sediment bed in a metal laboratory flume," Ph.D. thesis, California Institute of Technology, Pasadena, CA, 1994.

Field, R.A., M.E. Goldstone, J.N. Lester, and R. Perry, "The sources and behaviour of tropospheric anthropogenic volatile hydrocarbons," *Atmos. Environ.,* **26A**, 2983-2996 (1992).

Fischer, H.B., J. Imberger, E.J. List, R.C.Y. Koh, and N.H. Brooks, *Mixing in Inland and Coastal Waters,* Academic Press, New York, 1979.

Fogel, M.N., A.R. Taddeo, and S. Fogel, "Biodegradation of chlorinated ethenes by a methane-utilizing mixed culture, *Appl. Environ. Microbiol.*, **51**, 720-724 (1986).

Fogelqvist, E., "Carbon tetrachloride, tetrachloroethylene, 1,1,1-trichloroethane and bromoform in arctic seawater," *J. Geophys. Res.,* **90**, 9181-9183 (1985).

Freeze, R.A., and J.A. Cherry, *Groundwater*, Prentice-Hall, Englewood Cliffs, NJ, 1979.

Fuller, E.N., Schettler, P.D., and Giddings, J.C., "A new method for prediction of binary gas-phase diffusion coefficient," *Ind. Eng. Chem.,* **58**, 19-27 (1966).

Giger, W., C. Schaffner, F.G. Kari, H. Ponusz, P. Reichert, and O. Wanner, "Occurrence and behavior of NTA and EDTA in Swiss rivers," *Mitteilungen der EAWAG*, 32D, 27-31 (1991).

Gossett, J.M., "Measurement of Henry's Law constants for C_1 and C_2 chlorinated hydrocarbons," *Environ. Sci. Technol.*, **21**, 202-208 (1987).

Gschwend, P.M. and R.A. Hites, "Fluxes of polycyclic aromatic hydrocarbons to marine and lacustrine sediments in the Northeastern United States," *Geochim. Cosmochim. Acta*, **45**, 2359-2367 (1981).

Haag, W.R. and T. Mill, "Effect of subsurface sediments on hydrolysis of haloalkanes and epoxides," *Environ. Sci. Technol.,* **22**, 658-663 (1988).

Haderlein, S.B. and R.P. Schwarzenbach, "Adsorption of substituted nitrobenzenes and nitrophenols to mineral surfaces," *Environ. Sci. Technol.,* **27**, 316-326 (1993).

Haderlein, S.B., K. Weissmahr, and R.P. Schwarzenbach, "Specific adsorption of nitroaromatic pesticides and explosives to clay minerals," *Environ. Sci. Technol*., submitted (1995).

Hansch, C. and A. Leo, *Substituent Constants for Correlation Analysis in Chemistry and Biology*, Wiley, New York, 1979.

Harris, J.C. and M.J. Hayes, "Acid dissociation constant," In W.J. Lyman, W.F. Reehl, and D.H. Rosenblatt, Eds., *Handbook of Chemical Property Estimation Methods,* Chapter 6, McGraw-Hill, New York, 1990.

Hayduk, W. and H. Laudie, "Prediction of diffusion coefficients for non-electrolytes in dilute aqueous solutions," *AIChE, J.,* **20**, 611-615 (1974).

Hermens, J.L.M., "Quantitative structure-activity relationships of environmental pollutants," In O. Hutzinger, (Ed.),*The Handbook of Environmental Chemistry*, Vol. 2, Part B, Springer, Berlin, 1989.

Hine, J. and P.K. Mookerjee, "The intrinsic hydrophilic character of organic compounds. Correlations in terms of structural contributions," *J. Org. Chem.*, **40**, 292-298 (1975).

Horvath, A.L., *Halogenated Hydrocarbons. Solubility-Miscibility with Water,* Marcel Dekker, New York, 1982.

Howard, P. H., *Fate and Exposure Data for Organic Chemicals*, Lewis Publ., Chelsea, MI, 1989.

Hunkeler, D., P. Höhener, A. Häner, T. Bregnard, and J. Zeyer, "Quantification of hydrocarbon mineralization in a diesel fuel contaminated aquifer treated by *in situ* biorestoration," In K. Kovar and J. Krasny, Eds., *Groundwater Quality: Remediation and Protection*, IAHS Press, Wallingford, UK, 1995, in press.

Imboden, D.M. and R.P. Schwarzenbach, "Spatial and temporal distribution of chemical substances in lakes: Modeling concepts," In W. Stumm, Ed., *Chemical Processes in Lakes,* Wiley-Interscience, New York, 1985, pp. 1-30.

Imboden, D.M., and A. Wüest, "Mixing mechanisms in lakes," In A. Lerman, D.M. Imboden, and J. Got, Eds., *Physics and Chemistry of Lakes*, Springer, Heidelberg, 1995.

Jafvert, C.T., "Sorption of organic acid compounds to sediments: Initial model development," *Environ. Toxicol. Chem.,* **9**, 1259-1268 (1990).

Jafvert, C.T. and N.L. Wolfe, "Degradation of selected halogenated ethanes in anoxic sediment-water systems," *Environ. Toxicol. and Chem.,* **6**, 827-837 (1987).

Jafvert, C.T., J.C. Westall, E. Grieder, and R.P. Schwarzenbach, "Distribution of hydrophobic ionogenic organic compounds between octanol and water: Organic acids,"*Environ. Sci. Technol., **24***, 1795-1803 (1990).

Jeffers, P.M., L.M. Ward, L.M. Woytowitch, and N.L. Wolfe, "Homogeneous hydrolysis rate constants for selected chlorinated methanes, ethanes, ethenes, and propanes," *Environ. Sci. Technol.,* **23**, 965-969 (1989).

Johnson, C.A. and J.C. Westall, "Effect of pH and KCl concentration on the octanol-water distribution of methyl anilines," *Environ. Sci. Technol.,* **24**, 1869-1875 (1990).

Jung, R.F., R.O. James, and T.W. Healy, "Adsorption, precipitation and electrokinetic processes in the iron oxide (geothite)-oleic acid-oleate system," *J. Colloid. Int. Sci.,* **118**, 463-472 (1987).

Junge, C.E., "Basic considerations about trace constituents in the atmosphere as related to the fate of global pollutants," In I.H. Suffet, Ed., *Fate of Pollutants in the Air and Water Environments, Section I, Mechanisms of Interaction between Environments*, Wiley-Interscience, New York, 1977, pp. 7-26.

Kim, H.-Y., "Fate and transport of volatile organic compounds in the Aberjona Watershed," Ph. D. dissertation, Dept. of Civil and Environ. Eng., MIT, Cambridge, MA (1995).

Kim, H.-Y., H.F. Hemond, L.R. Krumholz, and B.A. Cohen, "In-situ biodegradation of toluene in a contaminated stream: Part 1. Field studies," *Environ. Sci. Technol.*, **29**, 108-116 (1995).

Kolpin, D.W. and S.J. Kolkhoff, "Atrazine degradation in a small stream in Iowa," *Environ. Sci. Technol.,* **27**, 134-139 (1993).

Könemann, H., "Quantitative structure-activity relationships in fish toxicity studies," *Toxicology*, **19**, 209-221 (1981).

Laha, S. and R.G. Luthy, "Oxidation of aniline and other primary aromatic amines by manganese dioxide," *Environ. Sci. Technol.,* **24**, 363-373 (1990).

Larson, R.A. and E.J. Weber, *Reaction Mechanisms in Environmental Organic Chemistry*, Lewis, Boca Raton, FL, 1994.

Lerman, A., *Geochemical processes. Water and Sediment Environments*, Wiley, New York, 1979.

Ligocki, M.P., and J.F. Pankow, "Measurements of the gas/particle distribution of atmospheric organic compounds," *Environ . Sci. Technol.*, **23**, 75-83 (1989).

Lincoft, A.H., and J.M. Gossett, "The determination of Henry's constant for volatile organics by equilibrium partitioning in closed systems," In W. Brutsaert and G.H. Jirka, Eds., *Gas Transfer at Water Surfaces,* Reidel, Dordrecht, 1984, pp. 17-25.

Mackay, D.M., D.L. Freyberg, P.V. Roberts, and J.A. Cherry, "A natural gradient experiment on solute transport in a sand aquifer, 1, approach and overview of plume movement," *Water Resour. Res.*, **22**, 2017-2029 (1986).

Millero, F.J. and M.L. Sohn, *Chemical Oceanography*, CRC Press, Boca Raton, FL, 1992.

Munz, C.H. and P.V. Roberts, "The effects of solute concentration and cosolvents on the aqueous activity coefficients of low molecular weight halogenated hydrocarbons," *Environ. Sci. Technol.,* **20**, 830-836 (1986).

Neely W.G., D.R. Branson, and G.E. Blau, "The use of the partition coefficient to measure the bioconcentration potential of organic chemicals in fish," *Environ. Sci. Technol.,* **8**, 1113-1115 (1974).

Neely, W.B., G.E. Blau, and J. Alfrey, Jr., "Mathematical models predict concentration-time profiles resulting from chemical spill in a river," *Environ. Sci. Technol.*, **10**, 72-76 (1976).

Othmer, D.F. and M.S. Thakar, "Correlating diffusion coefficients in liquids," *Ind. Eng. Chem.,* **45**, 589-593 (1953).

Paris, D.F., and N.L. Wolfe, "Relationship between properties of a series of anilines and their transformation by bacteria," *Appl. Environ. Microbiol.,* **53**, 911-916 (1987).

Patterson, S., D. Mackay, E. Bacci, and D. Calamari, "Correlation of the equilibrium and kinetics of leaf-air exchange of hydrophobic organic chemicals," *Environ. Sci. Technol.,* **25**, 866-871 (1991).

Reichardt, P.B., B.L. Chadwick, M.A. Cole, B.R. Robertson, and D.K. Button, "Kinetic study of the biodegradation of biphenyl and its monochlorinated analogues by a mixed marine microbial community," *Environ. Sci. Technol.,* **15**, 75-79 (1981).

Reichert, P. and O. Wanner, "Simulation of a severe case of pollution of the Rhine River," In *Proceedings of the Twelfth Congress of the International Association of Hydraulic Research;* Water Resources, Littleton, CO, 1987.

Reichert, P. and O. Wanner, "Enhanced one-dimensional modeling of transport in rivers," *J. Hydr. Div., ASCE,* **117(9)**, 1165-1183 (1991).

Roberts, P.V. and A.J. Valocchi, "Principles of organic contaminant behavior during artificial recharge". In W. van Dujvenbooden and P. Glasbergen, Eds., *Quality of Groundwater*, Elsevier, Amsterdam, 1981, pp. 439-450.

Roof, A.A.M., "Basic principles of environmental photochemistry," In O. Hutzinger, Ed., *The Handbook of Environmental Chemistry,* Vol. 2, Part B, Springer, Berlin, 1982.

Rutherford, J.C., *River Mixing*, Wiley, Chichester, 1994.

Schellenberg, K., C. Leuenberger, and R.P. Schwarzenbach, "Sorption of chlorinated phenols by natural sediments and aquifer materials," *Environ. Sci. Technol.,* **18**, 652-657 (1984).

Schwarzenbach, R.P., Assessing the behavior and fate of hydrophobic organic compounds in the aquatic environment - general concepts and case studies emphasizing volatile halogenated hydrocarbons," Habilitation Thesis, Swiss Federal Institute of Technology, Zurich, 1983.

Schwarzenbach, R.P., W. Giger, E. Hoehn, and J.K. Schneider, "Behavior of organic compounds during infiltration of river water to groundwater. Field studies," *Environ. Sci. Technol*., **17**, 472-479 (1983).

Schwarzenbach, R.P., R. Stierli, B.R. Folsom, and J. Zeyer, "Compound properties relevant for assessing the environmental partitioning of nitrophenols," *Environ. Sci. Technol.*, **22**, 83-92 (1988).

Seinfeld, J.H., *Atmospheric Chemistry and Physics of Air Pollution*, Wiley-Interscience, New York, 1986.

Small, H. "Hydrodynamic chromatography - a technique for size analysis of colloidal particles", *J. Colloid interface Sci.*, **48**, 147-161 (1974).

Stoistis, R.F., G.W. Poehlein, and J.W. Venderhoft, "Mathematical modeling of hydrodynamic chromatography," *J. Colloid Interface Sci.*, **57**, 337-344 (1976).

Suflita, J.M., J.A. Robinson, and J.M. Tiedje, "Kinetics of microbial dehalogenation of haloaromatic substrates in methanogenic environments," *Appl. Environ. Microbiol.,* **45**, 1466-1473 (1983).

Tables of Physical and Chemical Constants, Longman, London, 1973.

Tipping, E., "The adsorption of aquatic humic substances by iron oxides," *Geochim. Cosmochim. Acta*, **45**, 191-199 (1981).

Tolls, J. and M.S. McLachlan, "Partitioning of semivolatile organic compounds between air and Lolium multiflorum (Welsh Ray Grass)", *Environ. Sci. Technol.*, **28**, 159-166 (1994).

Tratnyek, P.G. and J. Hoigné, "Oxidation of phenols in the environment: A QSAR analysis of rate constants for reaction with singlet oxygen," *Environ. Sci. Technol.*, **25**, 1596-1604 (1991).

Ulrich, M., "Modeling of chemicals in lakes - development and application of user-friendly simulation software (MASAS & CHEMSEE) on personal computers," *Dissertation* ETH No. 9632, 1991.

Ulrich, M.M., S.R. Müller, H.P. Singer, D.M. Imboden, and R.P. Schwarzenbach, "Input and dynamic behavior of the organic pollutants Tetrachloroethene, Atrazine, and NTA in a lake: A study combining mathematical modeling and field measurements," *Environ. Sci. Technol.*, **28**, 1674-1685 (1994).

Veith, G.D., N.M. Austin, and R.T. Morris,"A rapid method for estimating log P for organic chemicals," *Water Res.*, **13**, 43-47 (1979).

Wanner, O., T. Egli, T. Fleischmann, K. Lanz, P. Reichert, and R.P. Schwarzenbach, "Behavior of the insecticides disulfoton and thiometon in the Rhine River: A chemodynamic study," *Environ. Sci. Technol.*, **23**, 1232-1242 (1989).

Zeyer, J. and H.-P. Kocher, "Purification and characterization of a bacterial nitrophenol oxygenase which converts *ortho*-nitrophenol to catechol and nitrate," *J. Bacteriol.*, **170**, 1789-1794 (1988).

R. INDEX

Note: *E* refers to "Illustrative Example"; *P* refers to "Problems"; *CS* refers to "Case Studies"; Compound Index comprises all chapters; Subject Index comprises only Chapters N, O, and P.

COMPOUND INDEX

SUBJECT INDEX